Smart Grids and Green Energy Systems

Scrivener Publishing
100 Cummings Center, Suite 541J
Beverly, MA 01915-6106

Publishers at Scrivener
Martin Scrivener (martin@scrivenerpublishing.com)
Phillip Carmical (pcarmical@scrivenerpublishing.com)

Smart Grids and Green Energy Systems

Edited by
A. Chitra
V. Indragandhi
and
W. Razia Sultana

Scrivener Publishing

WILEY

This edition first published 2023 by John Wiley & Sons, Inc., 111 River Street, Hoboken, NJ 07030, USA and Scrivener Publishing LLC, 100 Cummings Center, Suite 541J, Beverly, MA 01915, USA

For more information about Scrivener publications please visit www.scrivenerpublishing.com.

Wiley Global Headquarters
111 River Street, Hoboken, NJ 07030, USA

For details of our global editorial offices, customer services, and more information about Wiley products visit us at www.wiley.com.

Library of Congress Cataloging-in-Publication Data

ISBN 9781119872030

Cover images: Background, Disconnectors, Fedecandoniphoto | Transmission Tower, Roncivil | Dreamstime.com
Cover design by Kris Hackerott

Set in size of 11pt and Minion Pro by Manila Typesetting Company, Makati, Philippines

Printed in the USA

Contents

Preface

The ecological benefits of a greenhouse gas emission-free system are tremendous. Recent trends in energy are heading in that direction. In such circumstances, smart grids and green energy systems are emerging as two of the most important parts to this progress. The demand for renewable energy is increasing every day. Hence, the technology needs to be supported from an academic perspective to service industrial demands.

This book aims to bring together the research that has been carried out in the field of smart grids and green energy systems. Smart grids and green energy systems involve the expertise from various areas of research, such as electrical engineering, electronics, computer engineering, and mechanical engineering. This book will serve as a point of convergence wherein all these domains need to be addressed.

The various chapters are configured in order to address the challenges faced in smart grids and green energy systems from various fields and possible solutions. The outcome of this book can serve as a potential resource for industry professionals, engineers, and researchers working in the domain of smart grids and green energy systems.

The contributions for this book have been chosen from various esteemed national and international institutions. We would like to extend our sincere thanks to all the valuable contributors for the wonderful research contribution. Our sincere thanks to Vellore Institute of Technology, for providing all the support to make this book a success. We would like to extend our heartfelt gratitude to the Wiley-Scrivener team for providing endless support in making this book into a reality.

1

Studies on Enhancement of Battery Pack Efficiency Using Active Cell Balancing Techniques for Electric Vehicle Applications Through MATLAB Simulations

B. Akhila and S. Arockia Edwin Xavier*

Department of Electrical and Electronics Engineering, Thiagarajar College of Engineering, Madurai, India

Abstract

Battery management system (BMS) is one of the most vital parts of electric vehicles, and their precision and accuracy in every aspect are required. Balancing of cells in battery packs is an important task of BMS when it comes to increasing the lifetime of the whole battery pack, end of life (EOL), and the reliability. For balancing individual cells of a battery pack, there are two main methodologies: the passive balancing technique and the active cell balancing technique. In passive cell balancing, state of charge (SOC) is considered for equalization and voltage in case of active cell balancing. This work analyzes which of the two methods is efficient. Moreover, there are several methods that come under active cell balancing method and this work concentrates more on the usage of flyback converter technique and multi-winding transformer method. The proposed cell balancing is analyzed by simulations and results.

***Keywords*:** Active cell balancing, battery cell balancing, battery management system, passive cell balancing

**Corresponding author*: saexeee@tce.edu

A. Chitra, V. Indragandhi, and W. Razia Sultana (eds.) Smart Grids and Green Energy Systems, (1–16)

1.1 Introduction

Electric automotives were prevalent in the 19th century because of their ease of use and comfort. However, the invention of combustion engines dominated over their electrical counterpart. The EVs are reappearing in the automotive industries due advancement in the technologies and the consideration of environmental impacts.

Although the electric vehicles are slowly taking over the industries, still there are some uncertainties in the vehicle systems and dubiousness among the customers in purchasing. Charging time, range, safety, comforts, and many other factors are in the considerations list. One of the main systems manufacturers that are focusing more especially on electrical side is the battery management system (BMS) (Figure 1.1). Increasing their lifetime despite the surroundings where they are subjected to is the main objective. In this work, we are concentrating on one of the main tasks of BMS, the balancing [1].

1.2 Influence of Lithium Ion Batteries

Li-ion battery systems are becoming prevalent for the propulsion of electric vehicles because of their lighter weight, high energy density, extended lifetime, fast charge capability, low level of self-discharge, and eco-friendliness. Li-ion battery cell level and pack variables are needed to be maintained accurately for safe operation [15, 16]. BMS acts as a brain of a

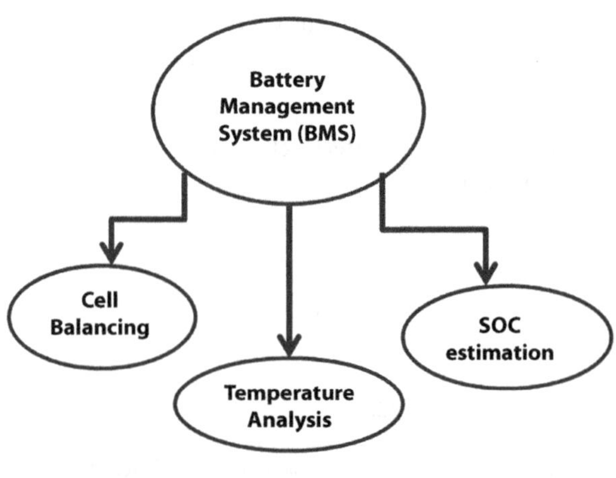

Figure 1.1 Typical BMS tasks.

battery pack that incorporates monitoring of temperature, voltage, current, and state of charge (SOC).

1.3 Cell Balancing

Not every cell is manufactured evenly. Researchers has found that lifetime of a battery pack in an EV is dependent on health of each cell, apart from the whole pack reaching its end of life (EOL). Therefore, it becomes important to monitor and control each and every cell. The cell that has lost its ideal characteristics of charging and discharging can be removed from the pack and replaced with new one. To maintain their shelf life, balancing of each cell becomes important [4].

1.3.1 Types of Cell Balancing

Cycling a battery pack eventually causes individual cells to become out of balance. Therefore, cell balancing becomes one of the important tasks of BMS [5]. The weakest cell limits the amount of charge that can be drawn and the strongest cell limits the extent to which it can be charged. Thus, balancing becomes important. There are two types of cell balancing techniques as shown in Figure 1.2 [2, 3].

1.3.2 Passive Cell Balancing

Passive cell balancing is a dissipative method in which the parallel resistors bleed excess charge from individual cell as shown in Figure 1.3. Bleeding here means that they dissipate the excess charge through this resistor.

1.3.3 Active Cell Balancing

Active cell balancing is a non-dissipative method in which the charge is redistributed equally among the cells. The cell with the higher charge will be redistributed to the cell with weakest SOC or voltage [8, 9].

The two methods of active cell equalization discussed in this work uses flyback converter and other method uses multi-winding transformer.

The operation of flyback converter is very simple. Here, the windings of inductor 1 and inductor 2 act individually rather than acting as a single transformer. This makes the flyback converter topology stand out from other transformers. When the current passes through the switch (SW), magnetic field is set up, which is stored in the core as energy. This reverse

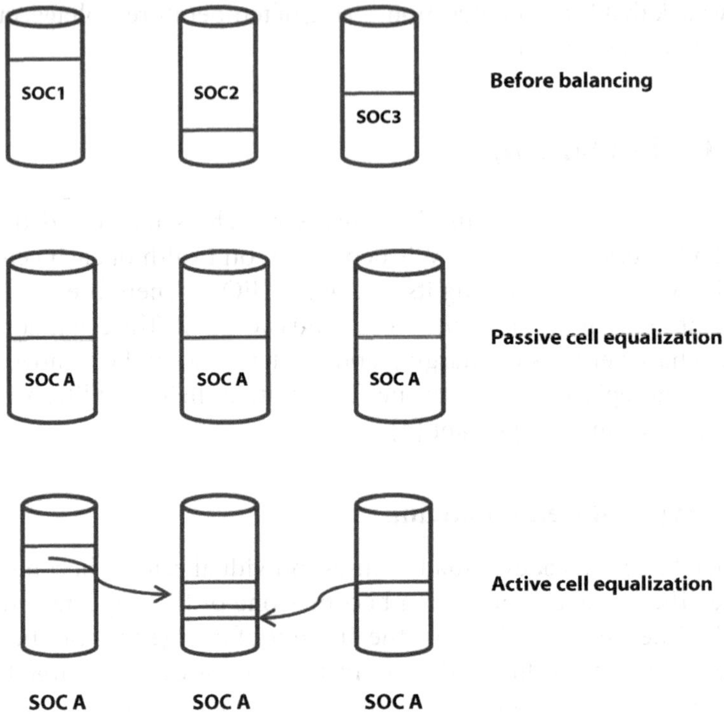

Figure 1.2 Types of cell balancing.

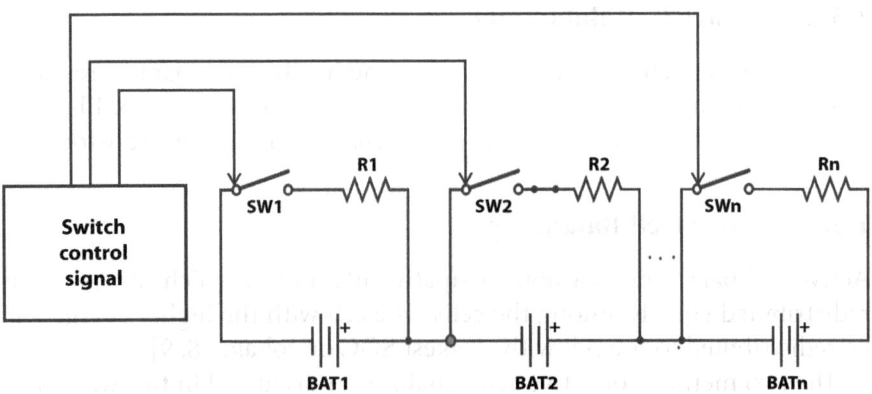

Figure 1.3 Circuit diagram for passive cell balancing.

biases the output diode. This magnetic field is collapsed when SW is in open condition and transfers the stored energy to the secondary winding that makes the flyback diode forward biased. Finally, the energy is eventually sent to the load as shown in Figure 1.4.

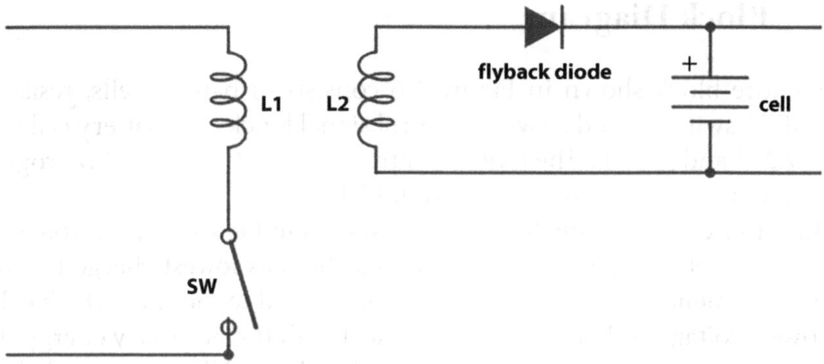

Figure 1.4 Circuit diagram for active cell balancing using flyback converter.

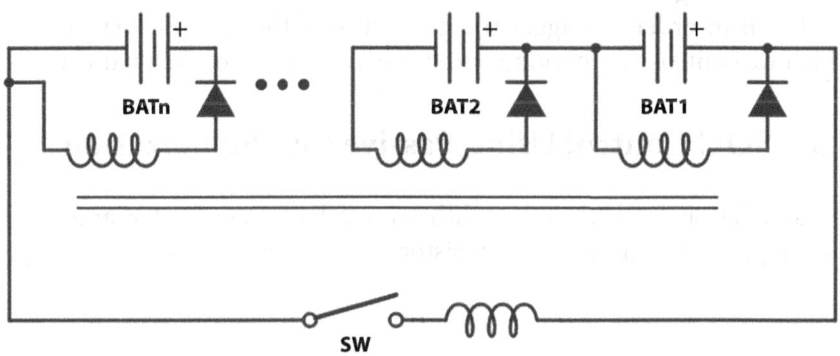

Figure 1.5 Circuit diagram for active cell balancing using multi-winding transformer.

While using the multi-winding transformer, the whole battery current is transferred to the transformer and the output of the transformer is rectified and brought inside the cell that has lowest SOC/voltage level through corresponding semiconductor switches as shown in Figure 1.5 [12].

1.3.4 Why Cell Balancing is Important

Cycling a battery pack eventually causes individual cells to become out of balance. Therefore, cell balancing becomes one of the important operations of the BMS. Not all cells behave alike because of columbic efficiency difference.

Weakest cell limits the amount of charge that can be drawn. Strongest cell limits the extent to which it can be charged.

1.4 Block Diagram

The entire block shown in Figure 1.6 consists of battery cells, resistors, MOSFET switches, and a switch control signal block. The battery cell used is of 7.2 V and 5.4 Ah. The switch control signal is the MATLAB program to trigger the gate terminal of the MOSFET switch.

In active cell balancing technique as shown in Figure 1.7, the voltage of the higher cell is redistributed to the cell that has lowest charge through storage element [11]. Switch control signal is used to monitor which cell is of lowest voltage and targets it to provide it with the necessary energy. The software used in this project is MATLAB/SIMULINK. With the help of this software output waveforms of SOC, current and voltage waveforms of the passive cell balancing and active cell balancing technique is obtained. MATLAB program is written which is fed to the function block of the MATLAB in order to trigger the gate pulse of the MOSFET switch with which we control the charging and discharging cycle of each battery cell.

1.5 SOC Control Using Passive Cell Equalization

The simulation model has three lithium ion batteries of 7.2 V and 5.4 Ah each in parallel with MOSFET- resistor pair as shown in Table 1.1. The gate

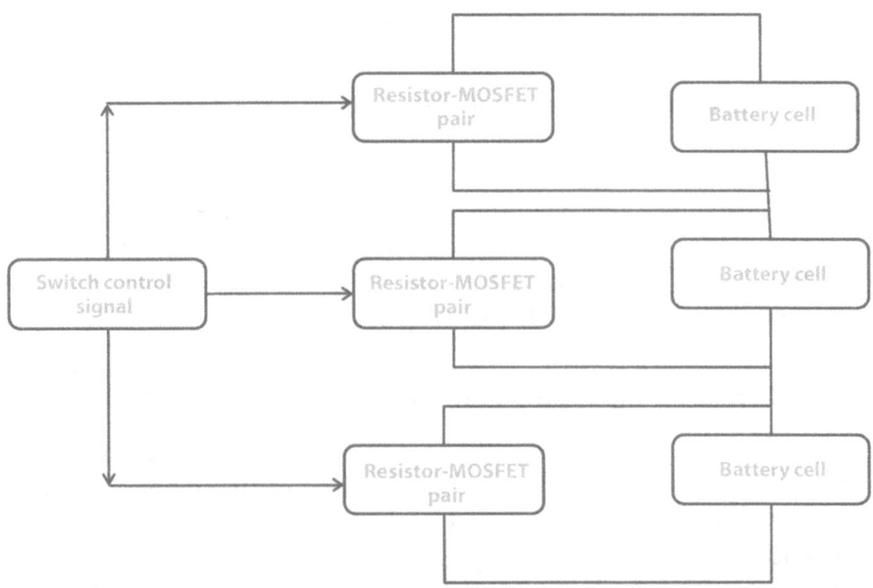

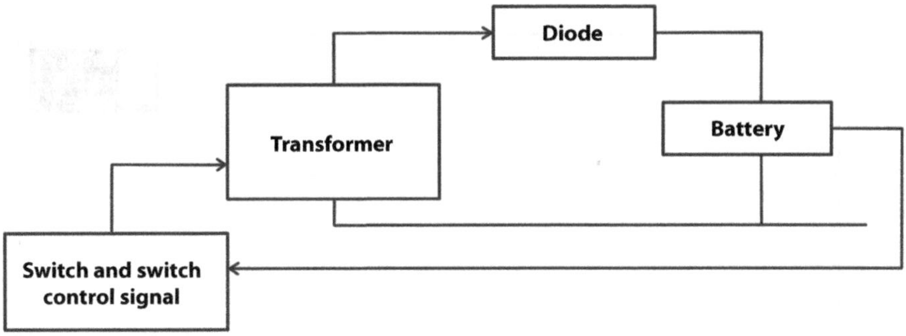

Figure 1.7 Basic block diagram for active cell balancing.

Table 1.1 Simulation specification of cells.

Specifications	**Cell 1**	**Cell 2**	**Cell 3**
Nominal voltage	7.2 V	7.2 V	7.2 V
Rated capacity	5.4 Ah	5.4 Ah	5.4 Ah
Initial SOC	15%	30%	60%

terminal of the MOSFET is triggered according to the MATLAB coding. Thus, we are monitoring which cell is to be drained and which should not. Each cells are set to different initial SOC for better comparative studies and analysis. Components and elements are used from the library browser of the simulink window. The codes are fed in the respective MATLAB function block. SOCs being the input for the code and output of the code will be fed to the gate terminal for triggering purpose.

The output of the code will be either 1 or 0 (high or low), and it will decide which of the three gate terminal should be turned on from which the discharging cycle will occur.

1.5.1 Equalization Results

In the simulation setup, we have three lithium ion cells in parallel with MOSFET-resistor pair. With the help of bus selector, SOC and current are

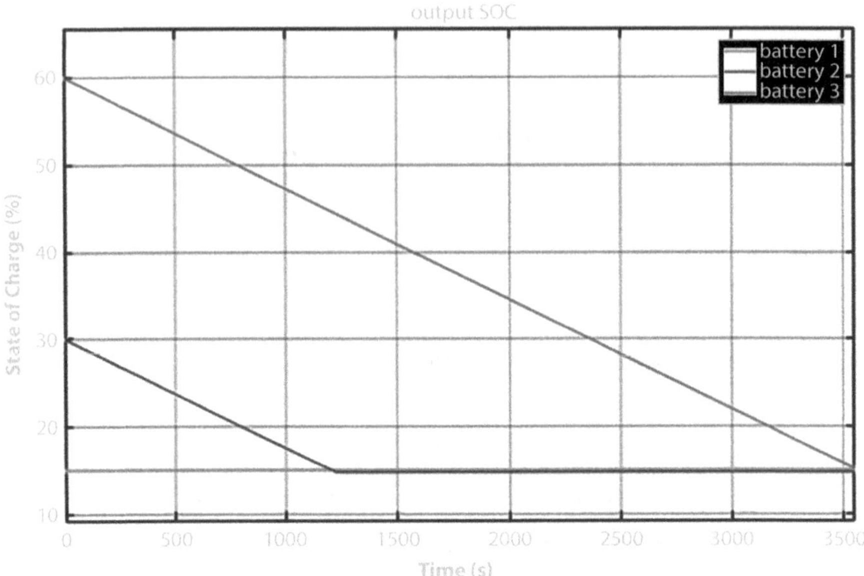
output SOC
battery 1
battery 2
battery 3
State of Charge (%)
60
50
40
30
20
10
0
500
1000
1500
2000
2500
3000
3500
Time (s)

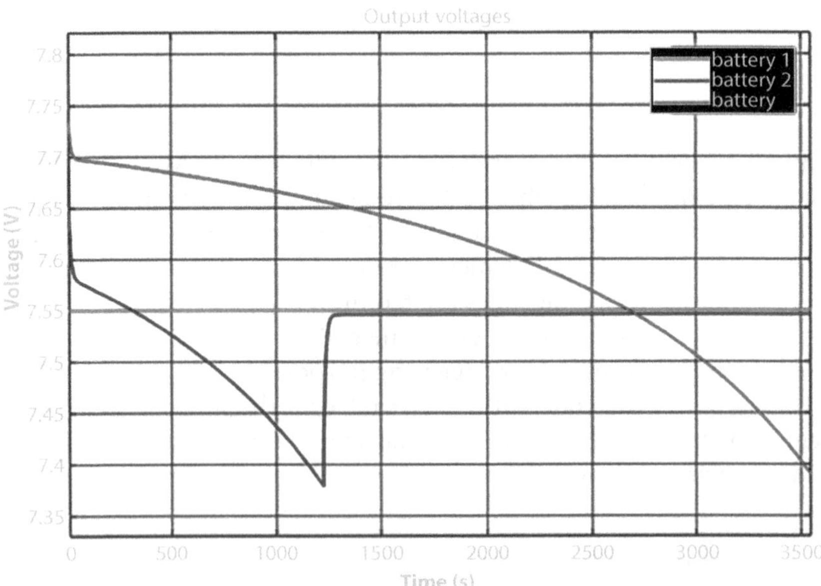
Output voltages
battery 1
battery 2
battery
Voltage (V)
7.8
7.75
7.7
7.65
7.6
7.55
7.5
7.45
7.4
7.35
0
500
1000
1500
2000
2500
3000
3500
Time (s)

measured. The SOC plot of 30% initial SOC is found to be balanced at 1,250 s and 60% initial SOC is found to be balanced at 3,550 s. Finally, the SOC of all three cells attains the balanced state of 15% SOC as shown in Figure 1.8. The ifelse statement continuously operates until it reaches the minimum SOC in each cell. Here, the SOC is taken as the input for the function block where set of nested ifelse code is available. The output of the program will be fed to the gate terminal of the MOSFET (1 or 0). The cell will be draining accordingly (Figure 1.9) until it reaches a preset value of SOC (minimum of all three cells).

1.6 Voltage Control Using Active Cell Equalization

1.6.1 The Flyback Converter Method

The model uses two capacitors as cells each of 5-F capacitance and the initial voltage of first cell is set to be 6 V and the second cell voltage is set to be 4 V [13]. Flyback diode id reverse biased when the MOSFET switch is closed and current passes through the primary winding as it sets up a magnetic field. When the MOSFET switch is opened, the magnetic field collapses thus forward biasing the diode in the secondary side as a result they transfer energy.

From the output voltage waveform as shown in Figures 1.10a and b of the flyback converter equalization topology, it is evident that the balancing of cells occurs only when needed, and the charges are redistributed accordingly (Table 1.2). In addition, there is no waste of energy through heat dissipation, since there is no bleeding resistor.

Table 1.2 clearly shows the cell that has been balanced to a certain point with little deviations and very less transients. From the results, it can be said that this is one of the reliable and comparatively less charge wastage method because of the effective redistribution of the voltage among every cells of the battery pack.

When it comes to on-road applications, the charge/voltage redistribution should be done carefully and effectively since a little amount of charge wastage causes significant impact on the efficiency of the whole vehicle. Thus, maintaining little to no charge wastage at the balancing system becomes a crucial and essential task.

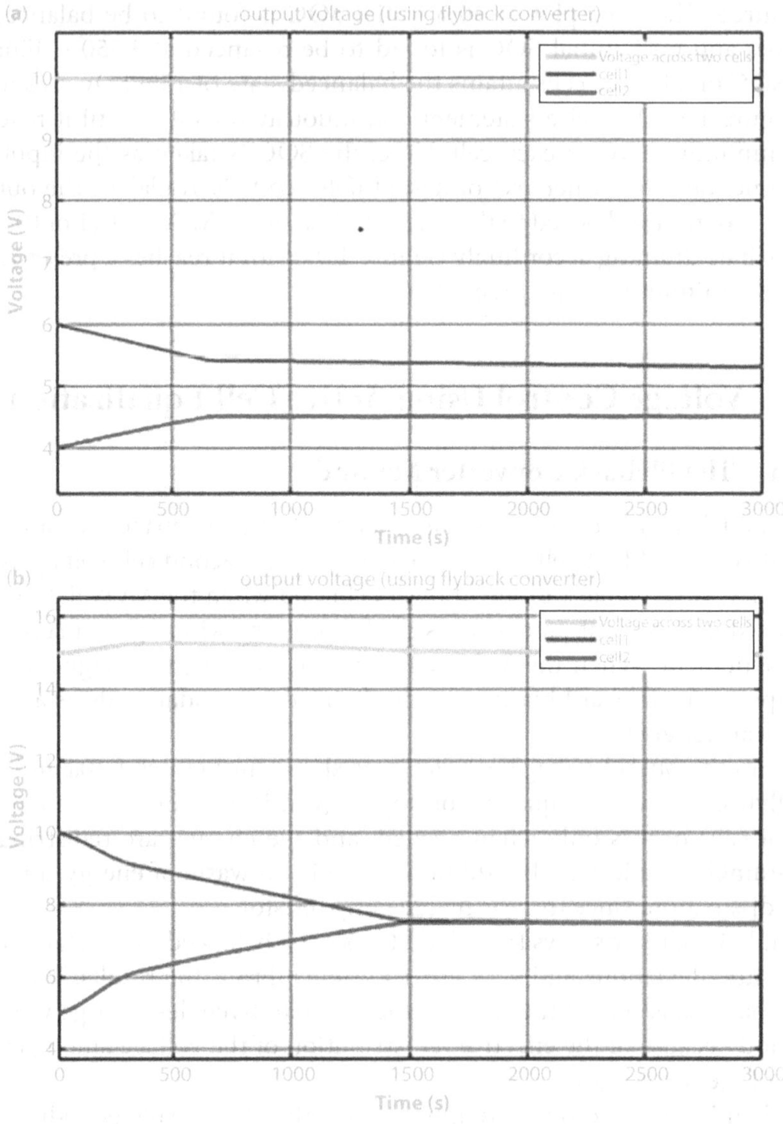
(a)
output voltage (using flyback converter)
Voltage across two cells
cell1
cell2
Voltage (V)
Time (s)
(b)
output voltage (using flyback converter)
Voltage across two cells
cell1
cell2
Voltage (V)
Time (s)

Table 1.2 Output voltages after cell equalization (using flyback converter).

	Cell no.	Capacitance (F)	Initial cell voltage (V)	Balanced voltage (V)
1	cell1	5	6	5.31
	cell2	5	4	4.502
2	cell1	5	10	7.472
	cell2	5	5	7.471

1.6.2 The Multi-Winding Transformer Method

The multi-winding transformer model is analyzed with two cells and three cells topology respectively to observe their behavior and deviations from ideal state when number of cells is multiplied as shown in Figures 1.11 a and b. The model uses capacitors of 5 F and diodes of 0.001-ohm resistance and 0.8-V forward voltage. Here, the entire battery current is transferred to transformer through the switch.

From the output waveforms, it is evident that the cell is getting balanced, but there are unnecessary discharges too (Table 1.3). Some of the cells need not be balanced since their voltages are nearly equal to the lowest charge cell. Despite that, they are being discharged and charged too. In addition, there are some transients in the beginning as shown in Figures 1.12a and b.

Table 1.3 Output voltages after cell equalization (using flyback converter—two-cell topology).

	Cell no.	Capacitance (F)	Initial cell voltage (V)	Balanced voltage (V)
1	cell1	5	6	1.161
	cell2	5	4	2.125
2	cell1	5	15	4.102
	cell2	5	10	4.721

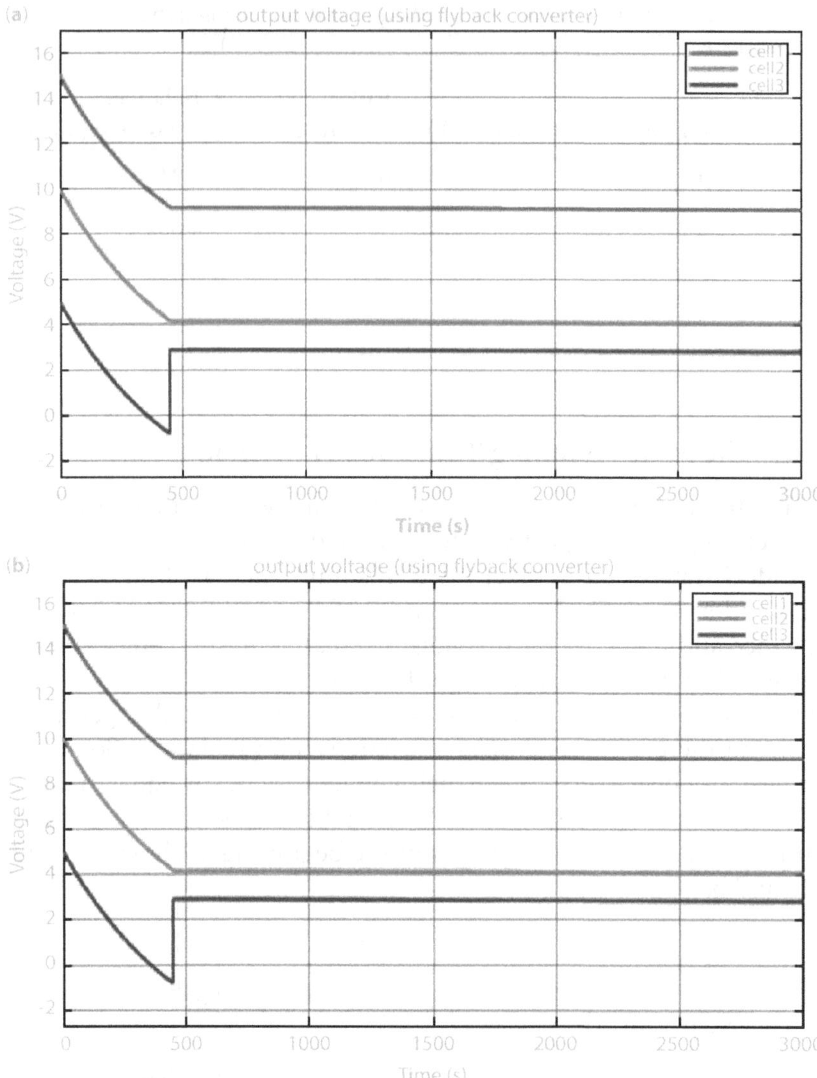
(a)
output voltage (using flyback converter)
cell1
cell2
cell3
Voltage (V)
16
14
12
10
8
6
4
2
0
-2
0
500
1000
1500
2000
2500
3000
Time (s)
(b)
output voltage (using flyback converter)
cell1
cell2
cell3
Voltage (V)
16
14
12
10
8
6
4
2
0
-2
0
500
1000
1500
2000
2500
3000
Time (s)

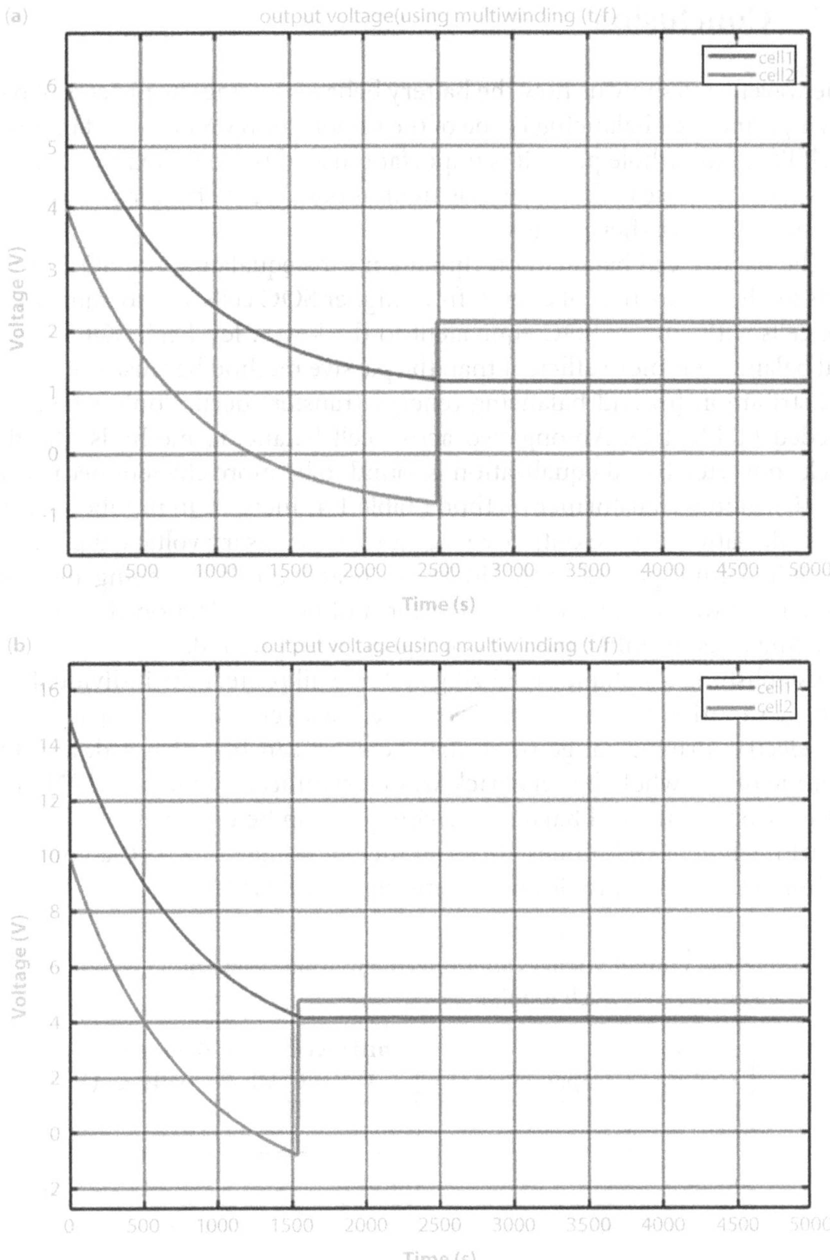
(a)
output voltage(using multiwinding (t/f)
cell1
cell2
Voltage (V)
Time (s)
(b)
output voltage(using multiwinding (t/f)
cell1
cell2
Voltage (V)
Time (s)

1.7 Conclusion

The waveforms show us how the battery behaves during the process of balancing. Since cell balancing is one of the important BMS tasks that increase shelf life of the whole pack, it is important to analyze their behavior in various circumstances. This work has figured out their behavior without the considerations of thermal aspects.

The passive cell balancing technique means equalizing the SOC of the cells by the dissipation of energy from higher SOC cells and formulates all the cells with similar SOC equivalent to the lowest level cell SOC. Active cell balancing is more efficient than the passive method because the charge redistribution [6] and balancing (energy transfer) occurs only when it is needed (Table 1.5). Among two active cell balancing methods, the flyback converter-based equalization is found to be more efficient because in multi-winding transformer method (Table 1.4) increase in number of cells cause deviations in the output voltage and unnecessary voltage dip during the initial time period [14]. However, in active cell balancing method, there is no wastage of energy in the form of heat dissipation through the bleeding resistor unlike passive cell equalization method.

Establishing a suitable equalizing system will protect the individual cell voltage variation that may vary over time, the overall battery capacity and the electric driving range of electric vehicles can be enhanced, and the lifetime of the whole battery pack will be significantly increased [7]. The future scope of the cell balancing techniques can be done with the help of smart intelligent techniques. This can include some of the soft computing techniques like fuzzy logic and genetic algorithm [10].

Table 1.4 Output voltages after cell equalization (using multi-winding transformer—two-cell topology).

	Cell no.	Capacitance (F)	Initial cell voltage (V)	Balanced voltage (V)
1	cell1	5	6	2.472
	cell2	5	4	0.4719
	cell3	5	2	1.89
2	cell1	5	15	9.072
	cell2	5	10	4.072
	cell3	5	5	2.815

Table 1.5 Comparative study of passive and active cell equalization.

	Passive cell balancing	Active cell balancing
Degree of difficulty in circuitry implementation	Easier implementation	Complex circuit
Energy	Some energy is wasted in the form of heat	Transfers the energy to neighboring cells instead of wasting
Components	Needs less components	Demands additional components
Cost	Cost effective	Increased cost
Efficiency	Less efficient due to heat dissipation through bleeding resistors	More efficient because the energy is transferred within the cells rather than wasting it

References

1. Omariba, Z.B., Zhang, L., Sun, D., Review of Battery Cell Balancing Methodologies for Optimizing Battery Pack Performance in Electric Vehicles. *IEEE Access*, 7, 129335–129352, 2019. https://ieeexplore.ieee.org/stamp/stamp.jsp?arnumber=8834858.
2. Gallardo-Lozano, J., Romero-Cadaval, E., Milanes-Montero, M., II, Guerrero-Martinez, M.A., Battery equalization active methods. *J. Power Sources, 246,* 934–949, 2014. https://dx.doi.org/10.1016/j.jpowsour.2013.08.026.
3. Baronti, F., Roncella, R., Saletti, R., Performance comparison of active balancing techniques for lithium-ion batteries. *J. Power Sources, 267,* 4, 603–609, 2014. https://doi.org/10.1016/j.egypro.2019.01.956.
4. Piao, C., Wang, Z., Cao, J., Zhang, W., Lu, S., Lithium-Ion Battery Cell-Balancing Algorithm for Battery Management System Based on Real-Time Outlier Detection. *Journal of Mathematical Problems in Engineering, Chong Qing University of Posts and Telecommunications,* 2015, 2015. https://doi.org/10.1155/2015/168529.
5. Cui, X., Shen, W., Zhang, Y., Hu, C., Zheng, J., Novel active LiFePO4 battery balancing method based on chargeable and dischargeable capacity. *Comput. Chem. Eng., 97,* 27–35, 2017. https://doi.org/10.1016/j.compchemeng.2016.11.014.

6. Han, W. and Zhang, L., Battery cell reconguration to expedite charge equalization in series-connected battery systems. *IEEE Robot. Autom. Lett.*, *3*, 1, 2228, 2018. DOI: 10.1109/LRA.2017.2728204.
7. Omariba, Z.B., Zhang, L., Sun, D., Review on health management system for lithium-ion batteries of electric vehicles. *Electronics*, *7*, 5, 72, 2018. https://doi.org/10.3390/electronics7050072.
8. Quinn, D.D. and Hartley, T.T., Design of novel charge balancing networks in battery packs. *J. Power Sources*, *240*, 2632, 2013. https://doi.org/10.1016/j.jpowsour.2013.03.113.
9. Min, G.-H. and Ha, J.-I., Active cell balancing algorithm for serially connected Li-Ion batteries based on power to energy ratio, in: *Proc. IEEE Energy Convers. Congr. Expo. (ECCE)*, pp. 2748–2753, 2017, DOI: 10.1109/ECCE.2017.8096514.
10. Mathew, M., Kong, Q.H., McGrory, J., Fowler, M., Simulation of lithium-ion battery replacement in a battery pack for application in electric vehicles. *J. Power Sources*, *349*, 94–104, 2017. https://doi.org/10.1016/j.jpowsour.2017.03.010.
11. Einhorn, M., Roessler, W., Fleig, J., Improved Performance of Serially Connected Li-Ion Batteries With Active Cell Balancing in Electric Vehicles. *IEEE T. Veh. Technol.*, *60*, 2448–2457, 2011. DOI: 10.1109/TVT.2011.2153886.
12. Li, S.Q., Mi, C., Zhang, M.Y., A high efficiency low cost direct battery balancing circuit using a multi-winding transformer with reduced switch count. *Applied Power Electronics Conf. and Exposition (APEC)*, Orlando, FL, USA, pp. 2128–2133, 2012, DOI: 10.1109/APEC.2012.6166115.
13. Shang, Y.L., Xia, B., Lu, F., Zhang, C.H., Cui, N.X., Mi, C., A switched coupling capacitor equalizer for series connected battery strings. *IEEE Trans. Power Electron.*, *32*, 10, 7694–7706, 2017. DOI: 10.1109/APEC.2017.7930884.
14. Federico, B., Roberto, R., Roberto, S., Performance comparison of active balancing techniques for lithium-ion batteries. *J. Power Sources*, *267*, 603–609, 2014. https://doi.org/10.1016/j.jpowsour.2014.05.007.
15. Kuh, B.T., Pitel, G.E., Krein, P.T., Electrical properties and equalization of lithium-ion cells in automotive applications. *IEEE Vehicle Power and Propulsion Conf*, Chicago, IL, USA, pp. 55–59, 2005, DOI: 10.1109/VPPC.2005.1554532.
16. Chen, Y., Liu, X.F., Fathy, H.K., Zou, J.M., Yang, S.Y., A Graph-Theoretic Framework for Analyzing the Speeds and Efficiencies of Battery Pack Equalization Circuits. *Int. J. Electr. Power Energy Syst.*, 98, 85–99, 2018. https://doi.org/10.1016/j.ijepes.2017.11.039.

2

Evaluation and Impacts of Minimum Energy Performance Standards of Electrical Motors in India

S. Manoharan[1]*, G. Sureshkumaar[2], B. Mahalakshmi[3] and V. Govindaraj[2]

[1]*Department of Electronics and Instrumentation Engineering, Karpagam College of Engineering, Coimbatore, India*
[2]*Department of Electrical and Electronics Engineering, Karpagam College of Engineering, Coimbatore, India*
[3]*Dept. of Physics, Hindustan College of Engineering and Technology, Coimbatore, India*

Abstract

The demand for electricity is rapidly increasing, posing a significant challenge to the entire power industry in meeting the unprecedented demand. The inefficiency of the low-quality motors leads to increased electric power consumption and the associated energy cost. As a result, improving the motor's efficiency must be a component of any overall energy saving strategy. Induction motors are widely employed as driving power in most industries, accounting for over 65% of total power utilization. These motors' energy-efficient design saves a lot of energy and lowers operating costs. The majority of energy consumed in industries is used to power various motors, which consumes a considerable amount of energy that may be greatly decreased by replacing the normal motor with a high efficiency motor. The goal of this study is to shed light on India's existing Minimum Energy Performance Standards (MEPS) for asynchronous motors by applying IS 12615: 2018, which is founded on the International Test Standard IEC 60034-2-1. This document also looks at the Indian government's energy-saving programs, such as the National Motor Replacement Program (NMRP), as well as the challenges that must be addressed and the path forward to implement this mission.

**Corresponding author*: manoish07@gmail.com

A. Chitra, V. Indragandhi, and W. Razia Sultana (eds.) Smart Grids and Green Energy Systems, (17–34)

***Keywords*:** Minimum Energy Performance Standards (MEPS), IE2, IE3, IE4, energy demand, efficiency, National Motor Replacement Program (NMRP)

2.1 Introduction

India's energy demand will grow faster than any other country over the next two decades, according to the International Energy Agency (IEA), with India overtaking the European Union (EU) as the world's third largest energy consumer by 2030. Under present policies, India's energy needs will grow three times faster than the global average, according to the IEA's India Energy Outlook 2021. India is currently the world's fourth largest energy consumer, trailing only China, the United States, and the EU. Industry's increased use of electrical motors and other machinery has also contributed to an increase in electricity demand [1].

During the COVID-19–induced countrywide lockdown, India's electricity consumption fell precipitously. India's peak electricity demand in fiscal year 2021 was 1.31% lower than the earlier fiscal year, despite emerging from the shadows of the COVID-19–induced slowdown. This suggests that robust growth is lacking, with demand rebounding to mid-2019 levels near the end of the fiscal year [2].

Monthly demand, on the other hand, has shown a significant improvement over peak lockdown months, owing primarily to thermal power. According to the National Load Dispatch Centre, electricity consumption in April 2020 was 18.1% and 15.1% lower than it was in February and March, respectively [3]. During the nationwide lockdown, electricity demand shifted from the commercial and industrial sectors to the residential sector. This, however, was insufficient to compensate for the drop in overall electricity demand. The shutdown of the industrial and commercial sectors has resulted in a drop in electricity demand, which may last for some time until the pandemic is over. In 2019, India's electricity consumption increased by 5% to 1,063 TWh. However, due to the pandemic's impact and reduced activity, electricity consumption patterns in 2020 are highly uncertain [4–6].

According to the 2019 Energy Statistics, the industry sector has always been a major consumer of electricity in India's consumption matrix. In 2017 to 2018, the industry sector accounted for 42% of total electricity consumption, followed by domestic (24%), agriculture (18%), and commercial sectors (9%), with others accounting for the remaining 7%. Motors are used in the industrial sector to power conveyors, and other materials handling and processing equipment. Ventilation fans, boiler forced draught fans, refrigeration compressors, and circulating water pumps are all examples of mechanical devices. Motors are employed within homes to power appliances for air conditioning and air handling units [7].

Figure 2.1 summarizes the sector-wise total electricity consumption in 2018–2019. The usual life expectancy of an electric motor is 15–20 years, depending on the quality of the raw materials used and the manufacturing process used. A motor's functioning cost accounts for roughly 95% of its cost of entire life span. With the expanding market penetration of electric motors, India has a significant opportunity to conserve energy. The operating cost of induction motors contributes significantly to total energy consumption and has been reported to be as high as 90%. In India, electric motors consume 70% of the electricity used in the industrial sector, which is encouraging the electric motor manufacturers to improve the motor efficiencies. There appears to be a significant opportunity to improve energy efficiency and reduce losses by up to 35% when compared to the current market scenario. In August 2011, IS 12615 was harmonized with universal IEC standard 60034-30 for efficiency classes (IE2 and IE3) and testing methodologies for electric motors made in India [8]. The evaluation of Energy Efficient Motor Standard IS 12615 is discussed in the following section.

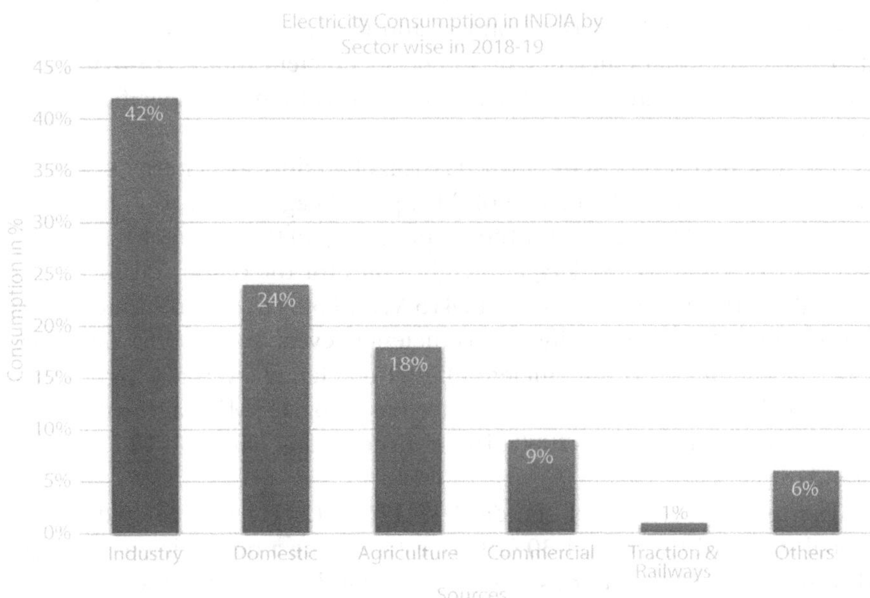

2.2 A Review of IS 12615 Evaluation

In response to energy shortages, carbon emissions reductions, and the availability of high efficiency motors (HEMs), a number of countries have taken initiatives requiring the withdrawal of lower performance class motors and the adoption of higher performance class motors in accordance with IEC 60034-30-1:2014, thereby defining minimum efficiency performance standards (MEPS). These regulations are expected to enforce technological challenges on all motor imported into their respective countries that are less efficient than the MEPS efficiency class.

The possibility of trade restrictions for exports is also addressed by updating the Indian Standard to meet international requirements IS 12615 was first published in 1989, with updates in 2004, 2011, and 2018 to bring it up to date with worldwide standards as well as to include all of the efficiency levels currently in use. MEPS was previously optional, but the Indian government has directed that the minimum efficiency of 3ϕ squirrel cage asynchronous motors that can be manufactured and marketed in the country must comply with IS requirements at the IE2 efficiency level. Figure 2.2 depicts the stages in the creation of the IS 12615 electrical motor efficiency standard.

The Bureau of Indian Standard (BIS) has adopted IS 12615 (1989) for energy efficient squirrel cage induction motor. IS 12615 specifies efficiencies at 75% and 100% at rated output and the ranges from 0.75 to 37 kW. It also specifies that efficiency of an energy efficient motor (EEM) at 75% and 100% of the rated output shall not be less than that specified in the standard [9, 10].

The first revision of standard IS12615:2004 addresses energy efficiency requirements and performance of 3ϕ squirrel cage asynchronous motors in two poles and four poles for the output rating of 0.37 to 160 kW, six-pole motors from 0.37 to 132 kW, and eight poles for the output rating 0.37 to 110 kW all frame sizes up to 315 L, 415 V, and 50Hz with continuous duty (SI) operation. It also specifies two efficiency levels, such as Eff2 (standard efficiency) and Eff1 (high efficiency). Eff1 efficiency levels are higher than those of Eff2. The efficiency levels are based on CEMEP agreement and adapted from IEEMA 19-2000. The test method is based on IEC 60034-2 Standard [11].

The second revision of standard IS 12615 (2011) is predicted on the global standard IEC 60034-30 (2008), establishing a new efficiency categorization for 3ϕ asynchronous motors. This standard specifies the necessities and enactment of, 3ϕ squirrel cage asynchronous motors in two, four, or six

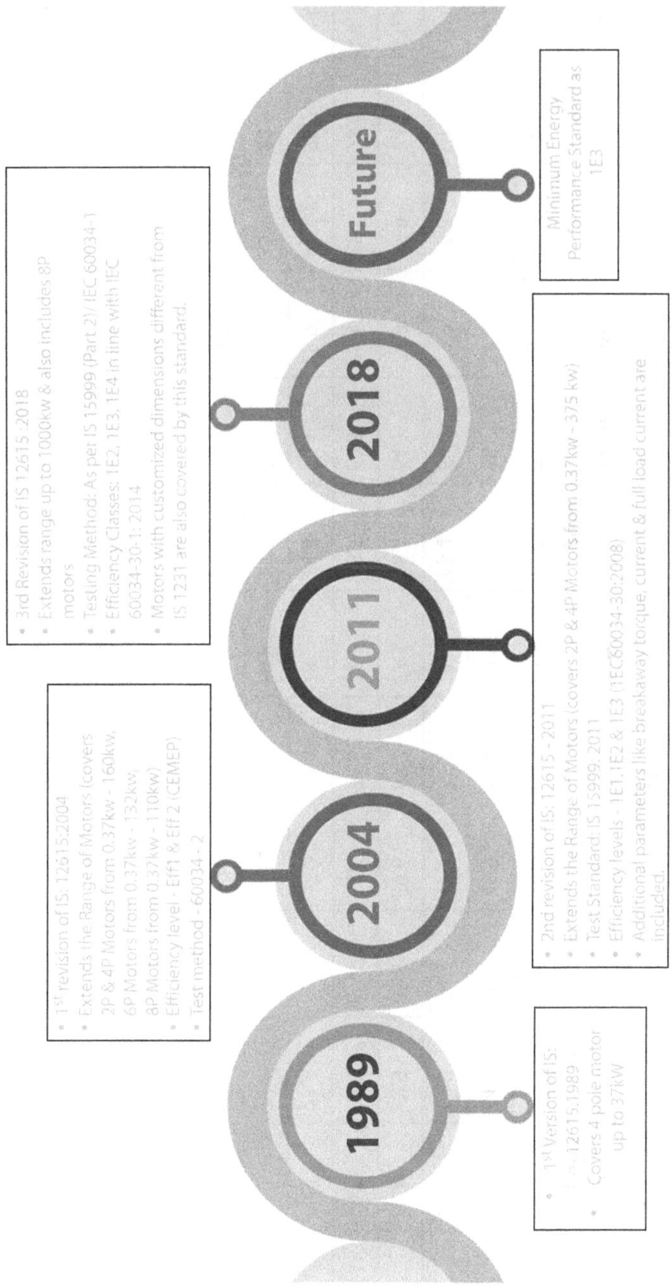
1989
2004
2011
2018
Future
1st Version of IS: 12615:1989
Covers 4 pole motor up to 37kW
1st revision of IS: 12615:2004
Extends the Range of Motors (covers 2P & 4P Motors from 0.37kw - 160kw, 6P Motors from 0.37kw - 132kw, 8P Motors from 0.37kw - 110kw)
Efficiency level - Eff1 & Eff 2 (CEMEP)
Test method - 60034 - 2
2nd revision of IS: 12615 - 2011
Extends the Range of Motors (covers 2P & 4P Motors from 0.37kw - 375 kw)
Test Standard: IS 15999: 2011
Efficiency levels - 1E1,1E2 & 1E3 (1EC60034-30:2008)
Additional parameters like breakaway torque, current & full load current are included.
3rd Revision of IS 12615 :2018
Extends range up to 1000kw & also includes 8P motors
Testing Method: As per IS 15999 (Part 2)/ IEC 60034-1
Efficiency Classes: 1E2, 1E3, 1E4 in line with IEC 60034-30-1: 2014
Motors with customized dimensions different from IS 1231 are also covered by this standard.
Minimum Energy Performance Standard as 1E3

Table 2.1 Comparison of various international efficiency (IE) classes.

Power (kW)	**Efficiency classes** (at 50-Hz frequency) IE2				IE3				IE4			
	Two poles	Four poles	Six poles	Eight poles	Two poles	Four poles	Six poles	Eight poles	Two poles	Four poles	Six poles	Eight poles
0.12	53.6	59.1	50.6	39.8	60.8	64.8	57.7	50.7	66.5	69.8	64.9	62.3
0.18	60.4	64.7	56.6	45.9	65.9	69.9	63.9	58.7	70.8	74.7	70.1	67.2
0.2	61.9	65.9	58.2	47.4	67.2	71.1	65.4	60.6	71.9	75.8	71.4	68.4
0.25	64.8	68.5	61.6	50.6	69.7	73.5	68.6	64.1	74.3	77.9	74.1	70.8
0.37	69.5	72.7	67.6	56.1	73.8	77.3	73.5	69.3	78.1	81.1	78.0	74.3
0.4	70.4	73.5	68.8	57.2	74.6	78	74.4	70.1	78.9	81.7	78.7	74.9
0.55	74.1	77.1	73.1	61.7	77.8	80.8	77.2	73	81.5	83.9	80.9	77
0.75	77.4	79.6	75.9	66.2	80.7	82.5	78.9	75	83.5	85.7	82.7	78.4
1.1	79.6	81.4	78.1	70.8	82.7	84.1	81.8	77.7	85.2	87.2	84.5	80.8
1.5	81.3	82.8	79.8	74.1	84.2	85.3	82.5	79.7	86.5	88.2	85.9	82.6
2.2	83.2	84.3	81.8	77.6	85.9	86.7	84.3	81.9	88	89.5	87.4	84.5
3	84.6	85.5	83.3	80.0	87.1	87.7	85.6	83.5	89.1	90.4	88.6	85.9
4	85.8	86.6	84.6	81.9	88.1	88.6	86.8	84.8	90	91.1	89.5	87.1

(Continued)

Table 2.1 Comparison of various international efficiency (IE) classes. (*Continued*)

Power (kW)	**Efficiency classes** (at 50-Hz frequency)											
	IE2				**IE3**				**IE4**			
	Two poles	**Four poles**	**Six poles**	**Eight poles**	**Two poles**	**Four poles**	**Six poles**	**Eight poles**	**Two poles**	**Four poles**	**Six poles**	**Eight poles**
5.5	87	87.7	86	83.8	89.2	89.6	88	86.2	90.9	91.9	90.5	88.3
7.5	88.1	88.7	87.2	85.3	90.1	90.4	89.1	87.3	91.7	92.6	91.3	89.3
11	89.4	89.8	88.7	86.9	91.2	91.4	90.3	88.6	92.6	93.3	92.3	90.4
15	90.3	90.6	89.7	88.0	91.9	92.1	91.2	89.6	93.3	93.9	92.9	91.2
18.5	90.9	91.2	90.4	88.6	92.4	92.6	91.7	90.1	93.7	94.2	93.4	91.7
22	91.3	91.6	90.9	89.1	92.7	93	92.2	90.6	94	94.5	93.7	92.1
30	92	92.3	91.7	89.8	93.3	93.6	92.9	91.3	94.5	94.9	94.2	92.7
37	92.5	92.7	92.2	90.3	93.7	93.9	93.3	91.8	94.8	95.2	94.5	93.1
45	92.9	93.1	92.7	90.7	94	94.2	93.7	92.2	95	95.4	94.8	93.4
55	93.2	93.5	93.1	91.0	94.3	94.6	94.1	92.5	95.3	95.7	95.1	93.7
75	93.8	94	93.7	91.6	94.7	95	94.6	93.1	95.6	96	95.4	94.2
90	94.1	94.2	94	91.9	95	95.2	94.9	93.4	95.8	96.1	95.6	94.4
110	94.3	94.5	94.3	92.3	95.2	95.4	95.1	93.7	96	96.3	95.8	94.7

(*Continued*)

Table 2.1 Comparison of various international efficiency (IE) classes. (*Continued*)

Power (kW)	Efficiency classes (at 50-Hz frequency)											
	IE2				IE3				IE4			
	Two poles	Four poles	Six poles	Eight poles	Two poles	Four poles	Six poles	Eight poles	Two poles	Four poles	Six poles	Eight poles
132	94.6	94.7	94.6	92.6	95.4	95.6	95.4	94	96.2	96.4	96	94.9
160	94.8	94.9	94.8	93.0	95.6	95.8	95.6	94.3	96.3	96.6	96.2	95.1
200	95	95.1	95	93.5	95.8	96	95.8	94.6	96.5	96.7	96.3	95.4
250	95	95.1	95	93.5	95.8	96	95.8	94.6	96.5	96.7	96.5	95.4
315	95	95.1	95	93.5	95.8	96	95.8	94.6	96.5	96.7	96.6	95.4
355	95	95.1	95	93.5	95.8	96	95.8	94.6	96.5	96.7	96.6	95.4
400	95	95.1	95	93.5	95.8	96	95.8	94.6	96.5	96.7	96.6	95.4
450	95	95.1	95	93.5	95.8	96	95.8	94.6	96.5	96.7	96.6	95.4
500–1,000	95	95.1	95	93.5	95.8	96	95.8	94.6	96.5	96.7	96.6	95.4

poles for frame sizes from 71 up to and including 315 L having 415 V, 50Hz, and output ratings from 0.37 to 375 kW for continuous duty (S1) operation with service factor 1.0 or S3 (intermittent periodic duty) with a cyclic duration factor of 80% or greater [12]. IE1 (standard efficiency), IE2 (high efficiency), and IE3 (premium efficiency) are the three efficiency classes described. In order to be categorized as "Energy Efficient," motors must satisfy at minimum IE2 efficiency levels, according to the IS 12615 [13].

The current Indian standard for EEM is IS 12615:2018, which covers efficiency levels and performance metrics for single-speed line-operated energy efficient, 3ϕ, 415 V, 50 Hz cage asynchronous motors in two, four, or six poles for output power ranging from 0.12 to 1,000 kW and frame sizes ranging from 56 to and including 315 L. Most of the motors that go within this specification are S1 type of service (continuous duty). Motors rated for duty cycles S2 and above that have a similar S1 duty cycle are also featured [14].

All imported and natively manufactured motors must comply with the revised standard, IS 12615:2018, published by the Bureau of Indian Standards (BIS), which specifies IE2 (high efficiency) as the minimum efficiency class [15]. Table 2.1 shows the values of efficiency for motors to classify as IE2, IE3, or IE4 in accordance with the IS 12615:2018 [16].

2.3 A Scenario of 'MEPS' for Electric Motors From Around the World

In March 2014, the International Electro-Technical Commission (IEC) published the standard IEC 60034-30-1. The IEC 60034-30:2008 standard, which established three efficiency levels, was superseded by this one for 3ϕ asynchronous motors. A new fourth efficiency level, or IE4, is included in the revised IEC 60034-30-1 standard. In addition to the IE4 rating, the standard now includes eight-pole motors and a wider power range. IEC has helped to develop EEM systems through the universally valid IEC 60034-2-1 test standard and the IEC 60034-30-1 rating framework consisting of four motor efficiency levels ("IE-Code"). The IE5 class, also known as ultra-premium efficiency, will be included in the future revision of the IEC standard. Between classes, energy losses are reduced by 20% or more, implying that an IE5 motor will have a 20% loss reduction compared to an IE4 motor. Although the ultra-premium efficiency class has yet to be established in standards, some manufacturers have already developed motors that will meet the requirements.

Table 2.2 MEPS-worldwide requirements for electric motors.

Country or region	Standard	Regulation	Power range	MEPS level
Brazil	Portaria nº 01/2017	ABNT NBR 17094-1	0.12 and 370 kW	IE3
Chile	NCh 3086 of 2008	IEC 60034-30-1	0.75 and 7.5 kW	IE2
Ecuador	RTE INEN 145	17 524 – 2017	0.746 up to 373 kW	IE2
Canada	NEMA MG-1	Energy Efficiency Regulations with Amendments	0.75 to 375 kW	Premium (IE3)
USA	NEMA MG-1	DOE 10 CFR Part 431	0.75 to 375 kW	IE3
Mexico	NEMA MG-1	NOM-016-ENER-2016	0.75 to 375 kW	Premium (IE3)
European union	Directive 2009-125-EC Regulation 640-2009	(EC) 640/2009 + (EU) 4/2014	0.75 and 375 kW	IE3 + IE2 for inverter duty
Ukraine	Decree N° 157, Resolution N° 804 and Resolution N° 1184	IEC 60034-2-1	0.75 up to 375 kW	IE3 + IE2 for inverter duty (From 09/2021 onwards)

(*Continued*)

Table 2.2 MEPS-worldwide requirements for electric motors. (*Continued*)

Country or region	Standard	Regulation	Power range	MEPS level
Great Britain	The Ecodesign for Energy-Related Products and Energy Information (Amendment) (EU Exit) Regulations 2019	(EC) 640/2009 + (EU) 4/2014	0.75 and 375 kW	IE3 + IE2 for inverter duty
Australia	GEMS Act of 2019	IEC 60034-30-1	0.73 up to <185 kW	IE2
New Zealand	GEMS Act of 2019	IEC 60034-30-1	0.73 up to <185 kW	IE2
Saudi Arabia	RTE INEN 14 BOD (Board of Directors) MEETING N° 1635	SASO 2893:2018	0.75 up to 375 kW	IE3
India	The Gazette of India S.O.178	IS 12615:2018	0.12 up to 1,000 kW	IE2
Japan	Energy Saving Act/Top Runner Program	JIS C 4034-30	0.75 up to 375 kW	IE3
South Korea	MKE-2017-206	KS C IEC 60034	0.75 up to 375 kW	IE3
Singapore	Energy Conservation Act (Cap. 92C)	IEC 60034-2-1	0.75 up to 375 kW	IE3

(*Continued*)

Table 2.2 MEPS-worldwide requirements for electric motors. (*Continued*)

Country or region	Standard	Regulation	Power range	MEPS level
China	Draft CEL 007:202	GB18613-2020	0.12 up to 1,000 kW	GB3 (IE3) (From 6/21 onwards)
Taiwan	Efficiency Standard and Benchmarks and BSMI Regulatory Inspection	CNS 14400	0.75 and 200 kW	IE3

Many developed countries around the world have tightened their MEPS levels to IE3 levels. MEPS for low-voltage motors have been critical in assisting countries in attaining their energy efficiency and CO_2 emission targets. MEPS have resulted in an overall improvement in motor efficiency and have made it easy to compare efficiency levels among manufacturers for motor users. The MEPS levels for electric motors in different countries and regions are listed in Table 2.2 [17, 18].

Few countries use their own testing methodologies while also referring to IEC 60034-2-1. MEPS for motors have now been set to the IE2 or IE3 level in nations that consume 76% of worldwide power generation. While noting that MEPS might vary based on market conditions in different countries, this is a substantial level of global synchronization for a universally traded product.

Some countries may become the target for low-quality motors that are not allowed anyplace else due to a lack of well-designed and implemented standards. The increased emphasis on green technology adoption is having a significant impact on global demand for EEMs. Consumer preferences are being influenced by the advantages that EEMs offer over regular motors, such as high-quality lamination, insulation, and minimal fan losses. Apart from that, environmental laws and regulations and nonrenewable energy conservation requirements, such as the Minimum Energy Performance Standards (MEPS), are driving up demand for EEMs around the world [19].

2.4 Government Initiatives to Improve the Energy Efficiency of Electric Motors

The Energy Conservation Act 2001 (EC Act) was enacted by the Indian government in August 2001, and Bureau of Energy Efficiency (BEE), a legislative organization under the Ministry of Power, was founded in 2002 to execute the EC Act. One of the key program areas for enhancing energy efficiency in the residential, commercial, and industrial segments identified in the act is Standards and Labeling (S&L). BEE started the S&L program in May 2006, and it now covers 26 product categories, with 10 of them being obligatory. In 2010, the BEE launched a voluntary labeling program for 3ϕ asynchronous motors.

BEE intends to update the existing 3ϕ asynchronous motor labeling program to reflect the most recent version of Indian and/or international standards, as well as migrate from the current optional to mandatory phase. In addition, BEE intends to broaden the program's scope to include single-phase induction motors in the S&L initiative. Energy Efficiency Services Limited (EESL), an agency, was established as part of the government's efforts to improve energy efficiency [20, 21].

One of the primary initiatives developed by EESL is the National Motor Replacement Program (NMRP). Consumers in India buy and use motors with lesser efficiency (IE1 or less) than those available on the universal market. The IE1 and below IE1 categories are expected to account for 90%–95% of the existing installed shares of motors. According to EESL, the market for EEM is expected to reach 10 million units by 2023–2024. Until 2023–2024, the investment potential for EEM is expected to be ₹ 125.34 billion [22].

2.4.1 National Motor Replacement Program

The NMRP is being carried out by the Institute for Sustainable Communities (ISC), a non-profit organization based in the United States, and EESL, a joint venture corporation under India's Ministry of Power. The ISC has been a key partner of EESL since the program's conception and implementation, and the two organizations are currently working together in Asia on the Energy Efficiency Alliance for Industry (E2 Alliance) Project [23].

The NMRP was created by EESL to encourage the usage of HEMs that complies with the IE3 standard. The motor replacement program includes technical criteria that address important consumer concerns, such as high starting costs, high expenditures for operations/maintenance and product quality. The NMRP's main goal is to make it easier and faster for the industry to adopt high-efficiency IE3 motors [24, 25].

Since its beginning, the E2 Alliance has facilitated 6,520 motor substitutions and developed a pipeline of 48,818 motor substitutions, according to a BW Online Bureau article dated April 7, 2021. The Alliance hopes to save 66.5 million kWh per year, or USD 62 million, by replacing old inefficient motors. It is expected that the replacement program will help in the development of additional qualified jobs in conceptual, financial, production, distribution, installation, after-sales, recycling, and other areas. To date, E2 Alliance has reduced greenhouse gas (GHG) emissions by 54,535 $MTCO_2$. The desired annual reduction is 4,62,864 $MTCO_2$.

2.4.2 Obstacles to Overcome and the Path Forward

Due to a variety of factors, the demand for IE2 and higher motors has been significantly increasing. To begin, raise awareness about the advantages of adopting HEMs in industries for energy and cost savings. Second, efforts made by bodies in charge of setting standards and policies, such as the BIS and the BEE, to mitigate climate change and conserve energy. BEE's very successful Perform Achieve and Trade (PAT) plan has forced many energy-intensive industries to adopt energy-efficient technologies more quickly in order to minimize their energy consumption and carbon emissions. Last year, the penetration of IE2 and higher motors was close to 65%, compared to only 15% in 2016 [26, 27].

Despite knowing about the advancement of HEMs, the industry has been slow to replace old motors with IE3 motors due to a number of obstacles. These barriers can be found at any point in the system, including the buyer, consultant, operations and maintenance employees, and suppliers. Buyers must be educated about the power and its associated cost savings, investment repayment period, excellent dependability and very low maintenance offered by IE3 motors. In India, the supply of IE3 motors has been patchy, owing to a combination of reduced demand and the high cost of setting up test and design facilities, both of which are too expensive for the micro, small, and medium-sized company (MSME) sector. As a result, it is critical to understand the constraints faced by motor manufacturers, particularly MSMEs, and how they can help accelerate the adoption of IE3 motors so that IE3 can become the MEPS in 2–3 years [28].

2.5 Conclusion

Power and energy sources include energy conservation as an important component. Indian government mandated the use of IE2 class motors beginning in October 2017, with the goal of exchanging old and inefficient motors with IE3 class EEM. However, there is still a large stock of low-efficiency motors with IE1 efficiency ratings or lower in use in Indian industries. India, as a developing country, is lifting the bar to IE3 (as worldwide markets are presently at IE3) through several legislations, with the goal of attaining IE4 by 2023. Energy efficiency is linked to a reduction in GHG emissions at both the global and national levels. As a result, energy efficiency has a lot to offer in terms of reducing energy consumption in motor and motor-driven systems. This will not only increase industry productivity but will also save energy and reduce GHG emissions.

References

1. *India Energy Outlook 2021*-International Energy Agency, France, 2021, https://iea.blob.core. windows.net/assets/1de6d91e-e23f-4e02-b1fb51fdd6283b22/India_Energy_Outlook_ 2021.pdf.
2. Business Standrad, COVID Impact: India Enters FY22 With Power Demand At Mid-2019 Level. Business Standard, New Delhi, 2020. https://www.business-standard.com/article/economy-policy/india-enters-new-fiscal-with-negative-power-demand-robust-growth-missing-121040100418_1.html.
3. Impact of lockdown on India's electricity sector. *Energya*, Energya, a study about climate change, energy and adapation, European Union 2020. https://www.energy-a.eu/impact-of-ongoing-lockdown-on-indias-electricity-sector-an-overview.
4. Beyer, R.C.M., Franco-Bedoya, S., Galdo., V., *Examining the Economic Impact of COVID-19 in India through Daily Electricity Consumption and Nighttime Light Intensity. Policy Research Working Paper; No. 9291. World Bank*, © World Bank, Washington, D. C, 2020, https://openknowledge.worldbank.org/handle/10986/33986.
5. Aruga, K., Islam, M.M., Jannat, A., Effects of COVID-19 on Indian Energy Consumption, *Sustainability*. MDPI, Open Access Journal. 12, 14, 1–15, 2020.
6. Pandey, V., Karthik, M., Makwana, M., Batra, P., Impact of Covid-19 on the Regional Power Demand and Generation in India. *IRADe*, Integrated Research and Action for Development, New Delhi, 2020. https://irade.org/Impact%20of%20Covid%2019%20on%20the%20Regional%20power%20power%20Grids%20in%20India.pdf.
7. Vasudha Foundation, *India's Power Outlook*, MDPI, Switzerland, 2020, https://www.vasudha-foundation.org/2020-indias-power-outlook-volume-1.

8. IEEMA, *The demand for electric motors is rapidly shifting from standard efficiency electric motors to high and premium efficiency motors*, Indian Electrical & Electronics Manufacturers Association, Mumbai, 2020, https://ieema.ore/division/rotating-machines/.
9. *Energy Efficient motor standard*, International Copper Association, India, Mumbai, 2019, https://docplayer.net/119085071-Energy-efficient-motor-standard-is-12615-2018.html.
10. *Energy Efficient Motors*, IIT, Kolkatta, https://www.iitk.ac.in/npsc/Papers/NPSC1998/p96.pdf.
11. *IE2 series TEFC SCR motors*, Bharat Bijlee, Mumbai, https://www.bharatbijlee.com/assets/downloads/prod/IE2motors.pdf.
12. Indian Standard, *Energy Efficient Induction Motors: THREE Phase Squirrel cage (Second Revision)*, Bharat Bijlee, Mumbai, 2011, https://law.resource.org/pub/in/bis/S05/is.12615.2011.pdf.
13. Havells, *IE2 & IE3 Energy Efficient 3 Phase LV Induction Motors*, Haveels, Noida, https://www.havells.com/content/dam/havells/brouchers/motors.Prima%20Series%20Motors/ Prima%20series%20motor%20Catalouge.pdf.
14. *Product Manual for line operated Three phase AC motors (IE Code), Efficiency classes and Performance specification, according to IS 12615:2018*, Bureau of Indian standards, New Delhi, 2019, https://bis.gov.in/wp-content/uploads/2019/05/PM_IS-12615_2.pdf.
15. National motor replacement programme India case study, http://united4efficincy.org/wp-content/uploads/2020/11/U4E_case-study_India-motor_2020-11-26.pdf, 2020
16. *Energy Efficiency classes for IEC motors*, Menzel Elektromotoren, Berlin, https://www.menzel-motors.com/iec-motors/.
17. *IEC-Electric motors-Increasing the Efficiency of Electric motors-One of the biggest energy efficiency opportunities*, International Electro Technical Commission, Switzerland, 2021, https://www.iec.ch/government-regulators/electric-motors.
18. *Examples by industry sector electric motors*, Peak Resources, Western Australia, 2018, https://www.peakresources.com.au/news/examples-by-industry-sector-electric-motors/.
19. UN Environment—Global Environment Facility, United for Efficiency (U4E), *Accelerating the Global Adoption of Energy-Efficient Electric Motors and Motor Systems, U4E Policy Guide Series*, UN Environment, Paris, France, 2017, http://www.united4efficiency.org.
20. CLASP, *Support for Development of Efficiency Policy for Induction Motors*, Collaborative Labeling and Appliance Standards Program, Washington, DC, USA https://www.clasp.ngo/rfps/support-for-development-of-efficiency-policy-for-induction-motors/.
21. ABB, India Bans Low Efficiency Motors to Save Energy and Cut Emissions. ABB, Mumbai https://new.abb.com/motors-generators/iec-low-

voltage-motors/articles/india-bans-low-efficiency-motors-to-save-energy-and-cut-emissions.
22. Power line Magazine, *Driving Efficiency-Reducing motor-driven electricity consumption for industries*, PLM, New Delhi, July 2020, https://powerline.net.in/2020/07/10/driving-efficiency/.
23. BW SMART CITIES, ISC & EESL partnership plans to save 66.5 MkWh of Energy, USD 7.3 million in a year on National Motor Replacement Program, Business world, Mumbai, 2021. http://bwsmartcities.businessworld.in/article/ISC-EESL-partnership-plans-to-save-66-5-MkWh-of-Energy-USD-7-3-million-in-a-year-on-National-Motor-Replacement-Program/07-04-2021-385792/.
24. Energy Efficiency Services Limited, A JV of PSUs Under Ministry of Power, *National Motor Replacement Program*, Energy Efficiency Services Limited, New Delhi, https://eeslindia.org/en/nmrp/.
25. *ISC Update: Motor Replacement Pilot Project in India*, Institute for Sustainable Communities, Montpelier, VT 05602, 2017, https://sustain.org/impact-update-10-2017/.
26. T&D India, T&D India, Mumbai, 2018. https://www.tndindia.com/ie-3-motors-can-help-build-low-carbon-future/.
27. Electrical India, Energy Efficiency in Motors and India's status. *Eng. Rev.*, Electrical India, Navi Mumbai, 2019. https://www.electricalindia.in/energy-efficiency-in-motors-and-indias-status/.
28. Use Of Energy Efficient Motors: The Need of Industry, ABB, Mumbai, 2019. https://engmag.in/use-of-energy-efficient-motors-the-need-towards-a-sustainable-and-resourceful-country/.

3

Smart Power Tracking and Power Factor Correction in a PV System

Karthika J.*, Santhosh B., Vallinayagam K., Thennavan S. and Narendran R.K.

Department of Electrical and Electronics Engineering, Sri Krishna College of Engineering and Technology, Coimbatore, India

Abstract

Design and simulation of a capable charger controller for a photovoltaic system are offered here. To control maximum power point tracking (MPPT) feature, the P&O (perturb and observe) algorithm is used as it offers a good MPPT efficiency and high accuracy and can even be applied in cheap digital devices. This offshoot is fed to the grid where the automatic power factor controller is designed to monitor the functioning of the system and to improve the power factor automatically whenever the power factor falls below the desired level.

Keywords: Photovoltaic, MPPT, power factor, capacitor bank, PV panel, cost effective

3.1 Introduction

The International Energy Agencies around the globe estimated that, by the year 2050, the renewable solar and wind energies would occupy a humongous share in the energy consumption globally, and many developed and developing nations have started installing momentous solar power capacity to supplement the traditional sources.

**Corresponding author*: karthika@skcet.ac.in

A. Chitra, V. Indragandhi, and W. Razia Sultana (eds.) Smart Grids and Green Energy Systems, (35–46)

Consequently, the number of domiciles being powered by solar energy, connected to the regional grid, is ascending day by day.

Such photovoltaic (PV) systems connected to the grids as shown in Figure 3.1 are incorporated with the solar panels that furnish most of the energy needs over the course of the day while being connected with the regional grid in the dark too.

Sometimes, more power is obtained than the predetermined output. At times like that, the profuse electrical energy is stored in the batteries or in the grid-connected PV setup which is fed back to the grid system. Depending upon the load value, the value of c varies accordingly.

The DC-DC-AC inverter design, which is also termed as the dual stage power conversion, makes up the solar PV system. To augment the voltage gain, stability, and design adaptability, a DC-DC converter is sandwiched, connecting the PV panel and the DC-AC converter.

A P-N junction fabricated semiconductor makes up the PV cell, which collectively constitutes the PV module. The yield of the band energy gap of the power-voltage and the current-voltage brings about the electron-hole pairs that fabricate current corresponding to the irradiation. Here, ambient conditions such as temperature, irradiance, and air mass are taken up as inputs, and power/voltage is given as output. As varying inputs has an influence in the output, maximum power point tracking (MPPT) is brought into play for drawing out the maximum power under variegated inputs.

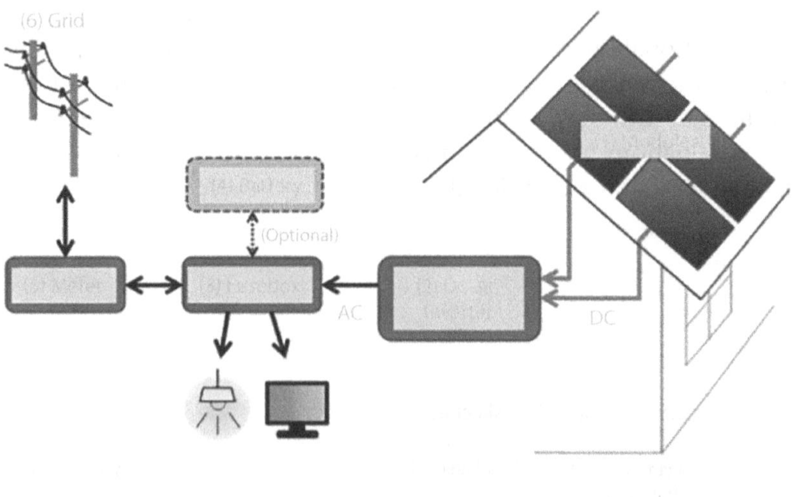

3.2 Literature Review

Characterization was once structured on calculations; several methods to observe MPP (maximum power point) for PV well-known shows have been examined by Abeyasekera *et al.*, [1]. A high-return solar pulse charger (PV-PC) for lead-acid battery (LAB) directed by using skill of a power increase that supported steady conductance MPP following (PIINCMPPT) was designed by means of Akorede *et al.*, [2]. The disadvantages of perturb and observe (P&O) technique have been examined, and it shows that the incremental conductance calculation has effectively followed the MPOP in any event, when air conditions change fastly. The work was completed through way of together reproduction and charts by Chen-Chi Chu *et al.*, [3]. Another strategy for MPPT called CVT (Constant Voltage Tracking) is introduced which dissected the trademark bend and actionable hypothesis of PV cluster. It includes a lift converter for elite SPTLE [4]. Mixture of obligation part for DC-DC support converter is executed on PIC16F876A RISC-microcontroller by Reetam Sen Biswas *et al.*, [5] A flufly-based irritate and notice (P&O) MPPT in sun-based board was introduced. Utilization of association dynamic channel is a convenient way to deal with Power Factor Correction (PFC) proved by Rifat Shahriar *et al.*, [6]. Contrasting and the well-known PFC, the projected strength aspect correction has lesser requirements of pressure machine appraisals, which prompts decrease cost, greater productivity, and decrease electromagnetic impedance as in Renuga Verayiah *et al.*, [7]. It likewise can dispose of the cumbersome inductor required in established PFC. When the records cutting-edge waveform is selected appropriately, electricity aspect limit leads to a limit in channel capacitor variation (and as a consequence to diminished framework cost), while assembly energy satisfactory prerequisites. Looking at that, the selection of waveform is free of the precise pressure transformation geography to be utilized. It applies in a similar way to support, flyback, buck, and one-of-a-kind geographies that have been utilized in energy issue correction applications.

3.3 Smart Power Tracking

As we have discussed, the first portion of our project is getting the maximum possible source of energy, and MPPT is one that we believe could shower what we expect, at its best [8]. MPPT is use to charge the use of controllers for the PV systems to obtain the utmost power output. Today, solar energy is believed to conquer a larger portion of the energy market in the upcoming years than other sources of renewable energy, so it is significant to obtain the utmost possible output from the solar panels and one such efficient tool is implementing MPPT. It improves the power output even when sunlight is less than ideal sunlight conditions. This algorithm makes sure the system continues to maneuver at the supreme power point even under diverse conditions like fluctuating solar irradiance, temperature, and the load by controlling the voltage. These are prototypical algorithms used for the controller designs of PV systems. The contemporary MPPTs that are equipped with inverters exist which serve threefold functions: the grid-connecting the solar power and ramify power for battery charging.

3.4 Perturb and Observe

The P&O mechanism, which is otherwise known as perturbation mechanism, is one of the few types of MPPT algorithms. As the name suggests, Figure 3.2 shows the concept behind this P&O process is tweaking the operational voltage or the current of the PV system based on the scrutinized power until a maximum possible output is obtained. This method is commonly pointed out as the hill-climbing method of the P-V characteristics of a PV solar systems. The curve upsurges on the left part of the MPP

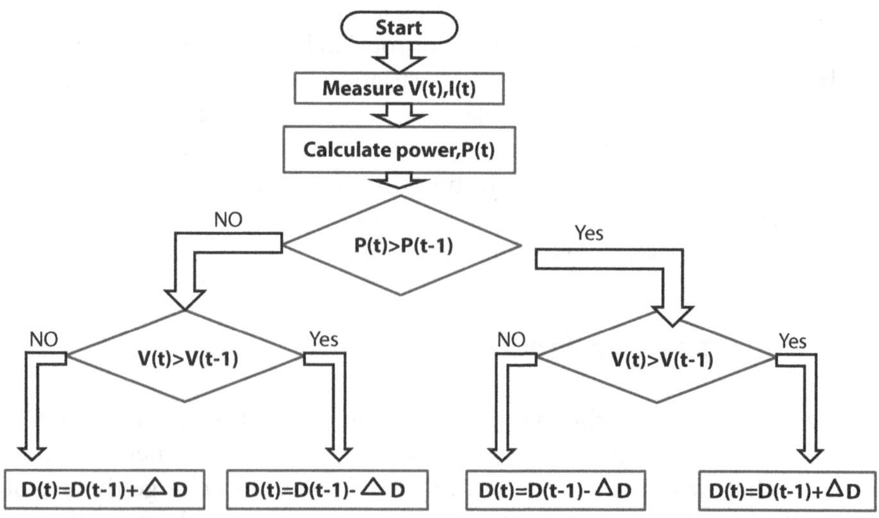

Figure 3.2 P-V characteristics for perturb and observe algorithm.

(dP/dV> 0), whereas the curve plunges on the right part of MPP (dP/dV< 0). Considering the ease of accomplishing it, the Perturb and Observation methodology is the most frequently used MPPT technique. Assuming that an appropriate and an accurate hill climbing strategy is backed, this method is something that might induce a very highly esteemed efficiency.

This is termed as the perturbation method as the power is varied by perturbing it. If the perturbation applied in a particular direction shoots the power up, then the perturbation is carry on with the same path. The power instantaneously falls after the maximum power is achieved and thus the perturbation is reversed. In addition, this oscillates around the maximum power once the steady state is achieved as given in Figure 3.3. The perturbation amount is kept small to maintain only a small variation of power.

The maximum power from the PV panels can be obtained by changeable the duty cycle of DC-DC converter and that is what an MPPT controller does. The converter is produced by combine a buck-boost dc/dc converter with a class-D series resonant paralleled loaded inverter. The impedance matching between the load and the PV system is achieved using the DC-DC boost converter.

In addition, now, after obtaining the maximum possible solar output from the PV panels, this colossal output could be fed to the next portion where the PFC could be streamlined.

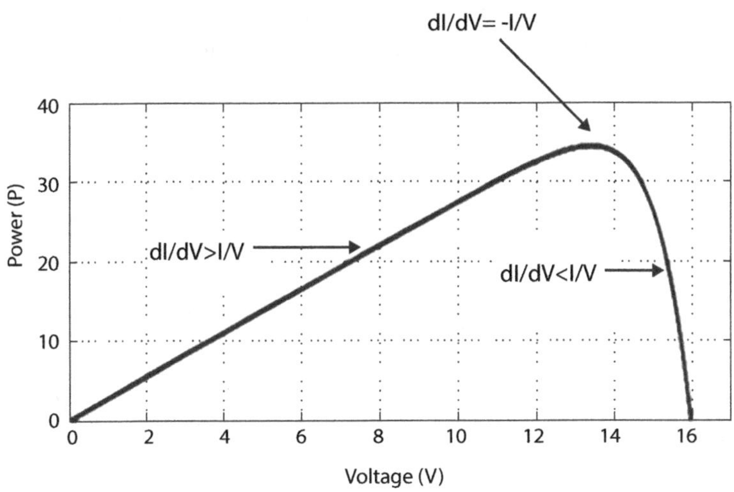

Figure 3.3 Flow chart of perturb and observation algorithm.

3.5 Need for Power Factor Correction

PFC is the technique of compensation of the lagging current via inducing a leading current by means of including capacitors to the supply. Power factor (PF) is termed as the ratio of the active power to the apparent power. The term apparent power can be explicated as the vector sum of active power and reactive power. Erratic current transferring returned and forth outcomes in reactive power that is set down in VA/VAR.

PF is also the term used in place of a move among the voltage and the current waveforms. The angular difference is directly proportional to reactive power as shown in Figure 3.4.

3.6 Correction Method

The major reason of the PFC is to make sure the load behaves like pure resistance if both the load voltage and the current can be sinusoidal. Correction of PF is as simple as wiring a few capacitors but, sometimes, an arduous one.

3.7 Capacitive Bank

A capacitive bank is the amalgamation of a number of capacitors that are connected either in parallel or series as shown in Figure 3.5. When there is a presence of inductive load, the PF capacitor bank can be switched on and can be switched off in case of no load. Whenever the PF falls because of the

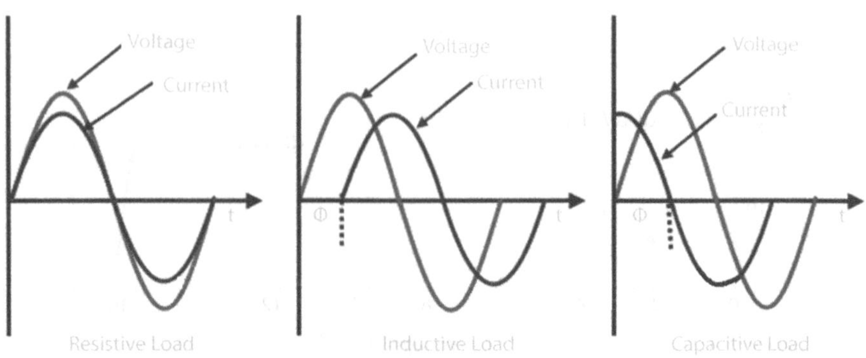

inductive loads it can be rectified by using the capacitors. The phase difference is laid between the current and the voltage by generating a magnetic field which is necessary for inductive load. However, a leading current can be brought forth by using the capacitors which eventually corrects the PF. These capacitors benefit us by reducing the burden and also it does not

Photo courtesy of Philippe Mertens

Figure 3.5 Capacitor bank.

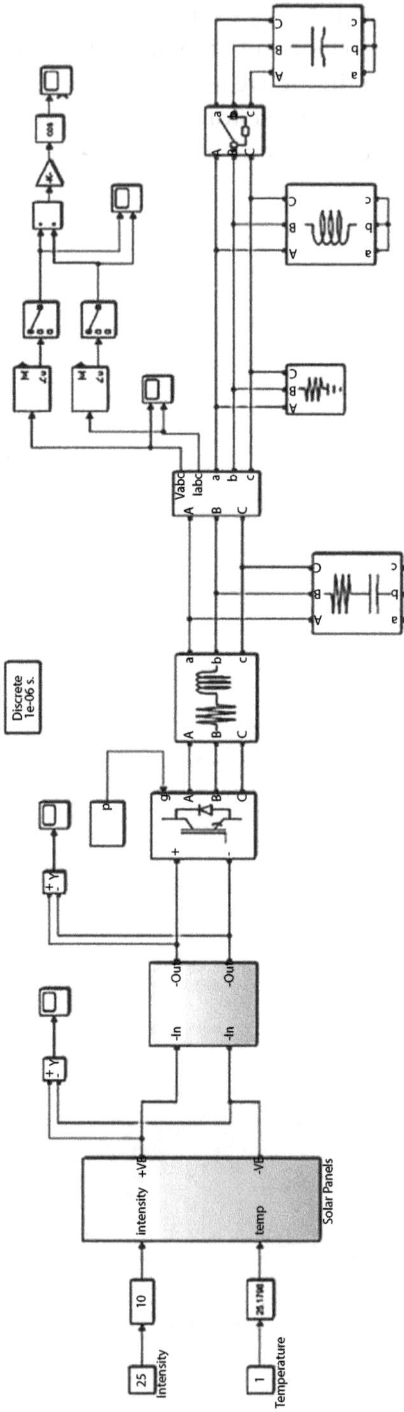

Figure 3.6 Simulation.

affect the load. Using these capacitors helps to reduce not only the losses but also, consequently, the electric bill of ours.

3.8 Simulation

Here, using Mat LAB, initially, a maximum power is generated using MPPT as used by Omar *et al.*, [9]. An MPPT controller with P&O algorithm is connected with the solar panels to obtain the utmost possible solar energy. The MPP controller is connected to the DC-DC converter, and the output acquired from the DC-DC converter is connected to the three-phase pulse generating inverter.

The generated DC voltage from the MPPT is given to the three-phase pulse generating inverter. The inverter is then purified by the L-C filters [10] and connected with the load in series as given in Figure 3.6. The PF starts falling below 0.8 when connected to the load. By introducing the capacitor bank, the PF becomes 1 in about 0.5 s and that is all about our project.

3.9 Result and Output

Figure 3.7 shows that the output voltage of the solar panel remains constant.

Figure 3.8 show that the yield of MPPT utilizing MPP is a result carried out in PV inverters to frequently change the impedance seen by the

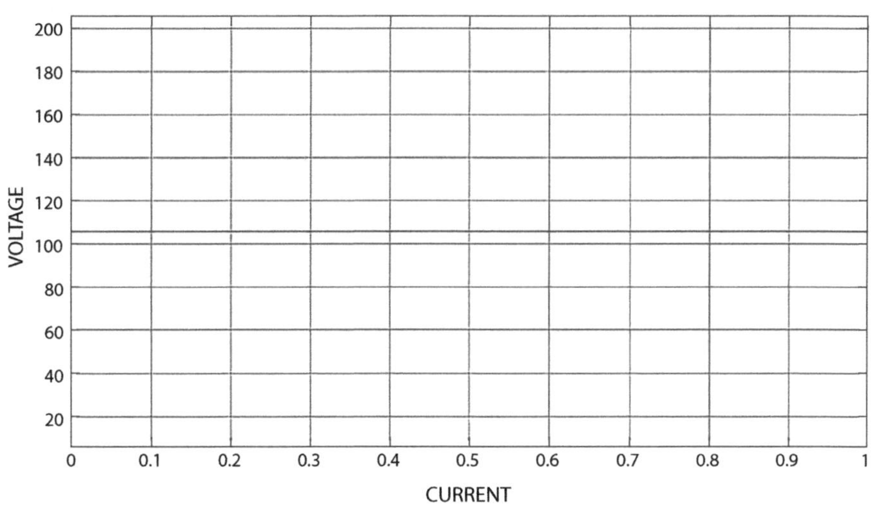

Figure 3.7 Output of solar panel.

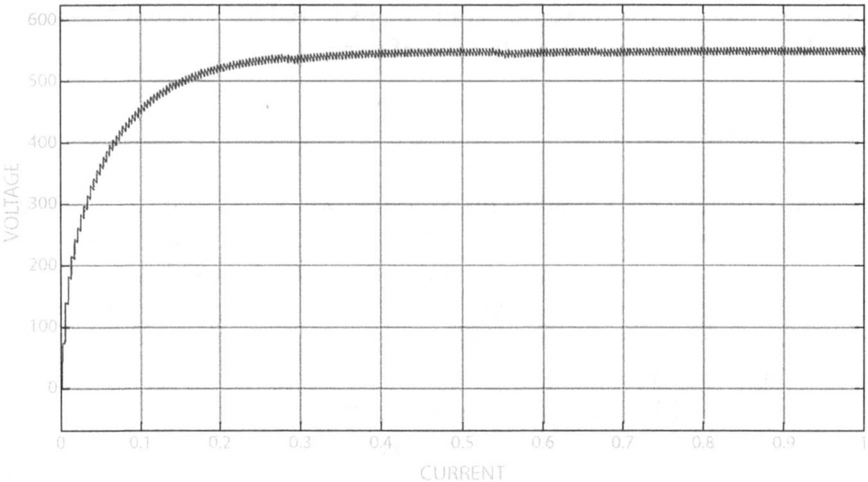
600
500
400
300
200
100
0
VOLTAGE
0
0.1
0.2
0.3
0.4
0.5
0.6
0.7
0.8
0.9
1
CURRENT

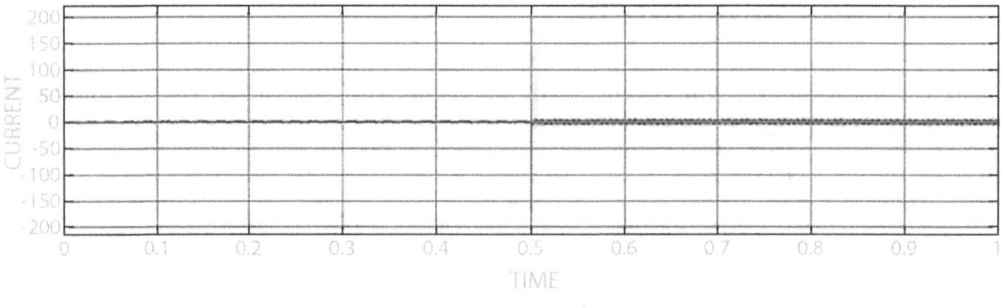
200
150
100
50
0
-50
-100
-150
-200
CURRENT
0
0.1
0.2
0.3
0.4
0.5
0.6
0.7
0.8
0.9
1
TIME

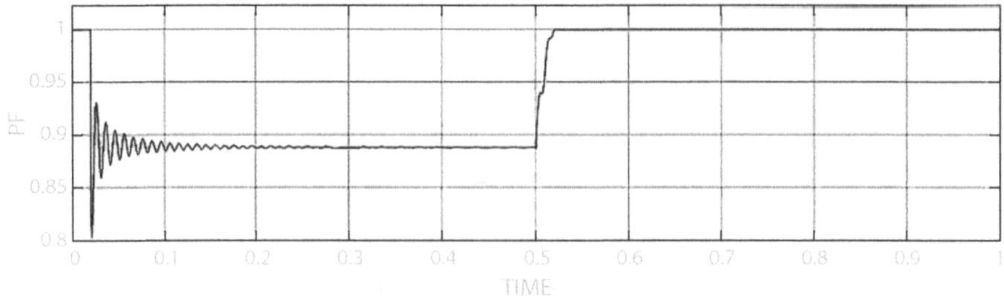
1
0.95
0.9
0.85
0.8
PF
0
0.1
0.2
0.3
0.4
0.5
0.6
0.7
0.8
0.9
1
TIME

sun-based cluster to keep the PV framework working at, or near, the peak power point of the PV board under fluctuating conditions, for instance, changing sunlight-based temperature, irradiance, and load.

Figure 3.9 shows the output of inverter by generating the DC voltage from the MPPT, given to the three-phase inverter. The three-phase inverter is getting purified by L-C filters.

Figure 3.10 shows output of PFC. On connecting the load with filters, PF is lagging below 0.8. Hence, we are connecting capacitor bank with the load, which makes the PF unity after 0.5 s.

3.10 Conclusion

MPPT algorithms are proven to be one of the most productive tools when it comes to electrical efficiency. Analyzing such algorithms of MPPT, we finally applied the cost-effective P&O design to get the maximum possible energy, and then, an automated capacitor bank is established to correct the PF by remunerating the reacting elements; whenever it falls below the level, we crave to improve the effectiveness of the system and to reduce the current bill which gives us a great sigh of relief.

References

1. Abeyasekera, T., Johnson, C.M., Atkinson, D., Amstrong, M., Suppression of line voltage related distortion in current controlled grid connected inverters. *IEEE Trans. Power Electron.*, 20, 1393–401, 2005.
2. Akorede, M.F., Hizam, H., Pouresmail, E., Distributed energy resources and benefits to environment. *Renew. Sust. Energ. Rev.*, 14, 724–734, 2010.
3. Chu, D.-C., Chieh-Li, C., Robust maximum power point tracking method for photovoltaic cells: A sliding mode control approach. *Solar Energy,* 83, 8, 1370–1378.
4. Enrique, J.M., Duran, E., Sidrach-de-Cardona, M., Andujar, J.M., Theoretical assessment of the maximum power point tracking efficiency of photovoltaic facilities with different converter topologies. 81, 1, 31–38, 2007.
5. Biswas., R.S., Mal, S., Automatic power factor improvement using microcontroller, in: *2015 International Conference and Workshop on Computing and Communication (IEMCON),* 2015. DOI: 10.1109/IEMCON.2015.7344500.
6. Shahriar, R., Alamgir, N., Hwang, K., Chong, U., A PWM-based scheme for power factor correction, in: *Proceedings of 2011 6th International Forum on Strategic Technology*, 2011. DOI: 10.1109/IFOST.2011.6021097.

7. Verayiah, R., Iyadurai, A., A comparison study on types of PV for grid connected photovoltaic power. *Indonesian Journal of Electrical Engineering and Computer Science (IJEECS)*, 6, 2, 349–356, 2017.
8. Abd Allah, B. and Djamel, L., Control of power and voltage of solar grid connected. *International Journal of Electrical and Computer Engineering (IJECE)*,. 6, 1, 26–33, 2016. ISSN: 2088-8708.
9. Benaissa, O.M., Hadjeri, S., Zidi, S.A., Modeling and simulation of grid connected PV generation system using Matlab/Simulink. *International Journal of Power Electronics and Drive System (IJPEDS)*, 8, 1, 392–401, 2017. ISSN: 2088-8694.
10. Zhiguo Pan, F.Z. Peng, S.W., Power factor correction using a series active filter. 20, 1, 148–153, 2005.

4

Grid Connected Inverter for PV System Using Fuzzy Logic Controller

Elam Cheren S.*, Sakthi Ganesh R., Vijay K., Surya V. and Venkatesha R.

Department of EEE, Sri Krishna College of Engineering and Technology, Coimbatore, India

Abstract
Solar photovoltaic (PV) system has widespread application within the grid connected system. This proposed system presents an optimization technique for the energy extraction in an exceedingly PV facility for grid connected PV system using fuzzy logic controller. The greatest force from a PV module will essentially rely upon change in temperature and radiation from the sun. To boost effectiveness PV frameworks, it utilizes a maximum power point tracking (MPPT) technique that extricates the maximum power generated by solar panel integrated with battery supplying load. The MPPT finds and keeps up activities at the most extreme outlet, utilizing a different tracking algorithm during change in climatic conditions. This work focuses on fuzzy-based control of MPPT technique for interlinking calculation with the grid. For this type of non-linear applications, fuzzy-based controller provides a better result rather than conventional one. In order to accomplish this concept, a PV module comprises MPPT framework, design of DC-DC converter, and fuzzy-based controller and simulated in MATLAB.

Keywords: Photovoltaic system, MPPT, fuzzy controller, grid connected system

4.1 Introduction

This chapter describes the grid connected inverters and how photovoltaic (PV) systems can be adopted with fuzzy logic control. Grid

**Corresponding author*: elamcheren@gmail.com

A. Chitra, V. Indragandhi, and W. Razia Sultana (eds.) Smart Grids and Green Energy Systems, (47–58)

connected technologies are discussed [2, 3]. The importance of solar traits in relations to temperature and irradiance and the manner the electric circuit voltage is designed are depicted inside the chapter [1]. Standards to style and setup practices of PV-grid related structures mentioned during this chapter play the various position on the cause of common coupling [4]. These standards helped within the development of the proposed PV system. This work describes the modeling and execution of MPPT technique build by fuzzy controller integrated with grid connected inverter for solar PV source [5]. The idea of executing the MPPT algorithm in fuzzy-based controller is arrived because of limitations of conventional controlling methods. There is an undeniable expansion in energy sector because of expansion in modern turn of events and increase in utilization. This situation excite interest in finding innovative and cost effective solution through research and related with the advancement of energy proficiency. In addition, the power generation should be feasible and from environmentally friendly power sources [6, 7]. During this indistinguishable time, the role of petroleum derivatives for the age of force is diminishing and furthermore getting costly.

To create and increase maximum energy factor monitoring and to manipulate a multi-supply fuel age frame work comprising of PV modules, wind mills and exclusive reasserts are the precept objective. PV force could be a sustainable power source that has as of late started huge interest and ought to inside the not so distant future supplants non-inexhaustible sources like petroleum derivatives. Be that as it may, to accomplish this alternative, we need to consider the cost of each unit of PV power generation with fuel sources [9, 10]. The raw materials utilized in the construction of PV cells and, furthermore, the innovation to be adopted in orchestrating the solar-based cells to make a solar module are viewed as main aspect that influence the productivity of solar module. On present scenario, PV modules have efficiency around 12%–26% to convert solar irradiance in to electrical energy. Higher range of efficiency can be achieved in Ga-As–based solar cells. Silicon sun-powered cells generate power with 12% to 14% efficiency [11]. Besides, productivity may drop because of burden situations, temperature of PV module, or diminishing of solar-based protection. To extract the most extreme evaluated energy from a PV source, it is important to operate in its ideal plug [12]. To solve this, a regulator called maximum power point tracker is needed. Accordingly, the capacity of the Maximum power point tracker is to make amend for the changing the electrical parameter of the cell. The MPPT technique improves the yield of PV module and provides the working values, which provides the extreme power from the PV module.

4.2 Methodology

The proposed approach exhibits the performance attributes of maximum power point tracking (MPPT), which is improved regarding motions about the MPP, speed, and affectability to parameter changes. Based on MATLAB simulation circuit for fuzzy logic controller, FIS code was modified and uploaded into a Simulink. The consequences display that lots of extra energy may be derived from a PV source by a fuzzy-based MPPT tracker as compared with different most energy tracker algorithm. The methodology of the algorithm is shown in Figure 4.1.

4.3 PV Module

PV energy is obtained as the outcome of immediate change of optical energy into electricity. The change is accomplished by means of semiconductor-based components referred as PV cells, which are likewise named as solar cells or PV cells. These cells are essentially flat and highly optical-sensitive diodes made of the similar materials which were utilized in semiconductors central processors, and related innovation. The working methodology of PV cell is as follows: The photons transfer their energy to these semiconductor diodes and in this manner it will create heat through recombining with the decidedly charged framed by the photons. A field will be set up between the intersections of a

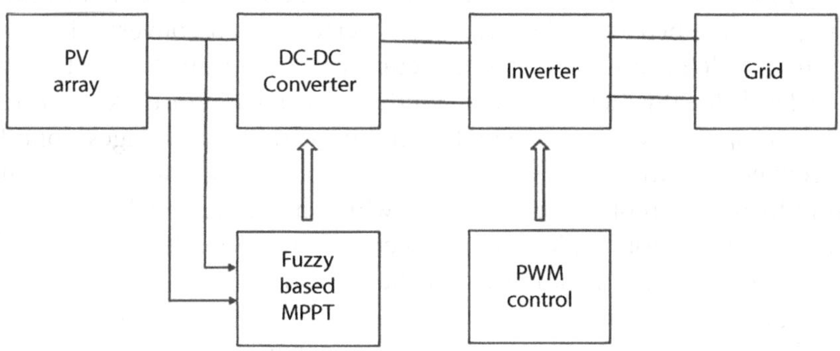

Figure 4.1 Block diagram.

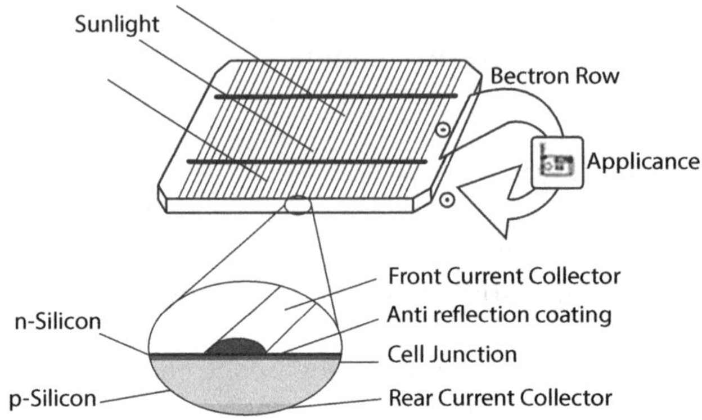

Figure 4.2 Solar PV module.

semiconductor. At the point when the electric field aligned the excited electrons aside of the PV cell, a distinction in voltage is in this manner shaped between the inverse A PV cells creates power as immediate flow. The electrical force made is restricted by the measure of got light irradiance, similar to the temperature of cells and the association of PV cells. The solar PV module is shown in Figure 4.2.

4.4 DC-DC Converter

Because of PV framework's nonlinearity nature, the yield power is variable steady with respect to difference in atmospheric conditions. Thus, the least simple apparatus can perform the work of enhancing the yield of PV module potential and current is named dc-dc converter. The output voltage is obtained with the help of converter, which converts the given DC input voltage, however, at a particular potential value higher than the input level. Ideally, here, the semiconductor capacities are like a control switch to apply the control signal, thereby making the changes done for the converter with low losses. At the point of reducing losses, capacitor and inductor are replaced by resistors, which has extremely low losses. In addition, several topologies handle electrical components effectively. The different topologies including boost, buck, and buck-boost are available with which it is possible to increase or decrease the voltage level. With the buck-boost topology, both step up and step down of potential are possible in single circuit.

4.5 MPPT

The power generated by the PV module shows severe non-linearity due to several conditions like variation in solar irradiance and ambient temperature, and the generation of maximum power under several climatic conditions is referred as a MPPT, which has greatest voltage and extreme output current. To achieve this, electrical tracking mechanism called MPPT is concocted and utilized. It advises that there should be single optimum voltage at PV terminal to perform in every condition. The DC-DC converter is shown in Figure 4.3.

A MPPT is an electronic framework that shifts the modules, which has electrical operating point by which the modules can convey extreme power.

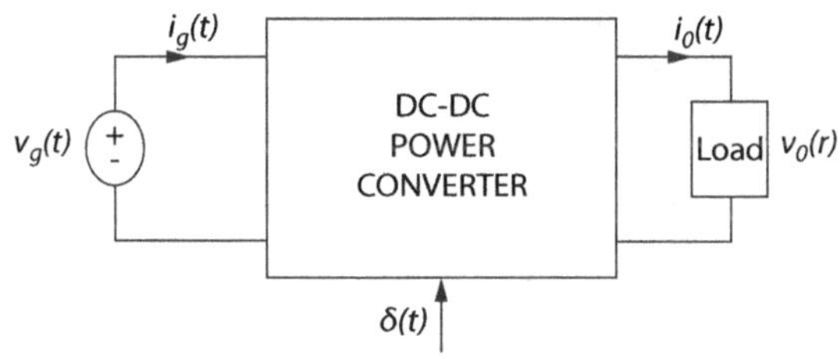

Figure 4.3 Structure of DC-DC converter.

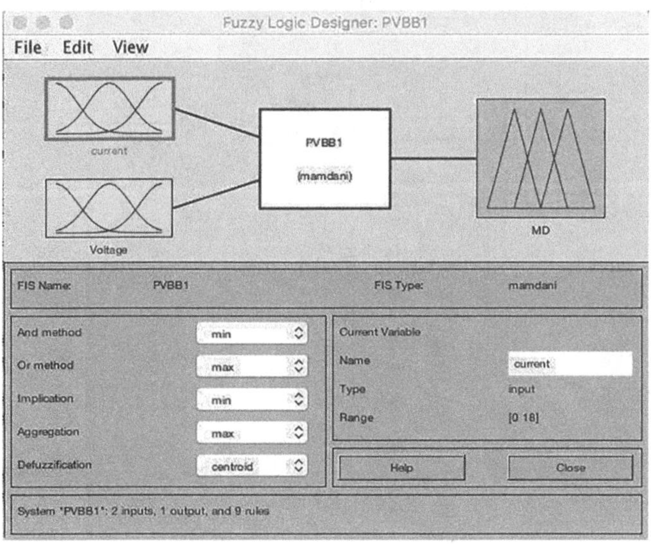

Figure 4.4 Fuzzy logic designer.

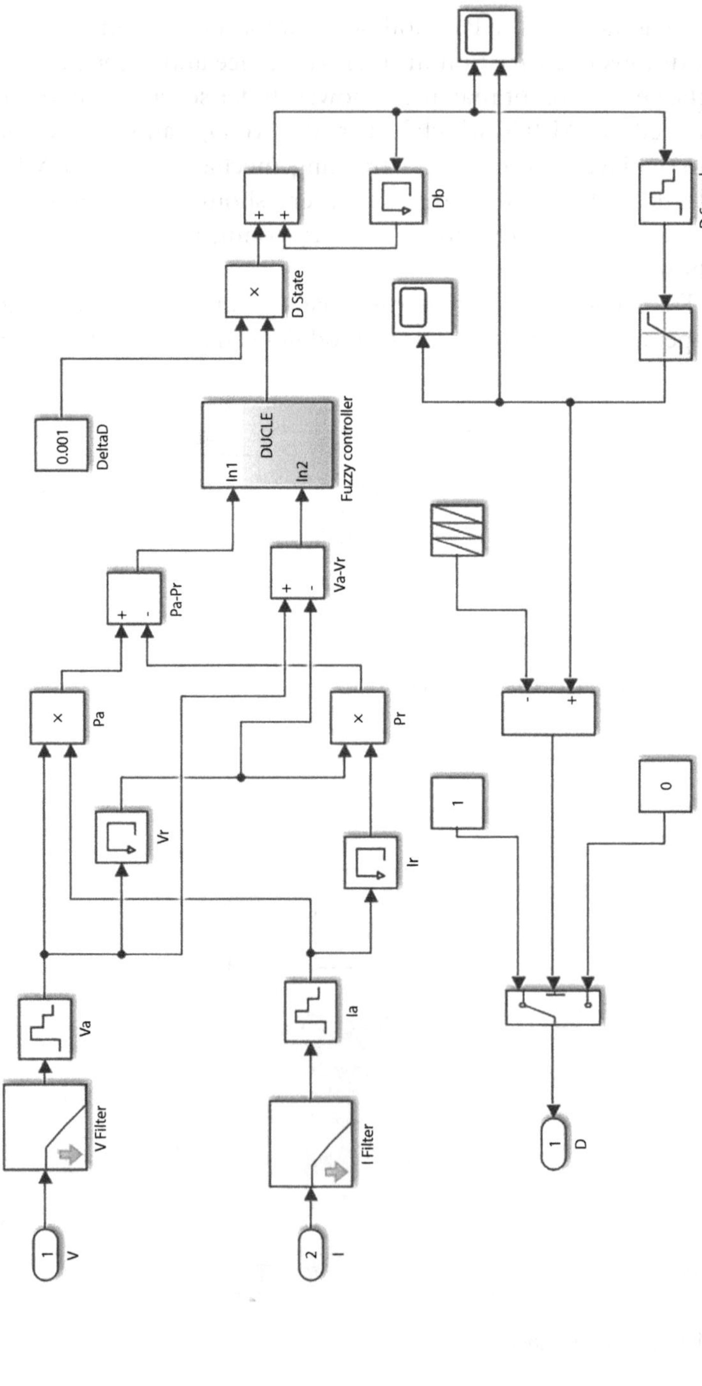

Figure 4.5 Fuzzy logic control design in MATLAB.

The essential technique for controlling charge is called pulse width modulation. It is an activity dependent on basically associating the modules straightforwardly to the battery. Fuzzy logic control characterizes two state variables as an input voltage and yield maximum voltage and maximum current (Im) [13]. The design of fuzzy logic controller is shown in Figure 4.4.

From measurements of fuzzy logic design, it gives a signal that is comparable to the duty cycle of the converter and pulse width modulator that transfers it to the converter. This pulse width modulator runs the estimation of duty cycle to achieve pulse width modulation. Where the control pulses are produced by the converter switch. In this simulation model, determination of fuzzy controller values are illustrated as a closed loop system [8]. The structure of fuzzy controller depends on fuzzy sets in which controller depends on fuzzy sets in which a variable is an individual at least one sets with a predefined level of enrollment. It is used to coordinate with the human activity measure in personal computers and evaluate uncertain data, settle on choice dependent on inconclusive data like a resistive load that is associated with the PV source by buck-boost converter and through inverter, grid [14]. The Simulink model of fuzzy logic controller is shown in Figure 4.5.

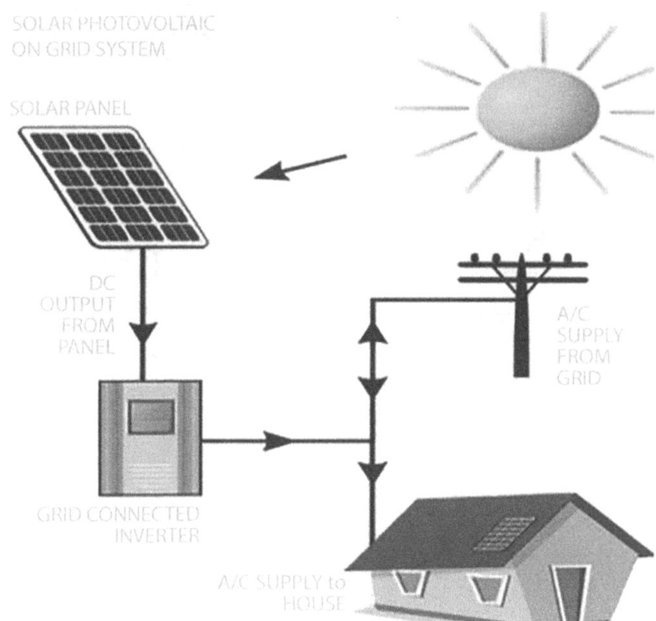

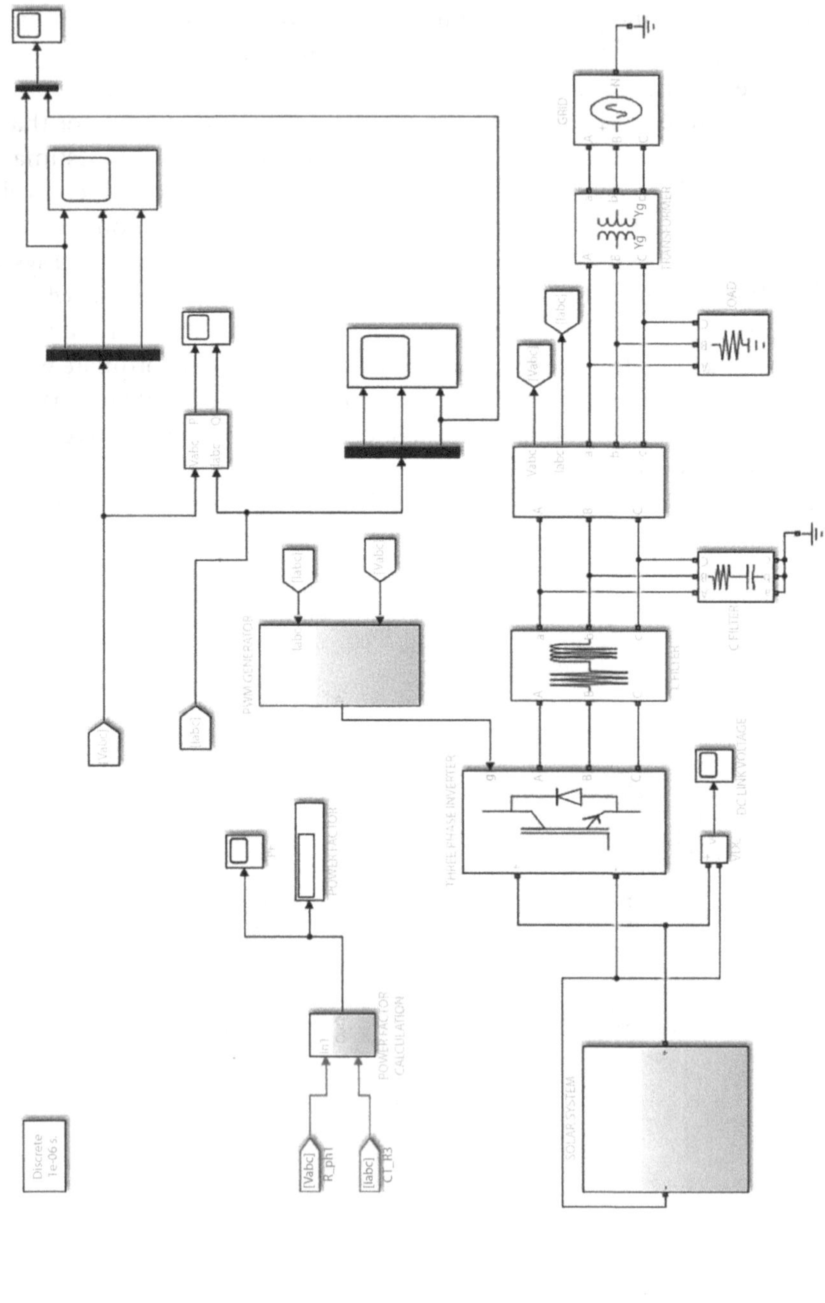
GRID
TRANSFORMER
LOAD
C FILTER
L FILTER
PWM GENERATOR
THREE PHASE INVERTER
DC LINK VOLTAGE
SOLAR SYSTEM
POWER FACTOR
POWER FACTOR CALCULATION
Discrete
1e-06 s.
[Vabc]
R_ph1
[Iabc]
CT_R3

4.6 Grid Connected PV System

As the power demand increases, PV power supply to the grid is growing rapidly. The integration of PV power generation to grid increase the power production rate, and it is eco-friendly too, which is rapidly growing. The whole setup is consisting of PV panel connected to DC-to-DC converter to boost up the voltage level, which, in turn, connected to the inverter for DC-to-AC power conversion, and, after this process, the AC power is integrated with the grid [15]. The grid connected PV system is shown in Figure 4.6.

4.7 Results and Discussion

The Simulink model in Figure 4.7 shows grid connected PV system with fuzzy logic control.

Figure 4.8 shows the output of voltage gained through three-phase inverter by using fuzzy-based MPPT algorithm.

Figure 4.9 shows the output of current gained through three-phase inverter by using fuzzy-based MPPT algorithm.

Figure 4.10 shows the output of power gained through three-phase inverter by using fuzzy-based MPPT algorithm.

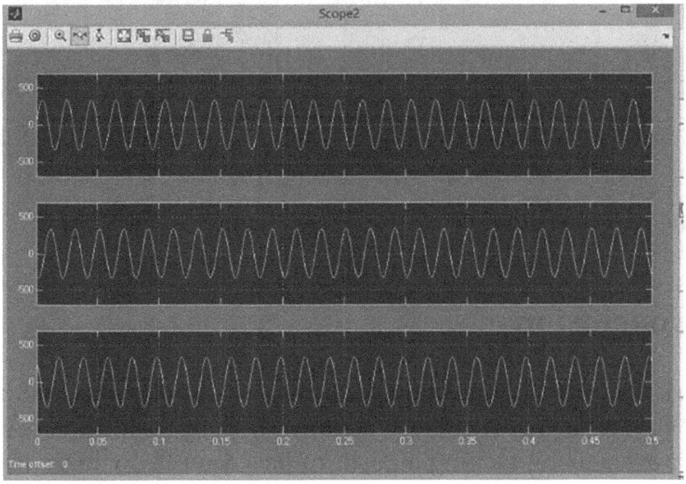

Figure 4.8 Voltage gain.

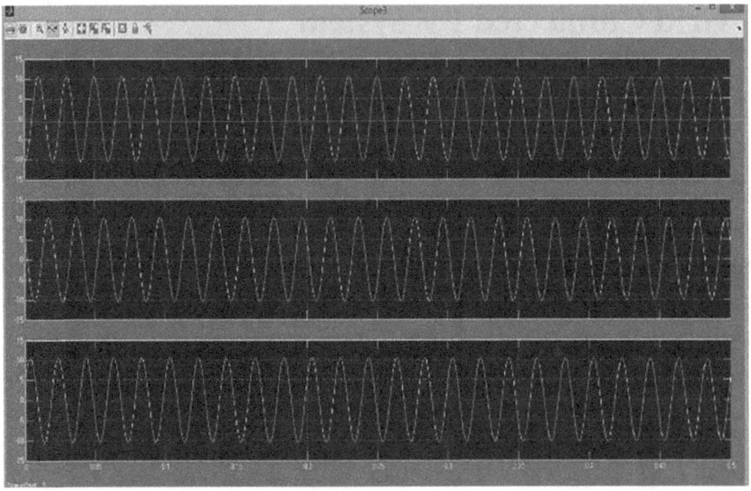

Figure 4.9 Current gain.

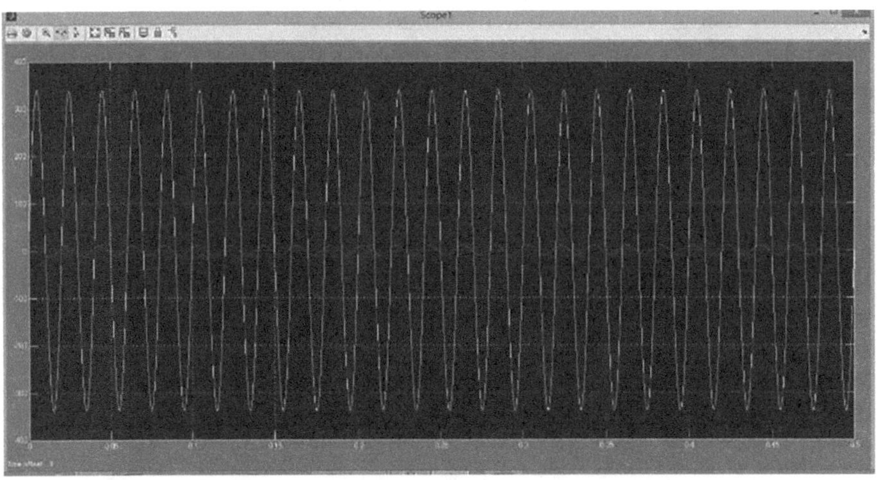

Figure 4.10 Power output.

4.8 Conclusion

This work focuses to conceive a technique to utilize maximum output from a PV system. We have referred many MPPT algorithms. We achieved optimal power using tacking algorithm based on fuzzy logic with combination of system model for algorithm progress and application. In our work, various subsystems and components are examined and modeled. These models are further evaluated for validation and integrate to achieve the prime power point tracking model. In order to create this algorithm, we utilize the fuzzy logic tools in the MATLAB. The integration of PV module, inverter with fuzzy logic controller, and MPPT model is utilized for tuning of membership functions and rules of fuzzy logic controller. We presented in our result that PV model is reasonably precise, and any solar panel model can be developed easily from manufacturer datasheets. In addition, the output results specify that suggested algorithm produces high efficiency under variable conditions like noise, climate, and other interferences.

References

1. Ocran, T.A. *et al.*, Artificial Neural Network Maximum Power Point Tracker for Solar Electric Vehicle. *Tsinghua Sci. Technol.*, 10, 2, 204–208, 2005.
2. Teodorescu, R., Liserre, M., Rodriguez, P., *Grid Converters for Photovoltaic and Wind Power Systems*, Wiley-IEEE Press, New Jersey, United States, 2007.
3. Carrsoco, J.M. *et al.*, Power-Electronic Systems for the Grid Integration of Renewable Energy Sources: A Survey. *IEEE Trans. Ind. Electron.*, 53, 4, 1002–1016, Aug. 2006.
4. Zedah, L.A., Outline of New Approach of the Analysis of Complex System and Decision Processes. *IEEE Trans. Syst. Man Cybern.*, 3, 28–44, Jan. 1973.
5. Spiers, D. and Royer, J., *Guidelines for the use of batteries in photovoltaic systems*, NESTE Advanced Power Systems, CANMET Energy Diversification Research Laboratory, Espoo, Finland, 1998.
6. Burger, B. *et al.*, Extreme high efficiency PV-Power Converters. *13th European Conference on Power Electronics and Applications*, 2009.
7. Kumar, H. and Tripathi, R.K., Simulation of variable incremental conductance method with direct control method using boost converter. *Engineering and Systems*, 2012.
8. Hannan, M.A. *et al.*, Real-time testing of a fuzzy logic controller based grid-connected photovoltaic inverter system. *2014 IEEE Industry Application Society Annual Meeting.*
9. Kondawar, S.S. and Vaidya, U.B., A comparison of two MPPT techniques for PV system in MATLAB Simulink. *Int. J. Eng. Res. Dev.*, 2, 7, 73–79, Aug. 2012.
10. Orozco, M., II *et al.*, Maximum power point tracker of a photovoltaic system using sliding mode control. *International Conference on Renewable Energies and Power Quality*, 2009.

11. Chouder, A., Guijoan, F., Silvestre, S., Simulation of fuzzy-based MPP tracker and performance comparison with perturb & observe method. *Rev. Des. Energies Renouvelables*, 11-4, 577, Dec. 2008.
12. Takun, P., Kaitwanidvilai, S., Jettanasen, C., Maximum power point tracking using fuzzy logic control for photovoltaic systems. *Proceedings of International Multi Conference of Engineers and Computer Scientists*, Hongkong, 2011.
13. Menadi, A., Abdeddaim, S., Betka, A., Benchouia, M.T., Real Time Implementation of a Fuzzy Logic based MPPT controller for Grid Connected Photovoltaic System. *Int. J. Renew. Energy Res.*, 5, 1, 236–244, 2015.
14. Abbadi, A., Hamidia, F., Morsli, A., Benbouabdellah, O., Chiba, Y., MPPT based Fuzzy Logic Controller for Grid Connected Residential Photovoltaic Power System. *Proceedings of International conference in Artificial Intelligence. In Renewable Energetic Systems*, 2019.
15. Bansal, B., Microgrid Controlling in Grid Connected Mode Using Fuzzy logic Controller With Optimization of DER's. *Int. J. Eng. Adv. Technol.*, 8, 5, 1128–1136, Jun. 2019.

5

An Experimental Investigation of Fuzzy-Based Voltage-Lift Multilevel Inverter Using Solar Photovoltaic Application

Gnanavel C.[1], Johny Renoald A.[2*], Saravanan S.[3], Vanchinathan K.[4] and Sathishkhanna P.[5]

[1]*Department of Electrical and Electronics Engineering, AMET Deemed to be University, Kanathur, India*
[2]*Department of Electrical and Electronics Engineering, Erode Sengunthar Engineering College, Erode, India*
[3]*Department of Electrical and Electronics Engineering, Muthayammal Engineering College, Namakkal, India*
[4]*Department of Electrical and Electronics Engineering, Velalar College of Engineering and Technology, Erode, India*
[5]*Department of Marine Engineering, AMET Deemed to be University, Kanathur, India*

Abstract

The new approach of the solar-fed symmetric voltage-lift multilevel inverter (SVLMLI) is proposed with a minimum number of components compared to the classical type of multilevel inverter (MLI). The comparison is made for open-loop and closed-loop controller for fuzzy logic control (FLC) and fuzzy-tuned proportional integral (PI) control. The FLC has significantly reduced the harmonic content by varying the modulation index. The main objective of this architecture is to design seven-level voltage-lift multilevel inverter (VLMLI), which is used for solar application with optimization control technique governed by fuzzy logical tuned PI controller approach delivers a major task. It has been for further analysis with the fast dynamic response and short settling time compared with PI controller. This circuit uses a single switch with minimizing harmonics levels, and the dv/dt and di/dt stress reduces across the load. The proposed MLI generates low distortion output uses through the minimum harmonics of the fast Fourier transform

**Corresponding author*: jorenoeee@email.com

A. Chitra, V. Indragandhi, and W. Razia Sultana (eds.) Smart Grids and Green Energy Systems, (59–74)

(FFT) framework done by utilizing MATLAB Simulink software, and the same circuit is associated with fuzzy-tuned Xilinx Spartan 3AN field program gate array (FPGA) prototype presented. Outcome of the proposed system is minimum total harmonic distortion (THD) valid through seven-level cases compared with prototype and simulation model of MLI presented.

***Keywords*:** Photovoltaic, symmetric voltage-lift multilevel inverter, proportional integral, field program gate array, fuzzy logic control

5.1 Introduction

Sunlight is directly converted to electricity and well suited to most locations; thus, solar photovoltaic (S-PV) are favored in comparison with other renewable energy sources. Various voltage-lift (VL) topologies are offered for DC/DC transformation in power electronics schemes. The VL has divided into four methods: 1. positive output voltage lift; 2. negative output voltage lift; 3. super lift; and 4. re-lifting approach introduced at 2003 [1]. The multilevel conversion techniques can be operated into two input environments: symmetric multilevel inverter (MLI) (SMLI) and asymmetric MLI (AMLI). Each type has become several circuits will apply the modulation for increasing number of levels, medium, and high output voltage applications. The various inverter topologies are proposed for MLI conversion in S-PV systems [2–4]. Three sorts of MLI constructions are diode clamped, flying capacitor, and cascaded H Bridge. The problem of this inverter structure includes an increased number of devices required and control complexity. Three types of traditional MLI structures such as cascaded H Bridge multilevel inverter (CHB), flying capacitor (FC), and neutral point-clamped (NPC) are particularly helpful to increase the output voltage level. They are connected for domestic and industrial applications like UPQC and FACTS devices. These drawbacks of this converter topology have many levels synthesized from several clamping diodes, clamping inductor, and DC sources. In the symmetrical technique, all the DC source voltages have equal quality, and it guarantees great modularity output of the circuit [5, 6].

In the proposed work, which can perform the necessary voltage lifting conversion to different load conditions. In general, the conversion approach is focused on multiple repeating parts, but the system design introduces the development of VL multilevel technique which can reduce the repeating components [7, 8]. The S-PV array module consists of S-PV1 to S-PV4 made a voltage level of (12*4) 48 V and 4.90 A. The parameters are open-circuit voltage and short-circuit current of 58.5 V and 4.9 A, respectively [9].

The ultimate aim of maximum power point (MPP) is tracking the parameters that are maximum voltage (V_{max}) and current (I_{max}) using the method of incremental conductance to determining the duty cycle and generate digital pulses to both converter (DC-DC) and multilevel inverter (DC-AC) switches. The incremental conductance of the solar is to be determining an optimum operating current and maximum output energy passing to next juncture of the converter in Figure 5.1. Here, VL multilevel converters are employed for the first time based on the series connection of the RL load, and it is commonly applied to interface renewable energy systems. The importance of staircase output voltage has been enhanced as the multiple voltages with the help of minimum power devices required. As a result, as proposed by authors, the development of the output voltage gain with variation of the modulation index limitation in fuzzy logic controller-based optimization was studied and compared [10, 11], which is relatively higher than that of the traditional multilevel inverters or reduced multilevel inverters [12, 13]. This system is designed with seven-level single-phase VLMLI, and optimization control techniques such as fuzzy logic and fuzzy-tuned PI-based control for reduce total harmonic distortion (THD) have been further analyzed to improve the multilevel output using MATLAB/Simulink and a single-phase FPGA hardware implementation using fuzzy-tuned Xilinx Spartan 3AN [14, 15].

5.2 Proposed SVLMLI

This report basically consists of only one MOSFET Switch, four parallel switching capacitor, and two freewheeling diodes to generates seven level output ($+Q_4$ 0 and $-Q_4$), ($+Q_3$ 0 and $-Q_3$), and ($+Q_2$ 0 and $-Q_2$) in Figure 5.1. Whereas classical MLI NPC, FC, and CHB [16, 17] require 12 switches to generate the seven level of output as shown in Table 5.1. The PV sources (PV voltage and current), V_{pv} and I_{pv}, are an output of PV array module. A double-pole double through switch (2P2T) and a four-point band switch are used in the MLI circuit. In the steady-state condition, $V_{C1} = V_{DC}$ (input). The load voltage V_{DC} (output) is extremely elevated from input source voltage by the use of capacitor (C_1) and inductor (L_1), where, during main switch off period, (1 – k) T decreases voltage level – (V_{output} – $2V_{DCinput}$).

The VL-MLI uses three capacitors: C_2, C_3, and C_4, which are used to divide the output voltage in seven levels. When duty cycle k = 0.5, the capacitor band switch (T_{on}) turn ON time starting from Q_2, Q_3, and Q_4 and Neutral, (T_{off}) turn OFF time, $-Q_4$, $-Q_3$, and $-Q_2$, respectively, in V_L-MLI: Therefore, the operation status is as follows.

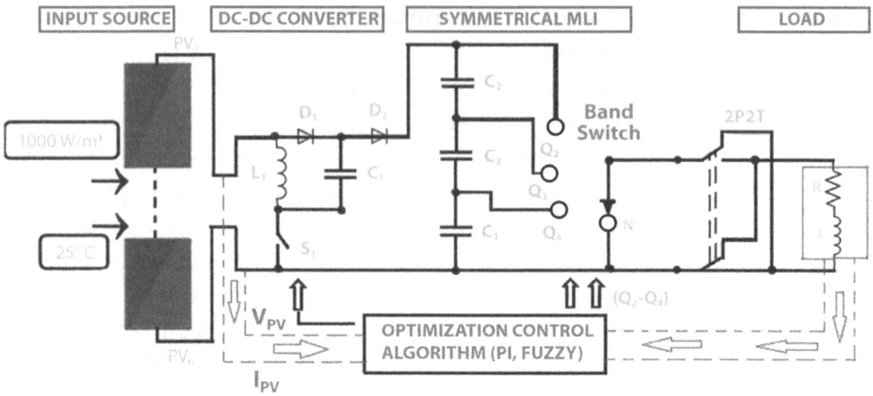

5.2.1 Trigger On State

Output voltage switch position Q_2 is a charging mode; here, V0 = +1/3V_{DC} is the double-pole double-through switch and the four-point band switch position is Q_2; the voltage across the capacitor V_{C2} is connected through change over (2P2T) switch in Figure 5.2. The voltage drop across the capacitor (D_2 and V_{C2}) has appeared at double-pole double-through switch. The precise positively charged half-cycle output voltage is appears in RL load.

Output voltage switch position in Q_3, here, V_0 = +1/2V_{DC}; the voltage drop across the capacitor (V_{C3} and V_{C2}) appeared to DPDT in Figure 5.2. The output voltage switch position Q_4 must be a charging mode. Here, V_0= +1/V_{DC}; the voltage drop across the capacitor (V_{C4} and V_{C3}) appeared to double-pole double-through switch. The precise charged positive half-cycle output voltage has appeared in RL load.

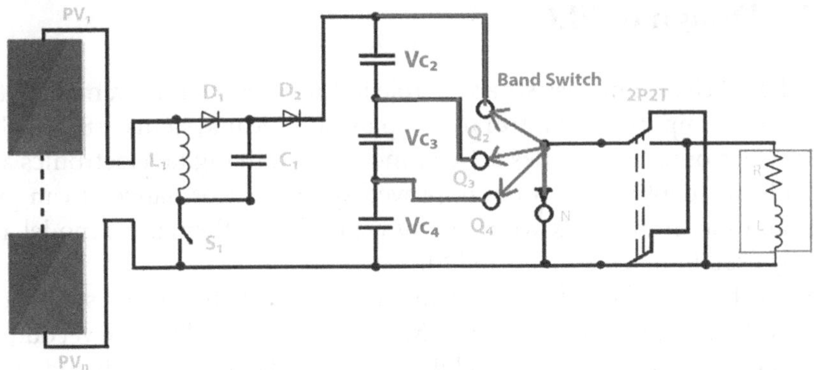

5.2.2 Trigger Off State

Output voltage switch position is neutral (N). Here, $V_0 = 0$. There is no voltage across the capacitor. In the next state, the output voltage switch position in $-Q_4$ should be a discharging mode; here $V_0 = -1/V_{DC}$; the voltage drop across the capacitor (N and $-VC_4$) has appeared at double-pole double-through switch. The precise discharged negative half-cycle output voltage appears in RL load. The output voltage switch position in $-Q_3$; here, $V_0 = -1/2V_{DC}$. The voltage drop across the capacitor ($-V_{C4}$ and $-V_{C3}$) appeared to double-pole double-through band switch. The output voltage switch position in $-Q_2$; here, $V_0 = -1/3V_{DC}$. The voltage drop across the capacitor ($-V_{C3}$ and $-V_{C2}$) appeared to double-pole double-through switch. The precise negatively charged half-cycle output voltage appears in RL load and inverter switching method, as shown in Table 5.2.

5.3 Design of FLC

The digital electronics assists a vital role in fuzzy logic strategy, making the process to simplify the control of complex nonlinear systems [18, 19]. The fuzzy electronics techniques are commonly used in digital electronics and provide an adaptive control of improved system performance. It can provide an exact framework without a usage of the mathematical model and widely used in machine control [20].

It has been explained in detail about control scheme of the symmetric voltage-lift multilevel inverter (SVLMLI) to get expected multilevel output. The FLC is divided into five modules: fuzzifier, database, rule-based, decision-maker, and defuzzifier, which are analyzed and presented as shown in Figure 5.3. These operator errors evaluate (E) = $V_{refence} - V_{output}$. The rate of conversion of error is calculated (COE) = En – En – 1 and the d/dt serves as the input attributes for the FLC, where the V_0 is end result output voltage of the multilevel inverter, the $V_{refence}$ is the preferred output voltage, δ_{mn} is changed over band switch position by the nth sampling instant, and the δ_{mn} is updated the switch position commanding signal generate to the FLC compared with $V_{reference}$ to generate the modulating signal (ms) required for PWM generation, which provides the appropriate gate signal to the multilevel inverter switch [21–23].

The change over switch-2P2T (mn) operation depends on the previous state of band switch position. The fuzzification is to the direct record highlights for knowledge and profit alongside standard utilization. The fuzzification values are called as crisp value, and it is used to develop the defuzzification procedure. The rules-based fuzzy systems are frequently connected in software. To obtain correct productivity, defuzzification is performed on fuzzy ruler view. The principles are made from an inference system using Mamdani algorithm presented [24, 25]. The contributions inputs are joined to the fuzzification block using a preprocessor. The RMS productivity voltage of seven level VLML inverter is developed from fuzzy control rules created based on the following criteria:

1. At the point, the fuzzy logic output is the reference signal made from the membership function in Figure 5.4a, the change of modulation index it must be larger. Hence, as for bringing, the output to the reference is quick. The membership function of an error signal is shown in Figure 5.4a.
2. The fuzzy logic is developed in fuzzy logic toolbox, and the different variables (e, Ce) are described by triangular

waveform; once in the reference, a tiny change in modulation index more improved the multilevel output waveform, as shown in Figure 5.4b.

3. Once the inverter output is higher than the reference, the change of modulation index should be negative and vice versa. This result of the saturation block limits the minimal values of THD.

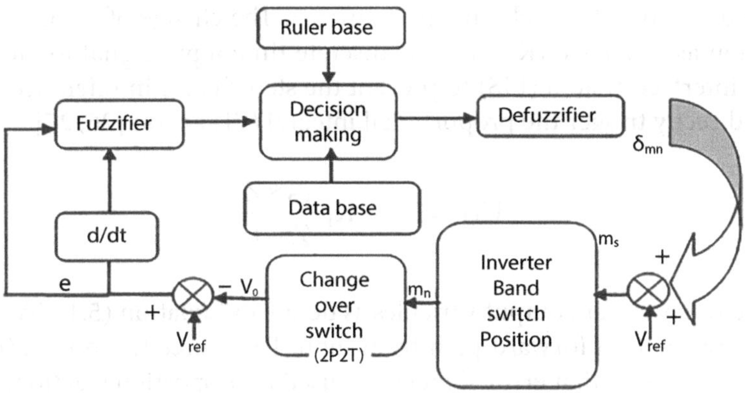

Figure 5.3 Structure of fuzzy logic controller scheme.

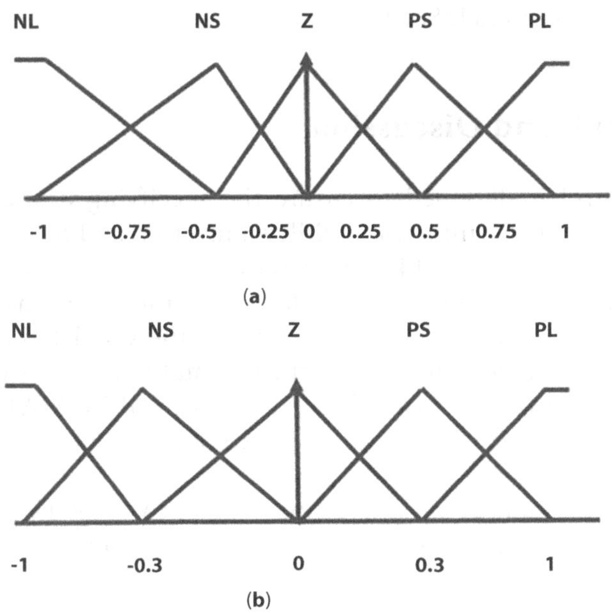

Figure 5.4 Membership function. (a) Reference signal. (b) Change of error signal.

5.4 FL Tuned PI Controller

A fuzzy-tuned PI controller in parallel form as shown in Figure 5.5 has been focused. The input and output variables are described, and the membership function of the controller is one of the most significant stages in a fuzzy logic controller design process. The multilevel inverter current is controlled, and change of error (Ce) and functions of errors are defined as input variables of the FLC. These two inputs have made non-linear membership function for each one in Figure 5.5. The change of error signal is mention as e (K) – e (K – 1). The discrete time input signal follows to a fuzzy interface system (FIS) to prevent the step change in reference signal from directly trigger the proportional integral (PI) action [26, 27].

$$U(t) = K_p + K_i \frac{Ts * Z}{Z - 1} \tag{5.1}$$

The K_p and Ki are output variables as define in Equation (5.1). Two gain blocks in the feed forward path from input PV source (r) to load (u) are applied to ensure that error signal (e) is used in proportional action, when the PI controller is linear. Therefore, the linear fuzzy-tuned PI controller is properly designed, and the fuzzy interface system to achieve better multilevel output is adjusted [28, 29].

5.5 Result and Discussion

The PI control is designed to ensure the specifying desired nominal operating point for symmetric multilevel inverter, and it is regulated at inverter circuit, such that it keeps nearer to a minimal functioning point in event of quick disturbances, collection point variations, modeling mistakes, and modules variations. It exposes the simulated closed-loop active fuzzy-PI tuned controller applied in multilevel inverter RL load. The response of output load voltage is performed by MATLAB using fast Fourier transform (FFT) evaluation to obtain a specific value, as in Figure 5.6.

The total harmonics analysis of fuzzy output voltage (V_{rms}) is 148 V, as shown in Figure 5.6. The seven-stage of VLMLI offers the maximum

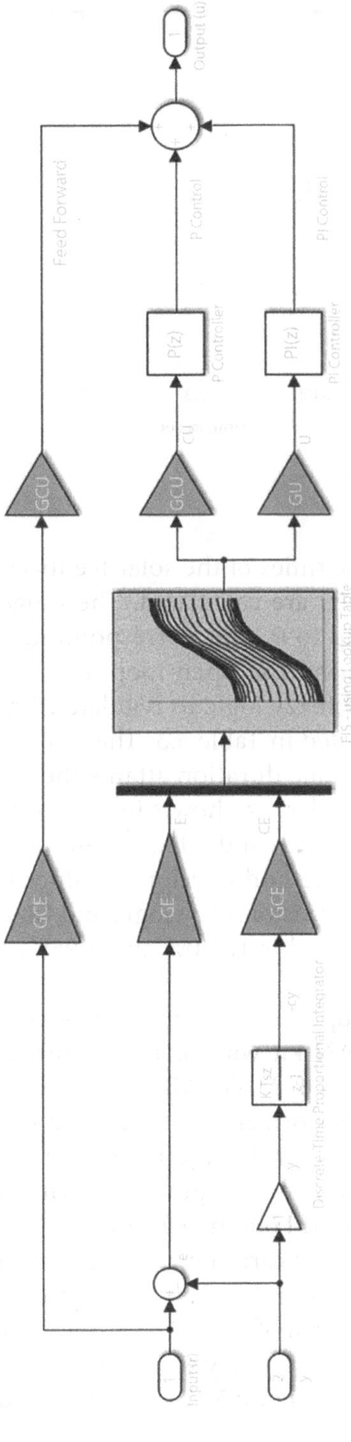
Input(r)
y
e
Discrete-Time Proportional Integrator
KTsz
z-1
GCE
GE
GCE
E
CE
FIS - using Lookup Table
GCU
GCU
GU
CU
U
P(z)
P Controller
PI(z)
PI Controller
Feed Forward
P Control
PI Control
Output (u)

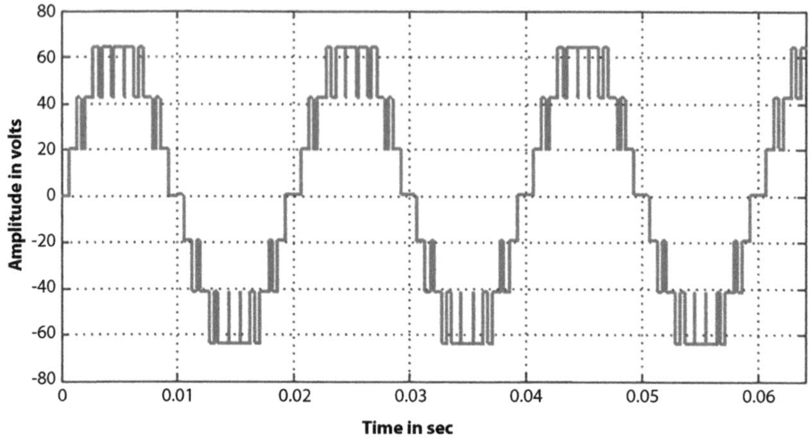

Figure 5.6 Output voltage waveform.

output result that is three times of the solar feedback voltage. The values of modulation index (MI) are calculated. The ranges from modulation index varied from 0.7 up to 0.1 for corresponding THD% and (output voltage) V_{rms} are evaluating the open-loop and closed-loop controller with PI, fuzzy logic, and fuzzy logic to regulate PI method achievement of THD values, as presented in Table 5.3. The step response of the fuzzy-tuned PI in a very short time duration attains the steady-state condition compared with PI controller, as shown in Figure 5.7. The PI operator controls proportional gain (K_p) and integral gain (K_i), where the applying Ziegler-Nicholas tuning method developed values that are 0.1 and 0.2, respectively. The FFT analysis of fuzzy logic and fuzzy-tuned PI controller output voltage THD level is 10.30% and 7.95%, as shown in Figures 5.8 and 5.9.

The experimental setup is operating and associated with the help of FPGA-Xilinx Spartan 3AN kit with battery improvement platform. The fuzzy logic PI-tuned VHDL-MATLAB code into the proposed FPGA model is presented. The performance of gate pulse and modulation signals is given through fuzzy-based rules and the operating parameters used in the simulation are the same for low-power experimental model in Figure 5.9. The system design VLMLI output is 110 V, 1.5 A, operated with an output frequency of 50 Hz, and a carrier switching frequency of 10,000 Hz, as shown in Figures 5.10 and 5.11. In order to generate a conventional multilevel output voltage waveform of seven levels, the VLMLI control with the relevant standalone mode is virtually sinusoidal, confirming the efficiency of this modulation method. Figure 5.12 shows the hardware configuration.

Table 5.3 Comparison between experimental and simulated result.

Modulation index	% THD Open-loop control (PWM)	Closed-loop PI control	Closed-loop fuzzy logic control	Closed-loop fuzzy assist PI control
0.7	50.21	38.40	17.39	8.99
0.6	48.13	38.31	16.20	7.95
0.5	45.31	38.13	15.09	6.01
0.4	44.35	37.99	14.09	5.89
0.3	42.01	35.90	12.57	4.69
0.2	41.96	34.02	11.00	5.10
0.1	39.56	32.16	10.81	3.81

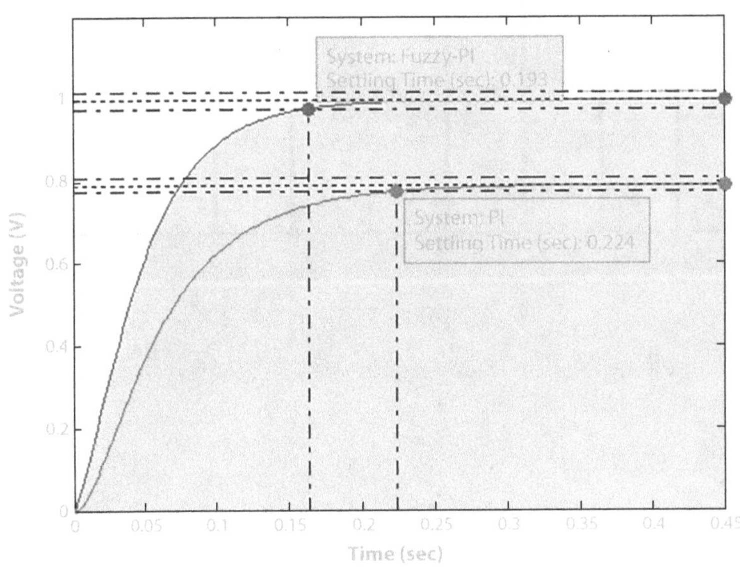

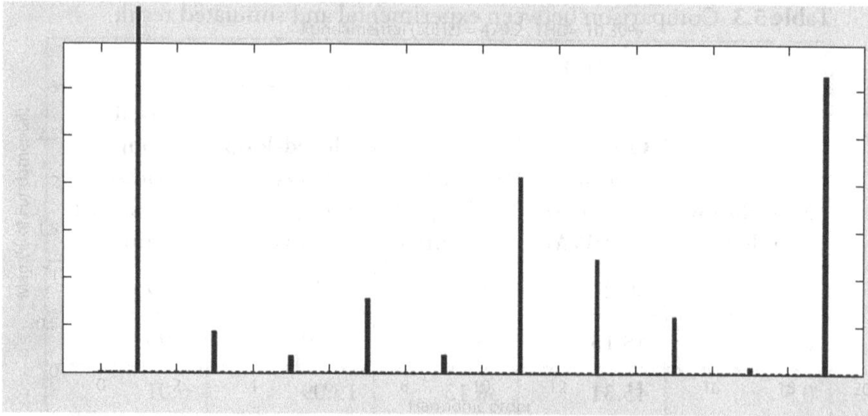

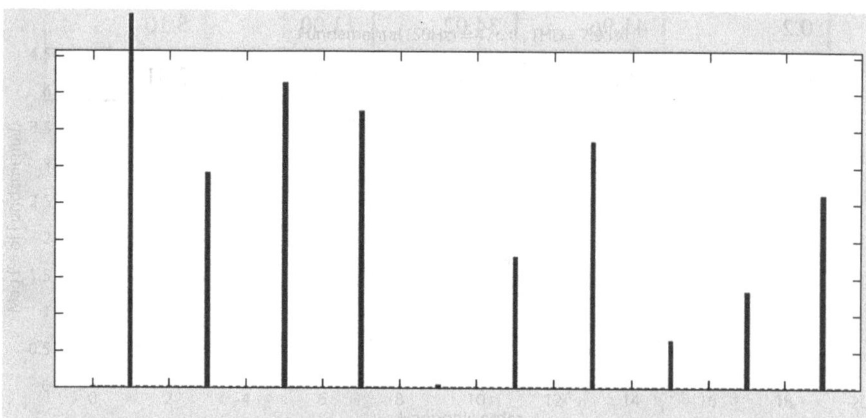

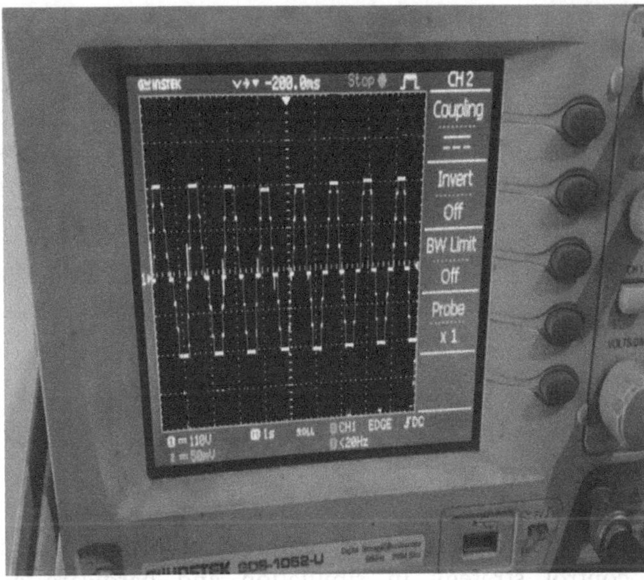

Figure 5.10 Output voltage of inverter.

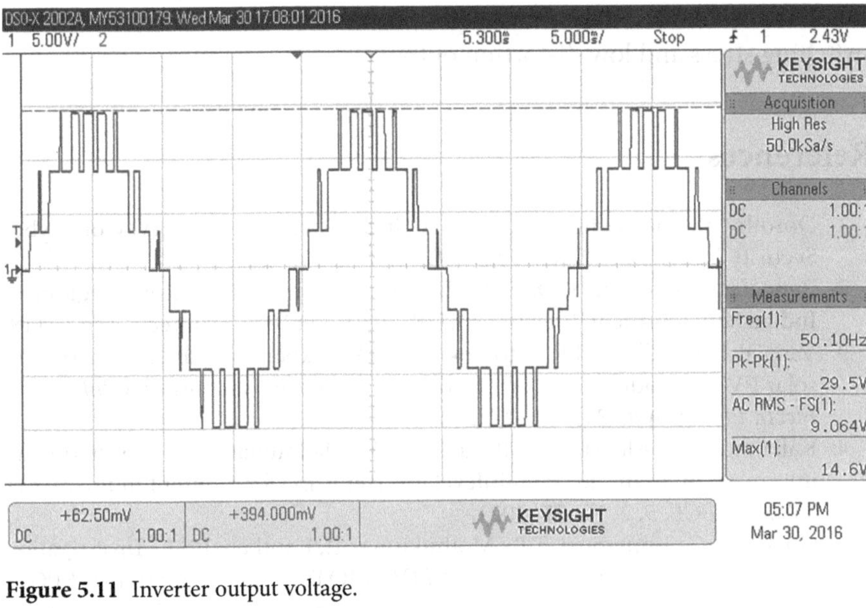

Figure 5.11 Inverter output voltage.

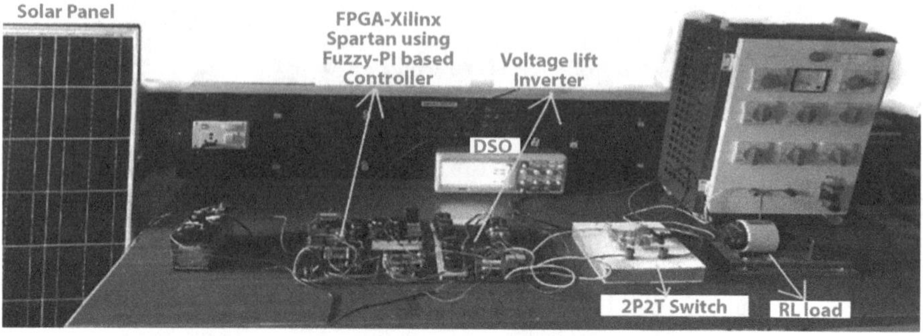

5.6 Conclusion

MLI is used in research seven-level symmetrical form of VL with fuzzy-tuned PI control strategy in simulation and hardware equipment. The multilevel output voltage with minimal distortion productivity voltage waveform and its performance factors, such as harmonic distortion values, are evaluated by 7.95%. The effectiveness of the experimental and simulated results was tested by single-phase seven-level inverter with less switching stress and low distortion output.

References

1. Dorothy, R. and Sasilatha, T., Smart Grid Systems Based Survey on Cyber Security Issues. *Bull. Electr. Eng. Inform.*, 6, 4, 337–342, 2017.
2. Renoald, J. and Vanaja, D., Solar Energy Assessment in Various Regions of Indian Sub-continent, Intechopen, UK, 2020, doi: 10.5772/intechopen.95118.
3. Vanaja, D.S., Albert, J.R., Stonier, A.A., An Experimental investigation of solar PV fed modular statcom in wecs using intelligent controller. *Int. Trans. Electr. Energy Syst.*, 31, 3, 1–27, 2021.
4. Kaliannan, T., Albert, J.R., Begam, D.M., Madhumathi, P., Power quality improvement in modular multilevel inverter using for different multi-carrier PWM. *EJECE*, 5, 2, 19–27, 2021.
5. Gnanavel, C., Immanuel, T.B., Muthukumar, P., Kanthan, P.S.L., Investigation on Four Quadrant Operation of BLDC MOTOR Using Spartan-6 FPGA. *International Conference on Soft Computing Systems*, pp. 752–763, 20182018.
6. Vanchinathan, K. and Valluvan, K.R., A metaheuristic optimization approach for tuning of fractional-order PID controller for speed control of sensorless BLDC motor. *J. Circuit. Syst. Comput.*, 27, 08, 1850123, 2018.

7. Albert, J.R. *et al.*, Investigation on load harmonic reduction through solar-power utilization in intermittent SSFI using particle swarm, genetic, and modified firefly optimization algorithms *J. Intell. Fuzzy Syst.*, 42, 4, 4117–4133, 2022. DOI: 10.3233/JIFS-212559.
8. Gnanavel, C., Muruganantham, P., Vanchinathan, K., Albert, J.R., Experimental Validation and Integration of Solar PV Fed Modular Multilevel Inverter (MMI) and Flywheel Storage System, *2021 IEEE Mysore Sub Section International Conference (MysuruCon)*, 147–153, 2021. Doi: 10.1109/MysuruCon52639.2021.9641650.
9. Rajavelan, M., Gnanavel, C., B. Immanuel, T., Muthukumar, P., Improved Sensitivity of SPR Instrument Using Multiple Reflection Technique. *Int. J. Eng. Technol.*, 7, 1310–1316, 2018.
10. Vanchinathan, K., Valluvan, K.R., Gnanavel, C., Gokul, C., Design methodology and experimental verification of intelligent speed controllers for sensorless permanent magnet Brushless DC motor. *Int. Trans. Electr. Energy Syst.*, 2021, e12991, 2021. https://doi.org/10.1002/2050-7038.12991.
11. Dhivya, M. and Albert, J.R., Fuzzy grammar hybrid split capacitors & split inductor applied in positive output luo converters. *Int. J. Sci. Res. Sci. Eng. Technol.*, 3, 327–332, 2017.
12. Gnanavel, C., Gokul, C., Vanchinathan, K., Test Scheduling of Core Based System-on-Chip Using Modified Ant Colony Optimization. *J. Eur. Systèm. Autom.*, 52, 6, 599–605, 2020.
13. Vanchinathan, K., Valluvan, K.R., Gnanavel, C., Gokul, C., Albert, J.R., An improved incipient whale optimization algorithm based robust fault detection and diagnosis for sensorless brushless DC motor drive under external disturbances. *Int. Trans. Electr. Energ. Syst.*, 31, 12:e13251, 2021. Doi: 10.1002/2050-7038.13251.
14. Gnanavel, C. and Stoner, A.A., Investigation and Validation of an Eleven Level Symmetric Modular Multilevel Inverter Using Grey Wolf Optimization and Different Evolution Control Algorithm for Solar PV Applications. *Circuit World*, 47, 1, 117–127, 2020.
15. Palanisamy, R., Govindaraj, V., Siddhan, S., Albert, J.., Experimental Investigation and Comparative Harmonic Optimization of AMLI Incorporate Modified Genetic Algorithm Using for Power Quality Improvement. 43, 3, 1–14, 1 Jan. 2022. DOI: 10.3233/JIFS-212668.
16. Vanchinathan, K. and Valluvan, K.R., Tuning of Fractional Order Proportional Integral Derivative Controller for Speed Control of Sensorless BLDC Motor using Artificial Bee Colony Optimization Technique, in: *Intelligent and Efficient Electrical Systems*, pp. 117–127, Springer, Singapore, 2018.
17. Hinago, Y. and Koizumi, H., A Single-Phase Multilevel Inverter Using Switched Series/Parallel DC Voltage Sources. *IEEE Trans. Ind. Electron.*, 57, 8, 2643–2650, Aug. 2010.

18. Martins, G.M., Pomilio, J.A., Buso, S., Spiazzi, G., Three-Phase Low-Frequency Commutation Inverter for Renewable Energy Systems. *IEEE Trans. Ind. Electron.*, 53, 5, 1522–1528, Oct. 2007.
19. Vanchinathan, K. and Valluvan, K.R., A study of sensorless BLDC motor drives and future trends. *Asian J. Res. Soc. Sci. Humanities*, 6, 9, 1863–1887, 2016.
20. Albert, J.R., Design and Investigation of Solar PV Fed Single-Source Voltage-Lift Multilevel Inverter Using Intelligent Controllers. *J. Control. Autom. Electr. Syst.*, 33, 3, 1–26, 2022. https://doi.org/10.1007/s40313-021-00892-w.
21. Sasikala, M., Albert, J.R., Hemalatha, V., Punitha, R., Sasikala, M., Solar roadways - the future rebuilding infrastructure and economy. *Int. J. Electr. Electron. Res.*, 4, 2, 14–19, 2016.
22. Santhiya, K., Devimuppudathi, M., Kumar, S., Albert, J.R., Real time speed control of three-phase induction motor by using lab-view with fuzzy logic. *J. Sci. Eng. Technol.*, 5, 2, 21–27, 2018.
23. Gnanavel, C., Nagarajan, L., Sivamani, D., Thirusenthilkumaran, P., Powering Smart Street Lighting Cities Using Visible Light Communication and Light Fidelity, in: *Biotechnological Communication*, vol. 6, pp. 1–5, 2020.
24. Albert, J.R., Keerthana, M.S., Design and Implementation of Super-Lift Multilevel Inverter using Renewable Photovoltaic Energy for AC Module Application. *Int. J. Sci. Technol. Eng.*, 2, 11, 617–624, 2016.
25. Ramaraju, S.K., Kaliannan, T., Androse Joseph, S., Kumaravel, U.D., Albert, J.R., Design and Experimental Investigation on VL-MLI Intended For Half Height (H-H) method to improve power quality using Modified PSO algorithm. *J. Intell. Fuzzy Syst.*, 43, 2, 1–14, 2022. DOI: 10.3233/JIFS-212583.
26. Thangamuthu, L., Albert, J.R., Chinnanan, K., Banu, G., Design and Development of Extract Maximum Power from Single-double Diode PV Model for Different Environmental Condition Using BAT Optimization Algorithm. 43, 1, 1–12, 1 Jan. 2021. DOI: 10.3233/JIFS-213241.
27. Albert, J.R., Dhivya, M., "Analysis on Super Lift Multilevel DC/AC Inverters using Photovoltaic Energy with AC Module Application, *IJSRD*, 5, 2, 479–481, 2017.
28. Gnanavel, C., Khanna, P.S., Rajavelan, M., Vanchinathan, K., Gokul, C., Real Time Monitoring and Controlling of Boiler Drum Parameters using Internet of Things (IoT). *Test Eng. Manage.*, 82, 2195–2198, 2020.
29. Albert, J.R. and Dhivya, M., Analysis of Super Lift Multilevel Inverter using photovoltaic energy with ac module applications. *IJSRD*, 5, 2, 479–481, 2017.

6

Potentials and Challenges of Digital Twin: Toward Industry 4.0

M. Baranidharan[1], Dattatraya Kalel[1] and R. Raja Singh[2*]

[1]*School of Electrical Engineering, Vellore Institute of Technology, Vellore, India*
[2]*Dept. of Energy and Power Electronics, Vellore Institute of Technology, Vellore, India*

Abstract
The ongoing global digital transformation, inspired by the Industry 4.0 initiative, has brought new concepts and emerging technologies. The digital twin (DT) technology is one of these innovative notions that has recently gained more attention. The DT idea attempts to allow smart factories by providing a link between real and virtual systems and digital reproduction or representation of the physical system. The data is gathered, analyzed, and fed into the simulated virtual model in order to forecast and enhance the physical system's performance. However, there are a number of obstacles to overcome in the development of the technology, including data analytics and data privacy. This article focused on the potential and challenges of DT technology that should possess in Industry 4.0. Exploration of the DT in the context of Industry 4.0 will aid future research and development.

Keywords: Digital twin, Industry 4.0, IoT, IIoT, artificial intelligence, cyber-physical system

6.1 Introduction

Currently, the world looks to GDP as an indicator of economic development, and GDP looks to Industry as a success indicator, because Industry is one of the global economy's pillars: In the European Union, manufacturing accounts for 2 million businesses, 33 million employees, and 60%

**Corresponding author*: rrajasingh@gmail.com

A. Chitra, V. Indragandhi, and W. Razia Sultana (eds.) Smart Grids and Green Energy Systems, (75–90)

of productivity growth. In a World Economic Forum report, from the last 30 years, an investment in digital technologies has grown GDP by $20, while a dollar invested in non-digital technology has only raised GDP by $3. By 2025, emerging innovations such as artificial intelligence and cloud computing are projected to account for nearly a quarter of global GDP, or 24.3%. Regardless, Business 4.0 has the capacity to raise global income level [1]. The industrial world is changing in a trend, which name as Industry 4.0. The smart industry, a concept borrowed from Industry 4.0 and smart manufacturing jargon, is often associated with its enabling technologies, which primarily revolve around the Industrial Internet of Things (IIoT), big data analytics, production processes, connected physical equipment (IoT), and sensors, as well as the previously described convergence of IT and OT, interoperable, and highly connected systems/platforms, as well as the analysis and use of data from a variety of sources, including computers, and the transformation of data into actions and decisions (automation and semi-automation), play a crucial role in this. The Smart Industry trend is causing a huge transformation in how factories and workplaces operate, making them cleaner, more efficient, more versatile, and environmentally sustainable, and accounting for 40% of global energy consumption [2]. Machines are transforming to be mindful of the humans around them and to have new interfaces for simpler and safer experiences, information communication tools that can be used in a range of smart business applications [3]. In the last two decades, the IoT has changed the way data is exchanged between various sources. Indeed, the widespread adoption of Internet-connected technology including sensors and actuators allows for a continuous flow of big data. More precisely, the digital twin (DT) is motivated by the convergence of AI with big data analytics for working IoT data, which is one of the most recent and perhaps the most significant advancements in the field of technology [4]. In the traditional approach, designers develop and forecast the life cycle using computer-assisted simulation and engineering techniques, as well as perform various physical testing methods. They optimize architecture to improve performance while lowering design costs. However, tolerances are limited in this approach, as are relationships between configurations, planning, and so on. DT technology, on the other hand, allows for real-time monitoring and digitization. A digital representation of a physical entity is known to as a DT. It uses IoT sensors and analytics to provide data-driven representations of physical structures. DT provides engineers with interactive tools for managing assets and resources while optimizing performance, in addition to inert computer-aided design (CAD) models. It simulates the behavior of physical objects by creating virtual models for them in a digital format. DT technology, which is basically a simulated simulation of a physical device, process, or product, offers a real-time view of how a physical asset is working, and it is rapidly gaining popularity in industries such as manufacturing, automotive, and construction. It provides an interface that permits tracking of past and present operations as well as future forecasts using a mixture of data and intelligence that reflects the design, meaning, and behavior of a physical device. As a result, DT combine software analytics, artificial intelligence, and machine learning data to create digital simulation patterns that upgrade and adapt in response to changes in their physical counterparts. This allows for real-time tracking, simultaneous updates from various sources, and the estimation of their remaining life and optimal maintenance intervals. The technology is dependent on internet connectivity; data protection is at risk when data is exchanged; and tools based on 3D CAD models rather than 2D drawings are needed across entire supply chains. Globalization and modern production methods are among the issues to be solved. It will be difficult to handle all of these project data for the DT between partners and sellers as the physical invention changes. Nowadays the digital twin-based application on various industries sector is there. The automotive and aerospace industries have adopted the digital twin concept. It is widely used not only in vehicle engineering and design customization but also in large transport and aircraft maintenance. Companies can simulate and verify every stage of growth in order to identify problems and potential failures before producing the final product using weight monitoring, aircraft tracking, precise weather stipulation, and DT. As a result, producers can improve operating processes while lowering overall engineering costs. According to a report on the future of industry, industrial firms of all sizes are focusing on improving manufacturing processes and improving relationships with customers. Manufacturers are embracing twin design customization to make the product customization process easier. This technology allows users to design and re-design items virtually before designing a physical product that meets all of the customer's needs.

6.2 Industry 4.0

Global competition and the need to adapt production quickly to the market's ever-changing demands are now driving industrial production. Only revolutionary developments in modern manufacturing technology will meet these requirements. Industry 4.0 is a promising strategy focused on the transformation of industry and manufacturing processes, and the addition of all investors in the company's value chain (suppliers and customers).

Figure 6.1 explains the digital transformation of industries. The application of the cyber-physical systems (CPSs) and IIoT standard concepts to industrial production systems is used to address technical aspects of these requirements. As a consequence, Industry 4.0's "execution process" is based on CPS building block connections. Industrial development is now driven by global competition and the need to adapt production rapidly to the market's ever-changing demands. In addition, to seamlessly integrate manufacturing and business processes, comprehensive technology support based on decentralized and adapted versions of manufacturing execution systems and enterprise resource planning is needed. Processing of vast amount of data from systems, computers, and goods is the third important factor. Usually, data is stored in the cloud [4, 5].

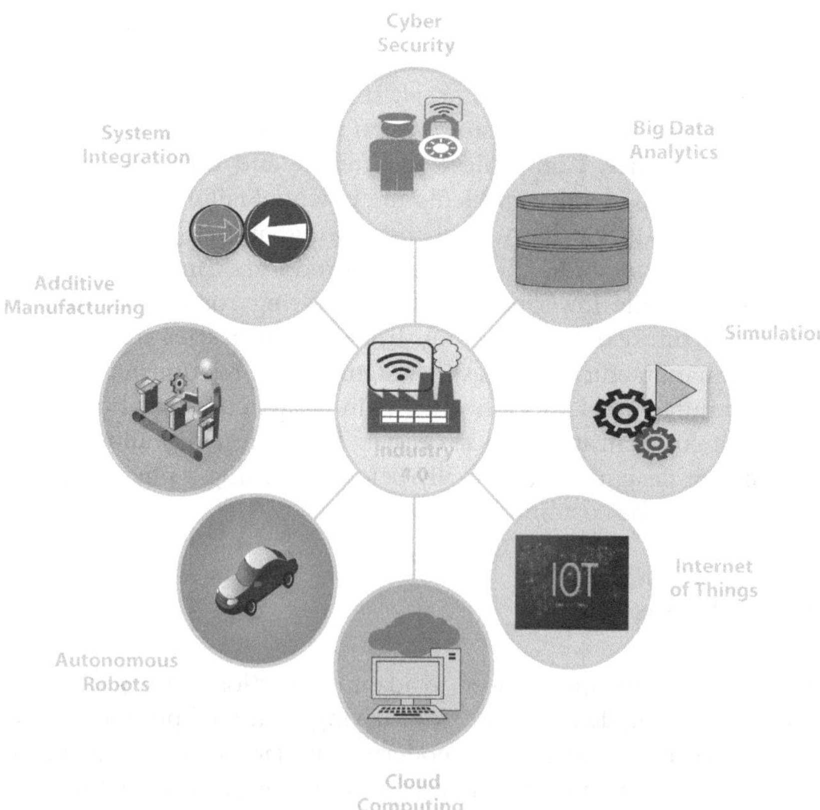

6.3 Digital Twin Technology

"A Digital Twin is an integrated multiphasic, multiscale, probabilistic simulation of an as-built vehicle or system that uses the best available physical models, sensor updates, fleet history, etc., to mirror the life of its corresponding flying twin." NASA 2012 in the AI and ML industry, there has already been a lot of progress, with new capabilities such as automation, prediction, and chats emerging. The DT technology is a virtual certainty that would primarily help companies in terms of productivity and competitive advantage. The world is moving toward smarter solutions, and DT satisfies the need for insights and research. This technology will continue as it matures and develops its model. The benefits of having a DT differ depending on how and where it is used. Using a DT to monitor currently items like a wind turbine or an oil pipeline, for example, will save millions of dollars in maintenance costs. Prototyping ahead of manufacture with DT will help minimize product defects and shorten time to market. Method enhancements include measuring staffing stages opposing production or positioning a supply chain with industrial or maintenance specifications [6]. Using supervision and simulation to improve achievement will result in improved efficiency and reliability. They can also reduce maintenance costs by anticipating failure before it occurs and ensure that arranging maintenance, repair, and replacement parts does not interfere with growth goals.

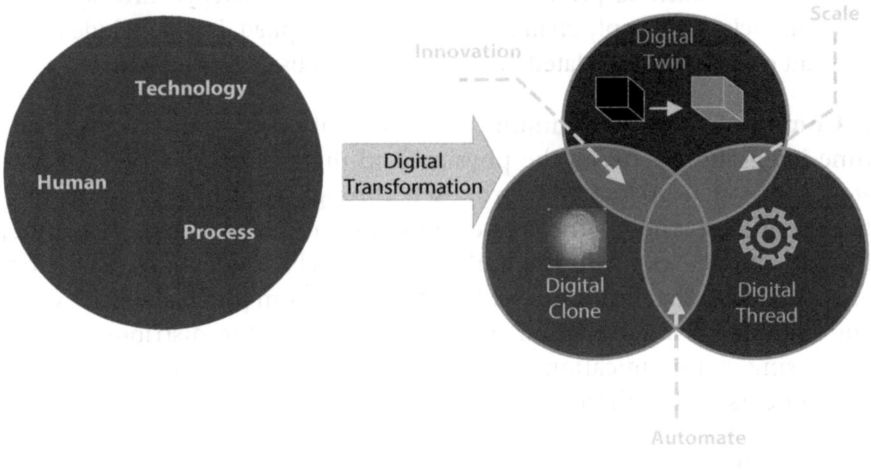

Through examining customization models and performing real-time performance monitoring, the DT will provide continuous improvements and ensure product consistency. Figure 6.2 shows the digital transformation of various physical systems.

6.3.1 Concept of Physical and Virtual Model of DTT

The concept of DT is presently more attention in research and with practitioners because of their massive array of application domains. The components of a DT of any product include the actual physical item operating in real space, a machine language-based version of the item—a twin, and real-time data obtained from sensors that link the physical item and its virtual twin. DT is one of the aiding technologies of Industry 4.0, and it links with real physical system and corresponding virtual system using simulation model, sensors data, and tools to analyze data. IoT (or) machine-to-machine communications are terms used to describe these intelligent interconnections and interoperability. Figure 6.3 shows the physical system and virtual model of robotics [7, 8].

1. Create: During the create phase, the physical process is outfitted with a slew of sensors that track serious inputs from the procedure and its surroundings. The sensors' data such as tensile force, displacement, torque, color unity, ambient temperature, atmospheric conditions, and moisture level are examples of external data that influence a physical asset's activities. The measurements can be converted into safe digital messages and sent to the DT using encoders. To the signals from the sensors, process-based knowledge from systems such as production implementation systems, ERP systems, CAD models, and supply chains systems can be applied. It will include a vast amount of constantly updated data for the DT to use as analysis input.

2. Communicate: The communicating stage allows for bidirectional, real-time integration between the physical and digital methods [9]. The six-step conceptual architecture proposed through for understanding how the DT works contains the create, communication, aggregation, analyze, insight, and act stages. In the build process, technologies such as sensors are installed in the physical asset to gather data from physical object. CAD and manufacturing execution system models were too distributed. Edge processing, communication interfaces, and edge protection are the three components of this phase.

A. **Edge processing:** The edge connection links sensors and historians, processes signals and data near the source, and sends data to the platform. By translating proprietary protocols into more simply understood data formats, this prevents network communication.

B. Communication interfaces: Via communication connection, data is moved from the sensor to integration. Given that, the sensor that generates the insight can be located nearly anywhere contingent on the DT configuration.

C. Edge security: A sensor and communication measurements have a slew of new security problems, which are still being ironed out. Firewalls, application keys, encryption, and system

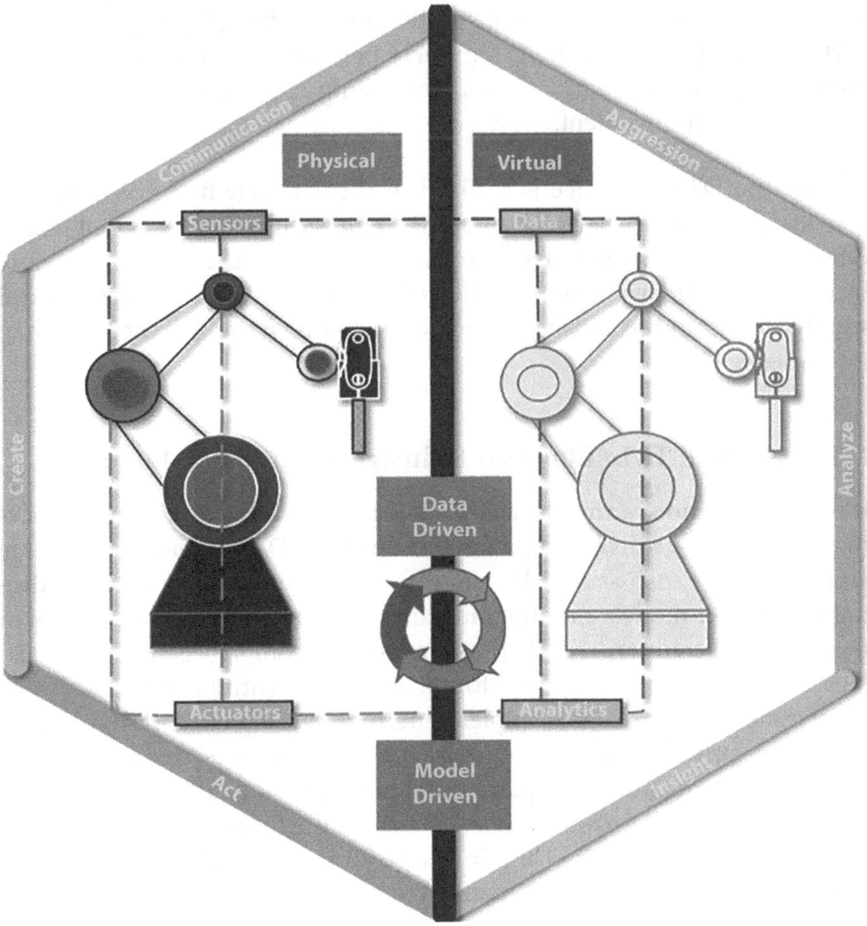

certificates are the most common security methods as more assets develop IP-enabled.

3. Aggregate: The aggregate stage will help with data ingestion into data store, processing, and analytics. Data aggregation and processing can take place on-premises or in the cloud

4. Analyze: Data is analyzed and visualized during the analyze stage. Advanced analytics tools and technologies can be used by data scientists and analysts to create iterative models that produce insights, suggestions, and decision-making guidance.

5. Insight: The findings from the analytics are provided through monitoring systems with visualizations in the insight phase, highlighting inadvertent discrepancies in the DT model's performance and the real-world comparison in one or more sizes, suggesting areas may require further analysis and improvement.

6. Act: The DT's influence is realized in the act stage by feeding actionable insights from earlier stages back into physical asset and digital phase. Observations are fed into asset process actuators, which control measure or control mechanisms, or are adjusted in back-end systems that control supply chains and gathering behavior, all of which are subject to human interference [10].

6.3.2 Digital Twin Effect on Industries—Industry 4.0

It is a combination of the physical and virtual worlds, with the goal of giving each industry a diverse digital representation. DT combines knowledge and technology such as AI, ML, and software analytics to create active digital simulation models that upgrade and adapt in response to changes in their physical counterparts [11]. As a result, the company gains the advantage of creating a computerized footprint of their entire product development cycle, from concept to implementation. This digital clone of physical properties, processes, and structures generates data on a continuous basis. By using reproductions, companies are able to provide early warnings, predict downtime, and grow new open doors for the future [12].

The DT is at the focus of the Industry 4.0 rotation, which encompasses automation, data sharing, and manufacturing processes to generate infinite growth opportunities for companies. The technology aids operators in understanding specific features, efficiency, and potential issues on the virtual simulated model by offering an exact digital replica of the machine. Since the technology allows for real-time monitoring of a physical plant using sensors connected to the whole setup, operators may be notified in advance of potential machine faults, downtime, or injuries [13]. By integrating AI, ML, and software analytics with data, the DT develops a simulation model that can upgrade a long side or in place of a physical complement [14].

The DT is vital to the growth of Industry 4.0 because it provides automation, data sharing, and joined-up production processes, as well as de-risking product deployment. Employees in the industry can track operations in real time, offering early notice of potential failures and allowing for real-time performance optimization and measurement with minimum productivity loss [15, 16].

6.4 Potential and Challenges in Applying Digital Twin Technology

This segment focuses on the problems that DT present. As the study progresses, it becomes apparent that the problems faced by data analytics, IoT, and IIoT are close to those faced by DT, with some examples mentioned below, and the DT with in Industry 4.0 ecosystem is explained in Figure 6.4 [17].

6.4.1 Information Technology Infrastructure

Accordingly, the existing IT infrastructure is a problem for both analytics and IoT. The digital twin requires infrastructure that enables IoT and data analytics to succeed; this will help in the efficient operation of DT. The DT would not be able to achieve its objectives without a well-connected and well-thought-out IT infrastructure [18].

6.4.2 Useful Data

The following task is gathering the information required for a DT. It must be noise-free, high-quality data, and a continuous, uninterrupted data stream. Since it is acting on bad and incomplete data, the DT can underperform if the data is bad and unreliable. For DT information, the quantity and quality of IoT signals are important. Preparation and review of system use are wanted to determine which data can be gathered and used to make the most of a DT [19].

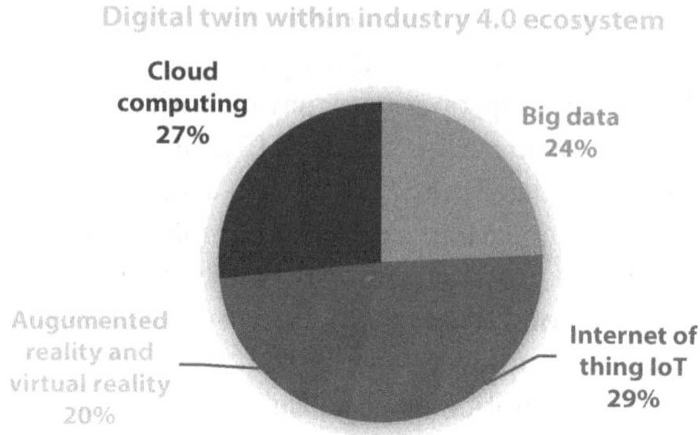

6.4.3 Trust

From the perspective of both the organization and the consumer, the problems associated with trust are numerous. To confirm that end users and organizations understand the benefits of a DT, it will help to overcome the problem of confidences, the technology must be discussed further and clarified at a fundamental level. It is critical to ensure user confidence by ensuring the DT work as intended. Trust in DT grows as more is known about them. The enabling technologies will provide further insight into the measures taken to ensure that privacy and security policies are followed during growth, thereby resolving confidence issues [21].

6.4.4 Expectations

Despite market leaders Siemens and GE accelerating DT adoption, care is necessary to illustrate the obstacles that exist for DT aspirations and the requirement for further understanding. The essential for strong foundations for IoT organization, as well as a better analysis of the system needed for analytics, will ensure that businesses use DTT. Fighting the notion that the DT should use because of existing movements is also challenging. The profits and drawbacks of DT must be addressed in order to take appropriate measures when designing DT systems [22].

6.4.5 Standardized Modeling

Since there is no standardized method to modeling such systems, it is the next problem in all aspects of DT development. There must be a consistent approach from the original design to the simulation of a DT, whether it is physics-based or designed-based. Standardized techniques ensure field and consumer awareness, as well as knowledge flow between stages of a DT's design and execution [23].

6.4.6 Domain Modeling

An issue arising from the need for standardization use is meaning that knowledge about area use is passed to individually of the implementation and operational periods of a DT modeling. It means that areas such as IoT and data analytics are consistent, enabling the DT to be effectively used in the future. Are these relevant in the future because they mean they'll be considered when making DTs, as well as using IIoT and data analytics [24].

6.5 Research and Development Challenges

In production, the DT has set a new goal: to model the dynamic actions of production processes, taking into account external factors such as human intervention and technological constraints. Data acquisition, data collection, data processing, and data storage are all steps in the data evaluation, acquisition, processing, and storage process. Modeling the actions of a machine uses a statically dependent decision-making algorithm. Modeling the improvisation of IoT systems in production sector digital twin was used for the to improve dynamic modeling. It is essential to develop efficient algorithms for achieving consensus, cooperative learning, and shared detection. It will be important to build a geometric data structure to fuse assembly features and sensor measurements, as well as quick search algorithms to adjust and compensate for complex changes in the real world. The increasing use of DT faces major obstacles. Complex assets and their actions can easily outstrip financial and computational resources, data governance capacities, and even business performance when matched digitally with precision and in real time [25].

6.5.1 Cost

DT necessitates a major investment in technology platforms, model creation, and high-touch support. Although the majority of these costs are declining, the decision to introduce a DT should still be balanced against alternative alternatives that which offer comparable value at a lower cost. If an organization is only involved in a few main variables, then an IoT system–based on sensors and a traditional database could be a more cost-effective way to gather information.

6.5.2 Precise Representation

For time being, no DT can be an exact replica of its physical equivalent. Combining a complex asset's physical, chemical, electrical, and thermal states is a difficult and costly task. As a result, engineers are forced to make assumptions and simplify their models in order to match the desired twin attributes with technical and economic limits.

6.5.3 Data Quality

Data is important for good models. It may be hard to verify in DT applications that rely on data from hundreds or thousands of various sensing, many of which are in harsh field conditions and communicate over insecure networks. Companies would need to establish mechanisms for detecting and isolating bad data, as well as managing discrepancies and anomalies in product data sources, at the very least [26].

6.5.4 Interoperability

Despite considerable progress in terms of transparency and optimization, there are still technological and economic obstacles to data sharing. Furthermore, if a DT depends on simulation or AI supported by a single vendor, then it may be difficult or impossible to duplicate the features using alternative suppliers, potentially locking businesses into long-term single-supplier partnerships.

6.5.5 Intellectual Property Protection

DT is a storehouse of information and intellectual property. Details of a product's design and performance are included in the templates and data that make up a twin. It may also hold confidential information about customer processes and use. This poses problems in terms of data ownership, identity security, data management, and data access governance by various user groups.

6.5.6 Cyber Security

Cyber criminals would be attracted to DT. Malicious entities attempting to interrupt an organizational strategy will use the data links that attach physical objects to their twins as a new point of entry. When DT performs a role in the management of their physical counterparts, manipulating one of them may have immediate and potentially disastrous consequences in the real world. Because of these characteristics, successful management of DT cyber security is a key priority for many organizations, and one that will pose challenges. DT has advanced functional supply chain with in technology, production, oil, and automotive sector since their modest beginnings in the aerospace and defense industries [27].

6.6 Future Scope of Digital Twin Technology

Due to advancement ICT tools, DT is seeing greatly increased use in industry. As a result, DT systems are becoming more reliable and affordable. The potential of DT is truly phenomenal. As the concept and the technology evolve, the collaborative and predictive impact of DT cannot be undermined. As an example, take the case of industrial chemical reactors. Let us say a company has erected a chemical reactor and also created its DT [28, 29]. However, due to operator negligence, the proportion of chemicals that need to be mixed has been incorrect. The DT's control system notices some unusual sensor input that it interprets as crossing the danger threshold. As a precautionary measure, it immediately shuts down the reactor and alerts the crew. Depending on the severity of the situation, the crew then takes further action. Not only does the DT save the company money, but if the chemicals are dangerous, it also prevents a possible explosion and saves precious lives. Sectors that benefit from DT include aerospace, defense, automotive, heavy machinery, consumer goods and electronics, and power and energy. In addition, while the concept of DT primarily focuses on manufacturing, it is finding use cases in areas as diverse as healthcare, supply chain, and even smart cities. As per estimates, the market for DT is set to grow at a rate of more than 30% per year from 2020 till 2025. An increase in demand for IoT and cloud-based platforms like Thing Worx and a desire to be globally competitive are set to propel the demand for DT technology. India and other countries are already investing into IIoT; DT is one more step in the right direction.

6.7 Conclusion

The fact that several new papers and patents have been written in the last two years attests to this. Furthermore, some market leaders are starting to integrate DTs into their product lines. Without the same development in AI, IoT, and IIoT, which are becoming key enablers for DT, it would not be possible. The capability of DT is described, highlighting how they can be interfaced with other software tools to build the systems. This paper discussed about implementation steps, application and potential of DT, in addition challenges of DT in the context of data privacy and security, research and development challenges.

References

1. Aitken, A., *Industry 4.0: Demystifying Digital Twins*, pp. 1–16, Lanner, Executive Briefing, 2018, Available: www.lanner.com.
2. Anderton, C.B., *Challenges and Benefits of Implementing a Digital Twin in Composites Manufacturing*, Cgtech.Com, [Online]. Available: https://www.cgtech.com/images/pdf/Challenges-and-Benefits-of-Implementing-a-Digital-Twin-in-Composites-Manufacturing.pdf.
3. Kaya, C. A. G. B. A. and Unal, S. B., *Industrial Engineering in the Digital Disruption Era*. vol. 2019, Springer, 2020. [online] http://link.springer.com/10.1007/978-3-030-42416-9
4. Bhatti, G., Mohan, H., Raja Singh, R., Towards the future of ssmart electric vehicles: Digital twin technology. *Renew. Sust. Energ. Rev.*, 141, February, 110801, 2021. https://doi.org/10.1016/j.rser.2021.110801.
5. Butt, J., Exploring the interrelationship between additive manufacturing and industry 4.0. *Designs*, 4, 2, 1–33, 2020. https://doi.org/10.3390/designs4020013.
6. Cohen, Y., Faccio, M., Pilati, F., Yao, X., Design and management of digital manufacturing and assembly systems in the Industry 4.0 era. *Int. J. Adv. Manuf. Tech.*, 105, 9, 3565–3577, 2019. https://doi.org/10.1007/s00170-019-04595-0.
7. Stief, P., Dantan, J., Etienne, A., Siadat, A., A new methodology to analyze the functional and physical architecture of existing products for an assembly oriented product family identification service reliability assessment in cloud. *Proc. CIRP*, 72, II, 940–946, 2018. https://doi.org/10.1016/j.procir.2018.03.074.
8. Döllner, J., Geospatial Artificial Intelligence: Potentials of Machine Learning for 3D Point Clouds and Geospatial Digital Twins. *PFG – J. Photogramm. Remote Sens. Geoinf. Sci.*, 88, 1, 15–24, 2020. https://doi.org/10.1007/s41064-020-00102-3.
9. Durão, L.F.C.S., Haag, S., Anderl, R., Schützer, K., Zancul, E., Digital twin requirements in the context of industry 4.0, in: *IFIP Advances in Information and Communication Technology*, vol. 540, pp. 204–214, 2018, https://doi.org/10.1007/978-3-030-01614-2_19.

10. Fuller, A., Fan, Z., Day, C., Barlow, C., Digital Twin: Enabling Technologies, Challenges and Open Research. *IEEE Access*, 8, 108952–108971, 2020. https://doi.org/10.1109/ACCESS.2020.2998358.
11. Paperback, B., *Optics in Industrial Manufacturing*. vol. 1998, pp. 1–52, 1998. [online] https://www.futurebridge.com/industry/perspectives-mobility/application-of-digital-twin-in-industrial-manufacturing/
12. Ghita, M., Siham, B., Hicham, M., Abdelhafid, A., Laurent, D., Digital twins: development and implementation challenges within Moroccan context. *SN Appl. Sci.*, 2, 5, 1–14, 2020a. https://doi.org/10.1007/s42452-020-2691-6.
13. Ghita, M., Siham, B., Hicham, M., Abdelhafid, A., Laurent, D., Digital twins: development and implementation challenges within Moroccan context. *SN Appl. Sci.*, 2, 5, 1–14, 2020b. https://doi.org/10.1007/s42452-020-2691-6.
14. Global, M., *The transformation of digital twins in manufacturing*, pp. 1–16, 2021, [online] https://manufacturingdigital.com/procurement-and-supply-chain/transformation-digital-twins-manufacturing.
15. I-Scoop, Digital transformation: online guide to digital business transformation. *I-Scoop Blog*, 41, 2016. [online] https://www.i-scoop.eu/digital-transformation/
16. IoT, I., Industrial IoT: Rise of Digital Twin in Manufacturing Sector, *Biz4intellia*, 1, 614, 1–11, 2021, [Online]. Available: https://www.biz4intellia.com/blog/rise-of-digital-twin-in-manufacturing-sector/.
17. Kaivo-oja, J., Kuusi, O., Knudsen, M.S., Lauraeus, T., Digital twins approach and future knowledge management challenges: Where we shall need system integration, synergy analyses and synergy measurements?, in: KMO 2019, *Communications in Computer and Information Science*, L. Uden *et al.* (Eds.), vol. 1027, pp. 271–281, Springer Nature Switzerland AG, 2019, https://doi.org/10.1007/978-3-030-21451-7_23.
18. Parrott, A. and Warshaw, L., *Industry 4.0 and the digital twin*, pp. 1–17, Deloitte University Press, 2017, [Online]. Available: https://dupress.deloitte.com/dup-us-en/focus/industry-4-0/digital-twin-technology-smart-factory.html.
19. Qi, Q., Tao, F., Hu, T., Anwer, N., Liu, A., Wei, Y., Wang, L., Nee, A.Y.C., Enabling technologies and tools for digital twin. *J. Manuf. Syst.*, 58, 3–21, 0–1, 2019. https://doi.org/10.1016/j.jmsy.2019.10.001.
20. Raja Singh, R., Yash, S.M., Shubham, S.C., Indragandhi, V., Vijayakumar, V., Saravanan, P., Subramaniyaswamy, V., IoT embedded cloud-based intelligent power quality monitoring system for industrial drive application. *Future Gener. Comp. Sy.*, 112, 884–898, 2020. https://doi.org/10.1016/j.future.2020.06.032.
21. Rasheed, A., San, O., Kvamsdal, T., Digital twin: Values, challenges and enablers from a modeling perspective. *IEEE Access*, 8, 21980–22012, 2020. https://doi.org/10.1109/ACCESS.2020.2970143.
22. Rathore, M. M., Shah, S. A., Shukla, D., Bentafat, E., Bakiras, S., The Role of AI, Machine Learning, and Big Data in Digital Twinning: A Systematic

Literature Review, Challenges, and Opportunities. *IEEE Access*, 9, 32030–32052, 2021, doi: 10.1109/ACCESS.2021.3060863.

23. Sharma, M. and George, J.P., *Digital Twin in the Automotive Industry : Driving Physical-Digital Convergence*, p. 8, 2018. Report [online] https://www.tcs.com/content/dam/tcs/pdf/Industries/manufacturing/abstrct/industry-4-0-and-digital-twin.pdf
24. []Softengi_WhitePaper_Digital-Twins_4.pdf. (n.d.).
25. Street, N.M. and Free, T., *Digital Twin Market Size By Application (Product Design & Development, Machine & Equipment Health Monitoring, Process Support & Service), By End-Use Manufacturing, Healthcare, Automotive, Aerospace & Defense, Energy Utility, Infrastructure buildings*, 2022, Report [online] https://www.marketresearch.com/One-Off-Global-Market-Insights-v4130/Digital-Twin-Size-Application-Product-12526459/
26. Us, A., Us, C., Us, J., Posts, R., *Digital Twin – Applications and Use Cases What is a Digital Twin ? Applications and Use Cases of Digital Twin*, 2021. Report [online] https://softengi.com/blog/use-cases-and-applications-of-digital-twin/#~:text=Applications20of%20Digital%20Twin%20have,and%20enhance%20overall%20business%20efficiency.
27. Twin, D., *Digital Twin and its Impact in Industry 4.0*, pp. 1–7, 2021, [online] https://vizexperts.com/blog/digital-twin-and-its-impact-on-industry-4-0
28. Wagner, R., Schleich, B., Haefner, B., Kuhnle, A., Wartzack, S., Lanza, G., Challenges and potentials of digital twins and industry 4.0 in product design and production for high performance products. *Proc. CIRP*, 84, 88–93, 2019. https://doi.org/10.1016/j.procir.2019.04.219.
29. Wanasinghe, T.R., Wroblewski, L., Petersen, B.K., Gosine, R.G., James, L.A., De Silva, O., Mann, G.K., II, Warrian, P.J., Digital Twin for the Oil and Gas Industry: Overview, Research Trends, Opportunities, and Challenges. *IEEE Access*, 8, 104175–104197, 2020. https://doi.org/10.1109/ACCESS.2020.2998723.

7

Real-Time Data Acquisition System for PV Module

Durgesh Kumar*, Ila Ashok, Sweta Kumari, Dipanjali and Lawrence Kumar

Department of Nanoscience and Technology, Central University of Jharkhand, Ranchi, Brambe, India

Abstract

This paper presents a simplified data acquisition system for the photovoltaic (PV) module. The proposed work deals with the measurement of PV module electrical parameters under real-time outdoor conditions. The developed prototype is based on open-source software and hardware, which makes it cost-effective. It uses *Arduino Uno* and *Python* for data acquisition and post-processing under real-time working condition. The developed data acquisition system displays the current-voltage and power-voltage curve in real-time condition. It also computes module parameters such as short-circuit current, open-circuit, fill factor, and maximum power from the acquired data. An electronic load using N-Channel Power MOSFET has been realized, which has been controlled by suitable gate-source voltage. The developed prototype is tested under uniform illumination and partial shading condition. The accuracy of the data obtained by the proposed prototype is validated by comparing experimental outcomes with digital storage oscilloscope data. This study will be helpful for the performance evaluation of PV module at real-time outdoor condition.

***Keywords*:** Solar photovoltaics, data acquisition, partial shading, I-V characteristics, MOSFET, electronic load

**Corresponding author*: Durgesh.mitm@gmail.com

A. Chitra, V. Indragandhi, and W. Razia Sultana (eds.) Smart Grids and Green Energy Systems, (91–104)

7.1 Introduction

In recent years, there is an enormous increase in worldwide energy consumption. The main driving force behind this energy demand upsurge is the rapid population growth and techno-economic advancement [1]. The growing demand for energy puts a lot of pressure on conventional energy sources that create complex environmental issues such as global warming and pollution [1–3]. Thus, the research and development of the renewable energy field have gained tremendous attention worldwide. Among the numerous renewable energy sources, solar energy is attractive choice to full fill the energy needs, and it is most abundant resource of energy available on planet earth associated with no potential damage to the environment [1]. The significant advantages of solar energy have strongly motivated the scientific community to develop an efficient solar energy technology for energy harvesting. For the utilization of solar energy, photovoltaic (PV) systems are most widely used, which directly converts light into electricity [4].

As of today, crystalline silicon (Si)–based PV module holds an outstanding position among all the numerous kinds of solar technology in the PV industry. It represents a maximum market share of commercial PV module production due to its low cost and high life expectancy. The abundance of silicon makes the crystalline silicon solar cell the most viable candidate in PV industry [5, 6]. The mature fabrication technology also reduced the cost of silicon-based PV modules [7–9].

The characteristic current-voltage (I-V) curve is essential to assess the PV performance for efficient harvesting of energy [10, 11]. The nonlinear I-V characteristic curve is used to extract electrical parameters (voltage, current, and power) of the PV module. These parameters predict and assess the performance of the PV module as well as are also used in mathematical modeling of the PV module. In addition to this, maximum power generated by the module under real-time outdoor can also be computed [12]. There are different techniques reported in the literature [12, 13] to acquire the characteristic curve of PV module. These different methods include variable resistor method [13], capacitive load method [14], bipolar power amplifier method, four-quadrant power supply method, DC-DC converter method, and electronic load method [11, 13]. Among these methods, the electronic load method has been widely used where a power MOSFET is used for testing the PV module [11, 15–18], which has several advantages over the other methods. Fast scan rate can be achieved by using MOSFET, which mitigates the effect of change in ambient conditions. At the same time, the hardware design is also robust and cost effective as it uses few electronic components for the data acquisition. Such advantages provide enough motivation in this proposed work to use power MOSFET as variable-load to record the I-V data of PV module.

Generally, to trace I-V curve, the data acquisition of PV module is carried out in laboratories by employing a continuous or a pulsed light source. However, it has some drawbacks. The continuous light sources have poor spatial uniformity and pulsed light sources require a fast scan rate. If sunlight is used as the source of light, then it provides uniform illumination over the PV module [17]. In addition, the new PV modules with the unidentified spectral response could be evaluated conveniently under natural sunlight [17–20]. Thus, in this work, real outdoor condition is chosen for the electrical characterization of PV modules. At the same time, the present study also investigates the module performance at partial shading condition, which is considered as a major real-time problem that significantly reduces energy generation and creates excess heating in the shaded module [21, 22]. To mitigate the effect of partial shading bypass diode is used [23, 24]. The effect of using bypass diode on characteristic curve is demonstrated and discussed in the result section.

In literature, many virtual instruments have been demonstrated to facilitate the data acquisition for PV module under real-time condition. Most of the reported work uses LabVIEW for data acquisition and post processing [25–30], some of work also uses MATLAB [31]. In the proposed method, the open source software *Python* and the open source hardware *Arduino Uno* are used. This makes developed data acquisition cost effective, and, at the same time, *Python* codes are platform independent, which can be ported to another system, e.g., Raspberry Pi. The acquired data accuracy has been verified through a digital storage scope (DSO) and found to be in good agreement.

Apart from the "introduction" presented in Section 7.1, the arrangement of the present work is augmented as follows: Section 7.2 deals with the description and design of the developed PV module instrumentation system. The experimental setup and data acquisition system are presented in Section 7.3. Experimental results are discussed in Section 7.4. Conclusion of this work has been presented in Section 7.5.

7.2 Description of Instrumentation Setup

The schematic diagram of the proposed data acquisition system is shown in Figure 7.1. The system mainly consists of electronic load, control circuit, and PV module under test.

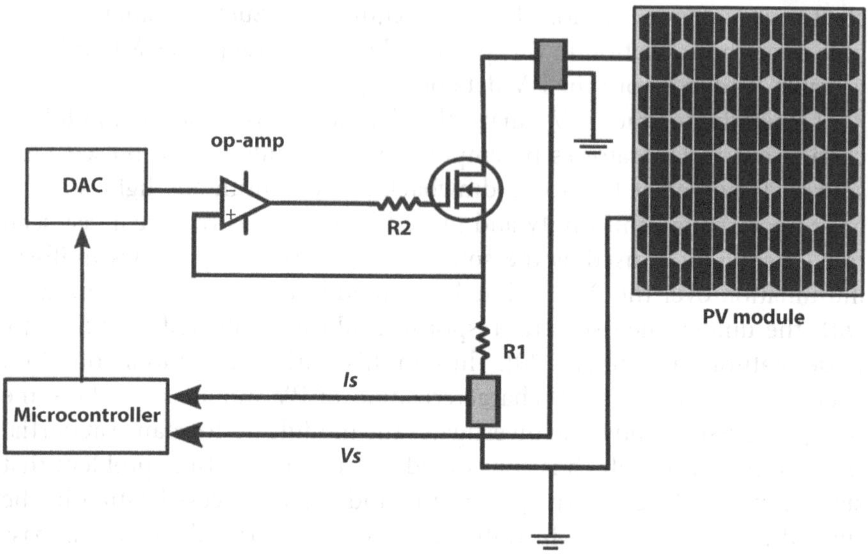

Figure 7.1 Schematic diagram of the data acquisition setup.

The PV module under the test has been connected to an electronic load realized with a power MOSFET and some inexpensive component. The MOSFET acts as a voltage controlled resistor. The resistivity of MOSFET is controlled by the gate voltage, which eventually controls the current drawn by the PV module. The operating point of the device is assessed by the characteristics of the PV module, test circuit configuration, and the characteristics of MOSFET [11]. The characteristics of the MOSFET in ohmic and constant current regions are expressed in Equations (7.1) and (7.2), respectively, as follows:

$$I_{DS} = K\left[(V_{GS} - V_{th}) \times V_{DS} - \frac{V_{DS}^2}{2}\right] \tag{7.1}$$

$$I_{DS} = K \times (V_{GS} - V_{th})^2 \tag{7.2}$$

where K is device constant, gate-source voltage is V_{GS}, V_{DS} is drain-source voltage, I_{DS} represents the drain current, and V_{th} is threshold voltage of MOSFET.

As seen from Figure 7.1, when the PV module is connected in this configuration, the PV current will be equal to drain current ($I_{DS} = I_{PV}$), which can be represented as follows [11]:

$$I_{DS} = I_{pv} = I_{sc} - I_{O} \exp (K_n V) \tag{7.3}$$

where V is the output voltage of the PV module, I_{sc} is short current of the module, I_{pv} is the PV current, I_o is the dark saturation current of the module, and K_n is constant. In Equation (7.3), series and shunt resistance of PV module are not taken into account. By applying the KVL, the following equation can be written:

$$V_{pv} = V_{DS} + I_{DS}R_1 \tag{7.4}$$

where V_{DS} is the voltage across MOSFET and R_I is the feedback resistance. The gate voltage controls the drain source resistance (R_{DS}), which eventually control the current drawn from PV module, thus the operating point of PV module could be changed by suitable gate-source voltage V_{GS} [11]. If the gate-source voltage V_{GS} is less than the threshold voltage V_{th}, then the MOSFET will be in the cutoff region and remains off. In cutoff region, MOSFET draw almost zero current from the PV module. When the MOSFET operates in ohmic region $((V_{GS} - V_t)>V_{DS})$, in this region, drain-source I_{DS} current varies linearly with gate-source voltage V_{GS}. In this region, Equations (7.1) and (7.4) are valid. When the MOSFET operates in constant current region $((V_{GS} - V_t) < V_{DS})$), Equations (7.2), (7.3), and (7.4) are valid. Thus, by changing the value of gate-source voltage V_{GS}, the operating point of the PV module maneuvers along its characteristic curve (I-V) between open-circuit voltage (V_{OC}) and short-circuit current (I_{SC}) [11].

The MOSFET gate is connected to the operation amplifier (opamp) as depicted in Figure 7.1. The inverting terminal of opamp is connected to the digital-to-analog converter (DAC) output. In this circuit configuration, opamp drives the MOSFET in such a way that voltage at the non-inverting terminal is made equal to voltage at the inverting terminal that is the same as DAC voltage. This is only possible when the voltage drop in the feedback resistance is equal to the DAC voltage. In the developed prototype, the value of feedback resistance R1 is 1 Ω, which makes DAC voltage the same as current drawn from the PV module.

The developed algorithm in *Python* starts the data acquisition process by initializing serial port and input/output port of the microcontroller board. *Python* is used for the data acquisition and post processing of the acquired data. The developed algorithm uses sequence of operations that are depicted in the flow chart shown in Figure 7.2.

The Arduino Uno serial ports appear as virtual serial COM ports for Python, which allows the data acquisition through different sensors. After initialization, Arduino Uno increments the DAC voltage and also

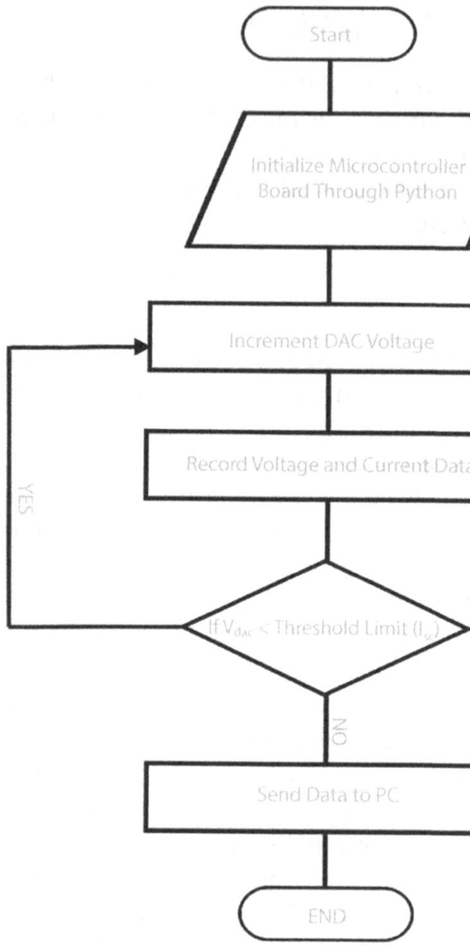

it records the current and voltage data from the sensors. When the DAC voltage becomes greater than short circuit of the module (i.e., $V_{dac} \geq I_{sc}$), it terminates the operation. After conversion of the recorded data, characteristic curves I-V and P-V and other parameters are displayed.

7.3 Experimental Setup and Data Acquisition System

The developed prototype to trace the I-V curve of PV module is depicted in Figure 7.3.

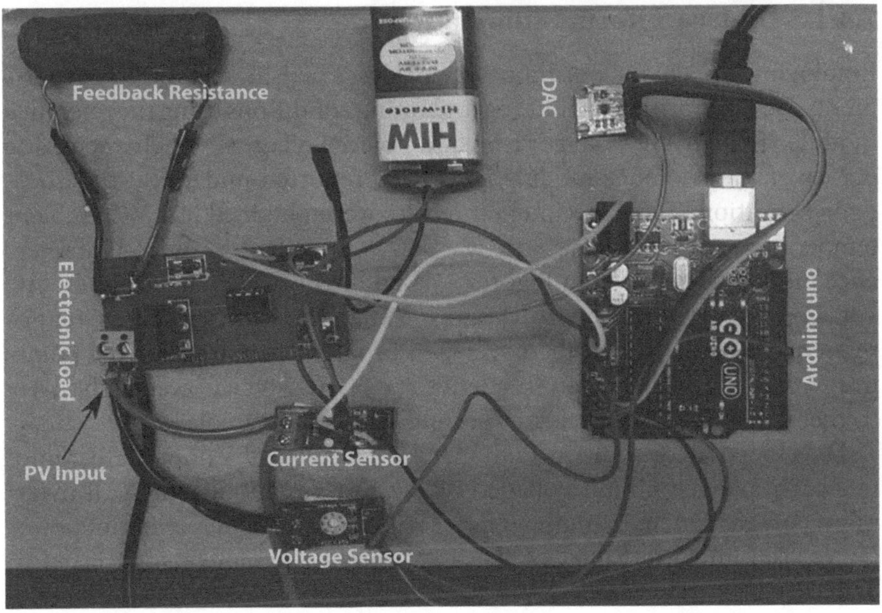

A *Microtek 10 W* polycrystalline silicon module is used as a reference for this work, which is placed under direct sunlight. An electronic load is realized using a power MOSFET IRFP460. To control hardware peripherals and acquiring the voltage and current data, ATmega328P microcontroller-based Arduino Uno board is used. A 10-bit I^2C-based DAC, MCP4725 is interfaced with Arduino Uno board, which is used to generate control signal. The current sensor (ACS 712) is used to measure the current drawn from the PV module, which can measure up to 5-A current. A voltage divider–based voltage sensor (KG045) is used for the measurement of module terminal voltage, which can measure up to 25 V. In addition to this, the setup consists of Operation Amplifier (LM358) resistor with 1-Ω feedback resistance.

7.4 Experimental Results

Using the developed prototype, data are acquired at uniform illumination and partial shading. Extracted curves are presented and discussed in following subsections.

7.4.1 Under Uniform Illumination

Employing the data acquisition system described in Section 7.2, experiments were performed on different operating conditions using the reference module. The current-voltage (I-V) and power-voltage (P-V) characteristic curves of the reference module have been reported under real-time outdoor condition. The experimental results were obtained at 22°C ambient temperature and solar irradiance was 950 W/m^2. The I-V and P-V of reference module is depicted in Figures 7.4a and b. In this configuration, DAC voltage was varied from 0 to 1 V, greater than the short-circuit current of reference module so that the developed prototype can trace all the points in I-V curve. The saw tooth frequency (i.e., DAC voltage) was set be 1 Hz so that it can trace one cycle in 1 second with 1,000 samples.

Developed algorithm (in Python) also calculates PV module parameters such as open-circuit voltage (V_{OC}), short-circuit current (I_{SC}), maximum current (I_m), maximum voltage (V_m), fill factor, and maximum power (P_{max}). The numeric values of these PV module parameters are listed in Table 7.1.

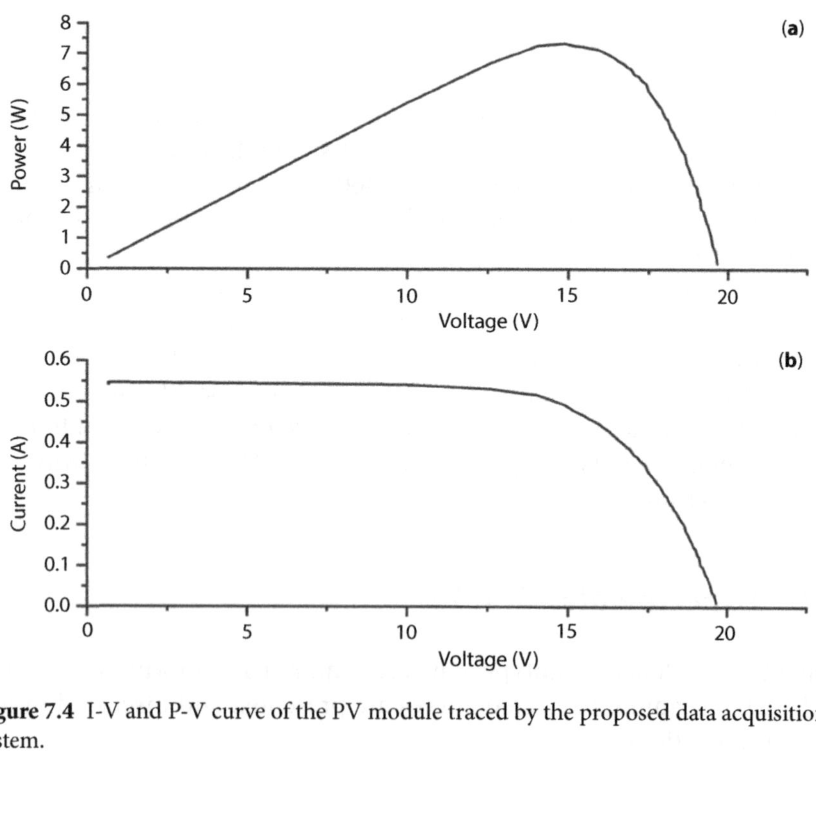

Figure 7.4 I-V and P-V curve of the PV module traced by the proposed data acquisition system.

Table 7.1 Calculated parameters of the reference module under uniform illumination.

I_{sc} (Short-Circuit Current)	0.5518
V_{oc} (Open-Circuit Voltage)	19.7510
P_{max} (Maximum Power)	7.3445
V_m (Maximum Voltage)	14.8926
I_m (Maximum Current)	0.4932

The above experimental data obtained through a virtual instrumentation system is validated by comparing it with digital storage oscilloscope waveforms. Here, the PV module, along with electronic load, has been connected to a digital oscilloscope to record the current and voltage waveform. The oscilloscope waveform of the voltage and current waveforms are depicted in Figures 7.5a and b.

A flat top is observed in the current waveform shown in Figure 7.5b acquired from the oscilloscope. As the DAC voltage is varied from 0 to 1 V,

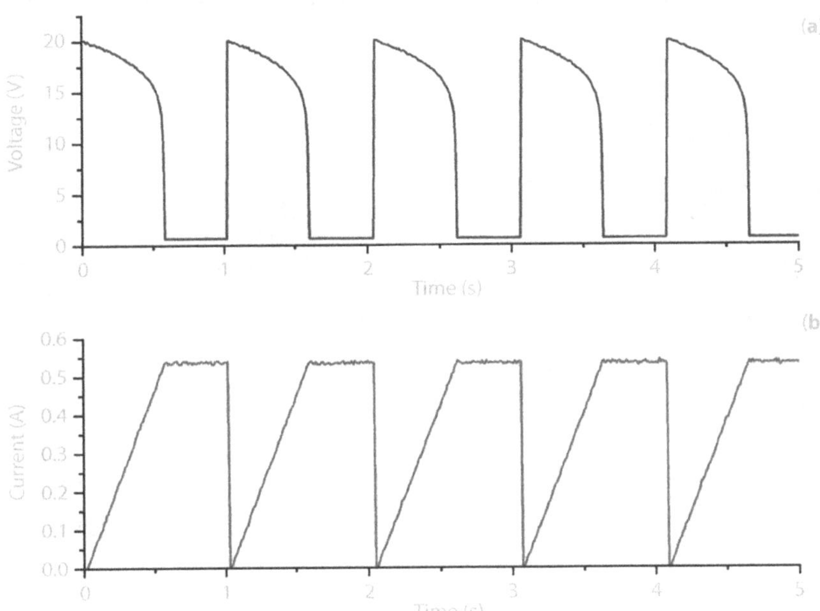

developed electronic load can draw up to 1A short current from the PV module. However, 0.5518 A was the value of short-circuit current drawn, which justifies the nature of the current waveform. The maximum power the reference module is around 7.3445 W. On comparison at several points, good agreement between oscilloscope data and acquired data from the developed prototype has been observed.

7.4.2 Under Partial Shading Condition

Performance evaluation of PV module at different partial shading conditions is also carried out. As an example, two reference modules connected in series connection are considered as depicted in Figure 7.6.

I-V and P-V, obtained under partial shading condition, are depicted in Figures 7.7a and b. During the shading, one module receives 820 W/m^2 irradiance and another module that is shaded receives 110 W/m^2 irradiance at ambient temperature 15°C.

Two power peaks that appeared are observed in the P-V curve, as depicted in Figure 7.7b. As the shaded module generates less photocurrent compared to module under uniform illumination, it behaves like a load (works in reversed bias condition), which significantly reduces generated output current. However, this effect on array is mitigated by using a bypass diode,

Figure 7.6 Partially shaded PV modules.

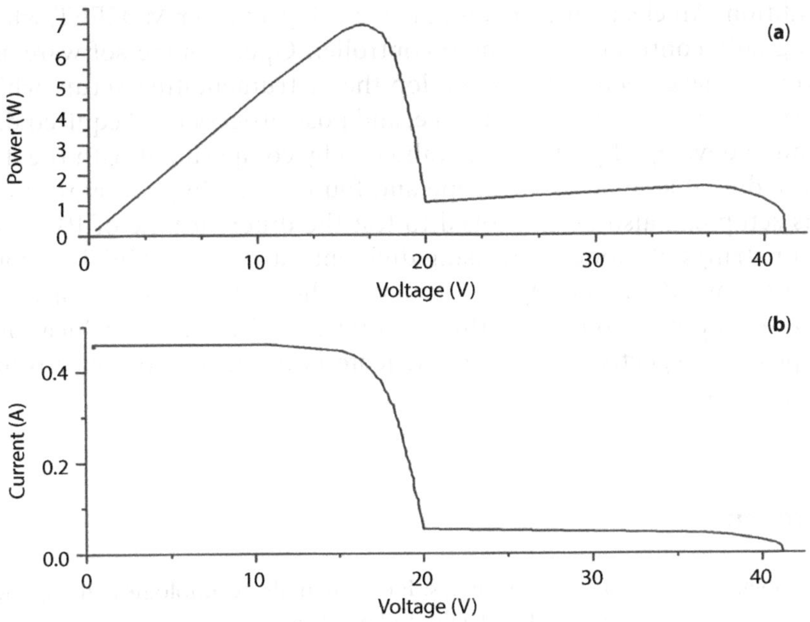

Figure 7.7 I-V and P-V curves of the PV module under the partial shading condition.

Table 7.2 Parameters of the reference module under partial shading condition.

I_{sc} (Short-Circuit Current)	0.4590 A
V_{oc} (Open-Circuit Voltage)	41.2301 V
P_{max1} (Global Power Peak)	6.9302 W
P_{max2} (Local Power Peak)	1.579 W
V_m (Maximum Voltage)	16.31 V
I_m (Maximum Current)	0.443 A

which justifies the two peaks in the P-V curve. Bypass diode also provides alternate path to the current to the shaded module. Different PV parameters are summarized in Table 7.2 extracted from curve shown in Figure 7.7.

7.5 Conclusion

A cost-effective standalone data acquisition system is developed to trace the characteristic curve of PV module under real-time operating condition. An electronic load is realized using a power MOSFET, which is digitally controlled by a microcontroller. Open-source software and hardware have been used to develop the instrumentation setup, which is used for data acquisition, storage, and post-processing. Acquired data through developed prototype is validated by comparing it with the data of the digital storage oscilloscope and found to be in good agreement. This setup can also be employed to test the different type of PV module or strings of modules by using different rating of MOSFET or combination of MOSFETs. This work may be helpful for the performance analysis of power converter. This prototype can be used for educational purpose, irrespective of having adequate knowledge about electronics engineering.

References

1. Solanki, C.S., Solar photovoltaics: Fundamentals, technologies and applications, Phi learning pvt. Ltd, Delhi, India, 2015.
2. Fahrenbruch, A. and Bube, R., Fundamentals of solar cells: Photovoltaic solar energy conversion, Elsevier, New York, USA, 2012.
3. Yadav, P., Pandey, K., Bhatt, V., Kumar, M., Kim, J., Critical aspects of impedance spectroscopy in silicon solar cell characterization: A review. *Renew. Sust. Energ. Rev.*, 76, 1562–1578, 2017.
4. Yadav, P., Tripathi, B., Pandey, K., Kumar, M., Investigating the charge transport kinetics in poly-crystalline silicon solar cells for low-concentration illumination by impedance spectroscopy. *Sol. Energy Mater. Sol. Cells*, 133, 105–112, 2015.
5. Saga, T., Advances in crystalline silicon solar cell technology for industrial mass production. *NPG Asia Mater.*, 2, 3, 96–102, 2010.
6. Yadav, P., Pandey, K., Tripathi, B., Kumar, C.M., Srivastava, S.K., Singh, P.K., Kumar, M., An effective way to analyse the performance limiting parameters of poly-crystalline silicon solar cell fabricated in the production line. *Sol. Energy*, 122, 1–10, 2015.
7. Lee, E., Lee, H., Choi, J., Oh, D., Shim, J., Cho, K., Lee, H., Improved LDSE processing for the avoidance of overplating yielding 19.2% efficiency on commercial grade crystalline Si solar cell. *Sol. Energy Mater. Sol. Cells*, 95, 12, 3592–3595, 2011.
8. Yin, W., Wang, X., Zhang, F., Zhang, L., 19.6% cast mono-MWT solar cells and 268 W modules. *IEEE J. Photovolt.*, 3, 2, 697–701, 2013.
9. Zhao, J., Wang, A., Green, M.A., Ferrazza, F., 19.8% efficient "honeycomb" textured multicrystalline and 24.4% monocrystalline silicon solar cells. *Appl. Phys. Lett.*, 73, 14, 1991–1993, 1998.

10. Humada, A.M., Hojabri, M., Mekhilef, S., Hamada, H.M., Solar cell parameters extraction based on single and double-diode models: A review. *Renew. Sust. Energ. Rev.*, 56, 494–509, 2016.
11. Kuai, Y. and Yuvarajan, S., An electronic load for testing photovoltaic panels. *J. Power Sources*, 154, 1, 308–313, 2006.
12. Van Dyk, E.E., Gxasheka, A.R., Meyer, E.L., Monitoring current–voltage characteristics and energy output of silicon photovoltaic modules. *Renew. Energy*, 30, 3, 399–411, 2005.
13. Duran, E., Piliougine, M., Sidrach-de-Cardona, M., Galan, J., Andujar, J.M., Different methods to obtain the I–V curve of PV modules: A review, in: *2008 33rd IEEE Photovoltaic Specialists Conference*, IEEE, pp. 1–6, 2008, May.
14. Muñoz, J. and Lorenzo, E., Capacitive load based on IGBTs for on-site characterization of PV arrays. *Sol. Energy*, 80, 11, 1489–1497, 2006.
15. Atia, Y., Zahran, M., Al-Hossain, A., Solar cell curves measurement based on LabVIEW microcontroller interfacing, in: *Proceedings of the 12th WSEAS International Conference on Automatic Control, Modelling & Simulation*, Catania, Sicily, Italy, pp. 29–31, 2010, May.
16. Zimmermann, U. and Edoff, M., A maximum power point tracker for long-term logging of PV module performance. *IEEE J. Photovolt.*, 2, 1, 47–55, 2011.
17. Jeevandoss, C.R., Kumaravel, M., Kumar, V.J., Sunlight based IV characterization of solar PV cells, in: *2011 IEEE International Instrumentation and Measurement Technology Conference*, IEEE, pp. 1–4, 2011, May.
18. Amiry, H., Benhmida, M., Bendaoud, R., Hajjaj, C., Bounouar, S., Yadir, S., Sidki, M., Design and implementation of a photovoltaic IV curve tracer: Solar modules characterization under real operating conditions. *Energy Convers. Manage.*, 169, 206–216, 2018.
19. Wagner, A., Peak-power and internal series resistance measurement under natural ambient conditions, in: *Proceedings EuroSun*, vol. 5, 2000, June.
20. Keogh, W.M. and Blakers, A.W., Accurate measurement, using natural sunlight, of silicon solar cells. *Prog. Photovolt.: Res. Appl.*, 12, 1, 1–19, 2004.
21. Dolara, A., Lazaroiu, G.C., Leva, S., Manzolini, G., Experimental investigation of partial shading scenarios on PV (photovoltaic) modules. *Energy*, 55, 466–475, 2013.
22. Kumar, D., Mishra, P., Ranjan, A., Dheer, D.K., Kumar, L., A simplified simulation model of silicon photovoltaic modules for performance evaluation at different operating conditions. *Optik*, 204, 164228, 2020.
23. Hemza, A., Abdeslam, H., Rachid, C., Aoun, N., Simplified methods for evaluating the degradation of photovoltaic module and modeling considering partial shading. *Measurement*, 138, 217–224, 2019.
24. Teo, J.C., Tan, R.H., Mok, V.H., Ramachandaramurthy, V.K., Tan, C., Impact of bypass diode forward voltage on maximum power of a photovoltaic system under partial shading conditions. *Energy*, 191, 116491, 2020.
25. El Hammoumi, A., Motahhir, S., Chalh, A., El Ghzizal, A., Derouich, A., Real-time virtual instrumentation of Arduino and LabVIEW based PV panel

characteristics, in: *IOP Conference Series: Earth and Environmental Science*, vol. 161, IOP Publishing, p. 012019, 2018, June.

26. Leite, V., Batista, J., Chenlo, F., Afonso, J.L., Low-cost IV tracer for photovoltaic modules and strings, in: *2014 International Symposium on Power Electronics, Electrical Drives, Automation and Motion*, IEEE, pp. 971–976, 2014, June.
27. Koutroulis, E. and Kalaitzakis, K., Development of an integrated data-acquisition system for renewable energy sources systems monitoring. *Renew. Energy*, 28, 1, 139–152, 2003.
28. Rezk, H., Tyukhov, I., Al-Dhaifallah, M., Tikhonov, A., Performance of data acquisition system for monitoring PV system parameters. *Measurement*, 104, 204–211, 2017.
29. Erraissi, N., Raoufi, M., Aarich, N., Akhsassi, M., Bennouna, A., Implementation of a low-cost data acquisition system for "PROPRE.MA" project. *Measurement*, 117, 21–40, 2018.
30. Martín, F.F., Llopis, M.V., Rodríguez, J.C., González, J.B., Blanco, J.M., Low-cost open-source multifunction data acquisition system for accurate measurements. *Measurement*, 55, 265–271, 2014.
31. Seo, S., Park, Y.S., Lee, Y.S., Development of a Cost-effective Data Acquisition System using an Open-source Hardware and Matlab/Simulink, in: *International Conference on Informatics in Control, Automation and Robotics*, vol. 2, SCITEPRESS, pp. 484–491, 2016, July.

8

Investigation of Controllers for "N" Input DC-DC Converters

A. Lavanya*, J. Divya Navamani, Nivas Jayaseelan and A. Geetha

Department of Electrical and Electronics Engineering, SRM Institute of Science and Technology, Kattankulathur, Tamil Nadu, India

Abstract

This paper discusses the role of controllers in "N" input DC-DC converters. Features like output voltage, current regulation, power flow management, and switching mode selection are vital in converters interfacing more than one input source and demanding while designing the controller. This challenge needs to be addressed in selecting an appropriate control scheme based on the application. The control structure should generate control signals for extracting the commanded power from the sources in the input side, supplying individually or simultaneously by maintaining regulated output voltage. This paper investigates the different types of controllers used in the N input converter, including the Proportional Integral Derivative controller, most commonly used and suitable for almost all types of topologies. The response of the controller for the N input converter is simulated and the results are compared. This paper also gives an overview of different types of controllers employed in the literature for various applications.

Keywords: Closed-loop control, controllers, DC-DC converter, multi-input converters, PI, PID

8.1 Introduction

Multiple input converters focus on researchers' attention due to their versatility in combining energy from different sources. The multi-input converter, as shown in Figure 8.1, allows the power from more than two sources to

**Corresponding author*: lavanyaa@srmist.edu.in

A. Chitra, V. Indragandhi, and W. Razia Sultana (eds.) Smart Grids and Green Energy Systems, (105–118)

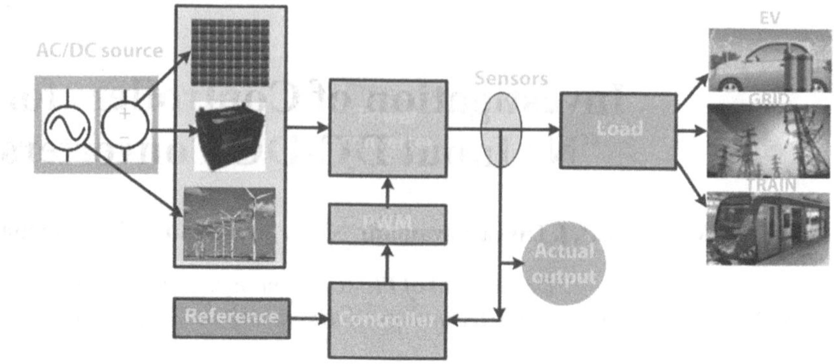

be combined with intermittent power and then transferred to a single load or a DC bus. Prabhala *et al.* [1] and Lavanya *et al.* [2] suggested integrating different energy sources with a standard single-input converter via a multi-input converter instead of a standard interface between each source and the DC bus. Affam *et al.* [3] gave a comprehensive review about the role of multi-input converters in applications like hybrid electric cars and renewable energy sources. For hybrid electric vehicles, the benefits of a high energy density source, for example, a battery, are coupled with the advantages of a power source, such as an ultra-capacitor, utilizing a multi-input converter as suggested by Angalaeswari *et al.* [4]. This enables the battery to provide consistent energy throughout the operation and the ultra-container can offer the high load current if required. A comparative study is made by introducing different controller tuning methods.

"N" input DC converters employed in power supplies play an important role in many applications such as telecommunications, grids, automotive, and robotic systems. These applications need to be controlled output and current voltage regardless of load and supply interruptions.

8.2 Role of Control Technique in Multivariable System

In general, voltage control and current control are two methods used to regulate the output voltage fluctuations, as illustrated in Figures 8.2 and 8.3. The controller reduces the discrepancy between the actual output and the intended value used in the system and decreases the error. In switched-mode converters, the controller is the critical aspect of stability and precision.

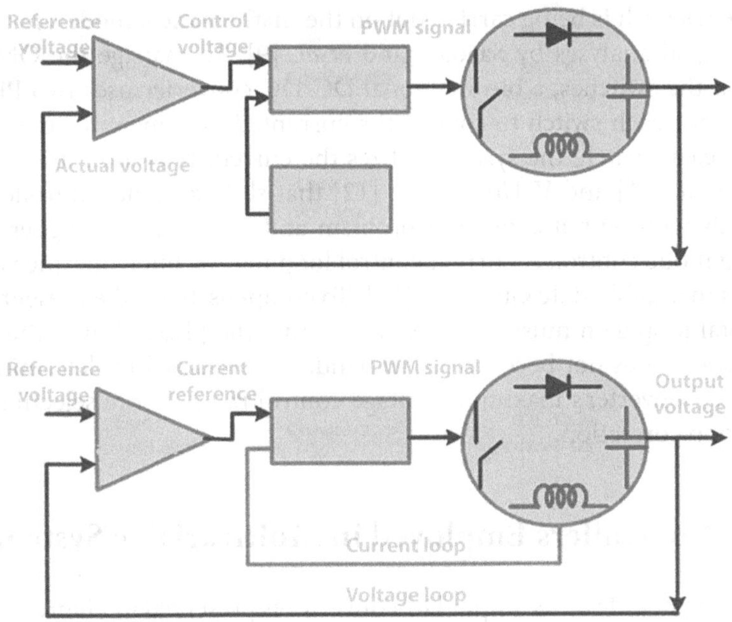

Voltage mode control was the main form of control due to load fluctuations and loop gain variation, coupled with input voltage as given by Naga Sai Kalyan *et al.* [10]. The alternative solution mode monitoring has been developed for this purpose. The PWM signal is produced by putting the control voltage on one input of the comparator and the transport voltage at the other input in a voltage mode control method. The PID controller is a mixture of proportional, integral, and derivative control. The PID controller can remove a continuous error of state Pirasteh-Moghadam *et al.* [5]. The proportional, integrated, and derivative control is changed to obtain a better solution to improve transient responsiveness as discussed by Wang *et al.* [6]. Changing one parameter will impact two other variables' effectiveness Sushita *et al.* [7], discuses about the effectiveness of PID controller compared to PI controller for BLDC motor drive. The generic PID family controllers, which are widely employed, need accurate linear mathematical models as given by Nobile *et al.* [8]. In parameter fluctuation, non-linearity, and charge disturbance, the PID controllers have failed to function adequately. In the context of a single switching cycle, DC-DC converters have three potential configurations. In this instance, state-space averaging techniques are the most often utilized way to construct a mathematical model as cited in Nobile *et al.* [8]. Only small signals may utilize the models generated using these techniques.

The research is being carried out on the mathematical models for large-scale signal analyses by Saadatmand *et al.* [9]. To manage the ON/OFF state of the switches, a bi-directional DC/DC converter uses two PI controllers for each switch to control the current. To compute the duty cycle (D) of each switch, the system utilizes the current reference. It is clear in Ding *et al.* [11] and V. Utkin *et al.* [12] that sliding mode controller easily deals with output constraint problem and advantages of higher order sliding mode control. A current control loop may be built after the voltage loop is in a stable state Guo *et al.* [13]. To compensate for the present loop, the total loop gain must be decreased so that the phase shift at the unity gain point does not become out of hand. It is essential to determine the DC-DC converter's maximum voltage control range before selecting current-loop correction.

8.3 Controllers Employed in Multivariable System

The basic module of "N" input DC converter is presented in Figure 8.4. The various types of linear and non-linear controllers are utilized in these converters to achieve the desired specification and performance. Since these converters have more than one switch, more than one converter are also employed as shown in Figure 8.4. The controllers like PI, sliding mode, intelligent controllers like fuzzy, and advanced controllers like model predictive and hybrid controllers are tested in similar converters.

By using the sliding mode, a design may be implemented that is unaffected by changes in parameters and external influence Markkassery *et al.* [18]. There are two ways of approaching the situation. The reaching mode in which the trajectory shoots from any beginning position and

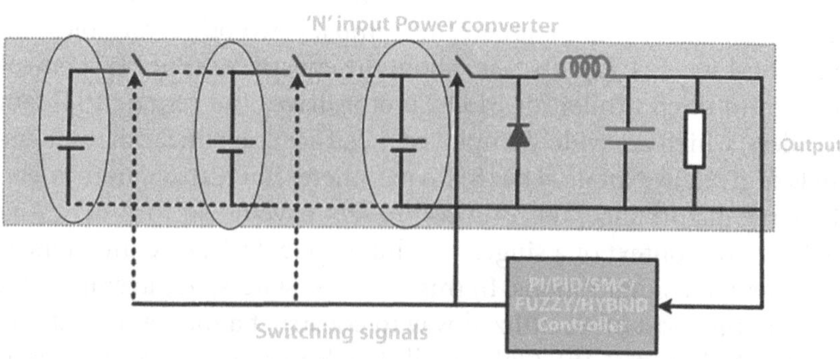

travels toward the switching surface is more important while designing the controller. It is called a sliding mode, which has the trajectory slide down the switching surface. As the surface equation is changing, the system remains constant in this mode as discussed by Santhosh *et al.* [19] and Noor-ul-Ain Hanif *et al.* [20]. In design, control law is applied to the converter to design the sliding regime and select a sliding surface, whose properties meet asymptotically the required dynamic. Therefore, the control law is constructed so that it gets the desired surface. An important control technique has to control the sources based on input cells. The current and voltage control cells are used as the to obtain the controlled output current needed for the proposed converter. For a consistent output current, the use of multiple voltage and current sources is recommended.

The switching structure has a substantial part in determining the power utilization of the different sources connected, i.e., if power is varied simultaneously or independently. This suggests the importance of the selection of controller. Especially when the converter has PV sources as input there is a compulsion to draw maximum power from the constantly varying source. Therefore, the switching sequence is based on the controller and the algorithm used to meet the desired objectives. Without controller this objective cannot be achieved. The hybrid type control technique is mostly found in this converter structure in order to overcome the drawbacks in conventional controllers. In applications like electric vehicle to utilize the energy sources effectively, a power electronic interface with suitable controller is required. A hybrid fuzzy PI-based control scheme is proposed to control and track the predefined speed profile in electric vehicle as illustrated in Figure 8.5.

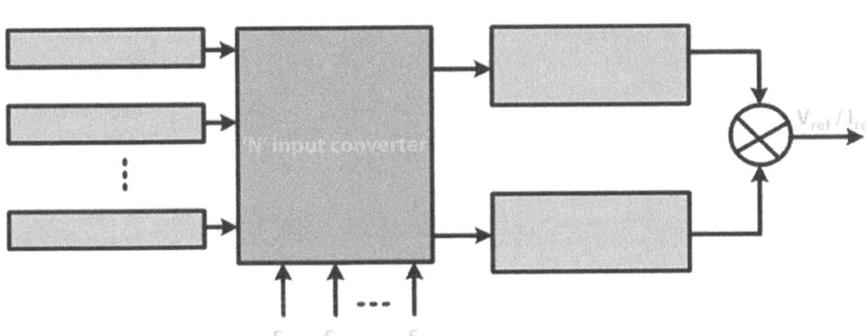

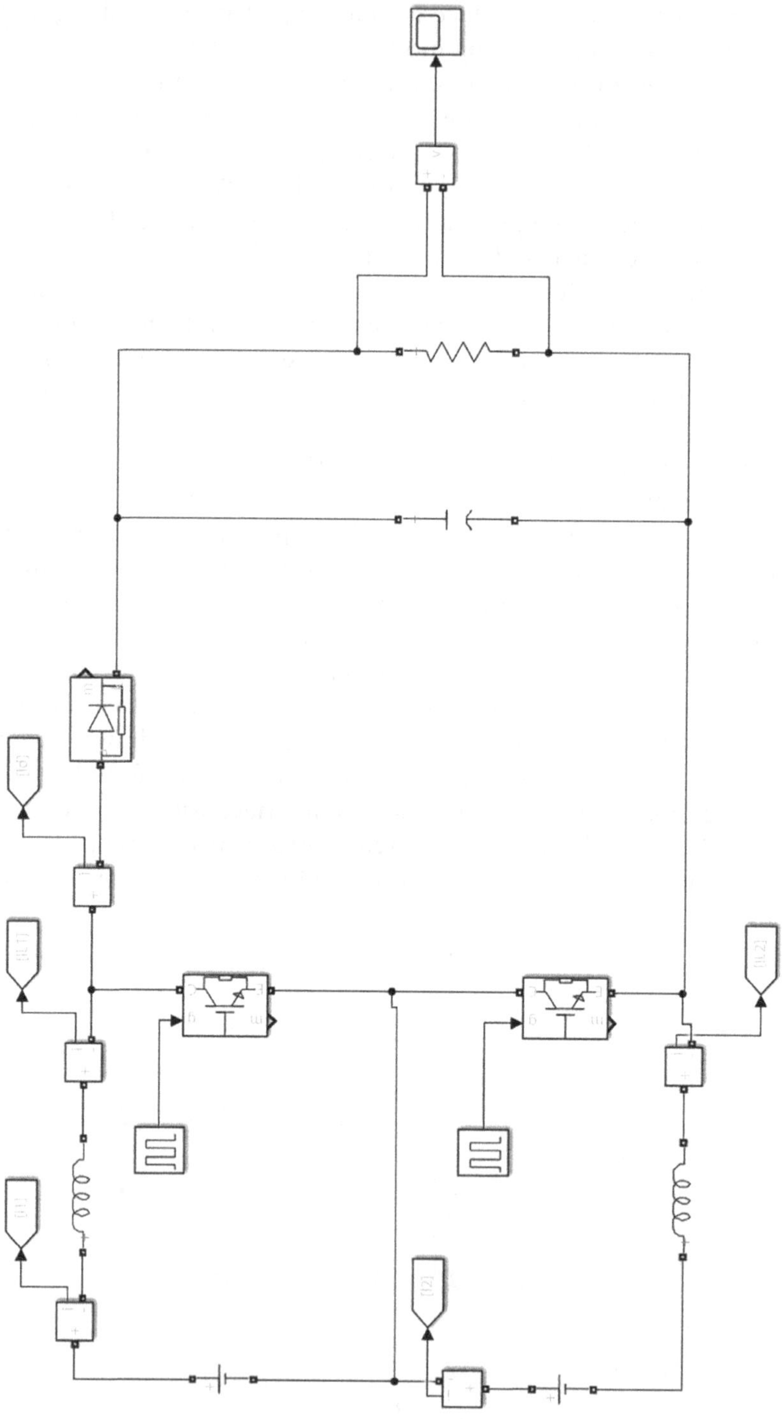

(a)

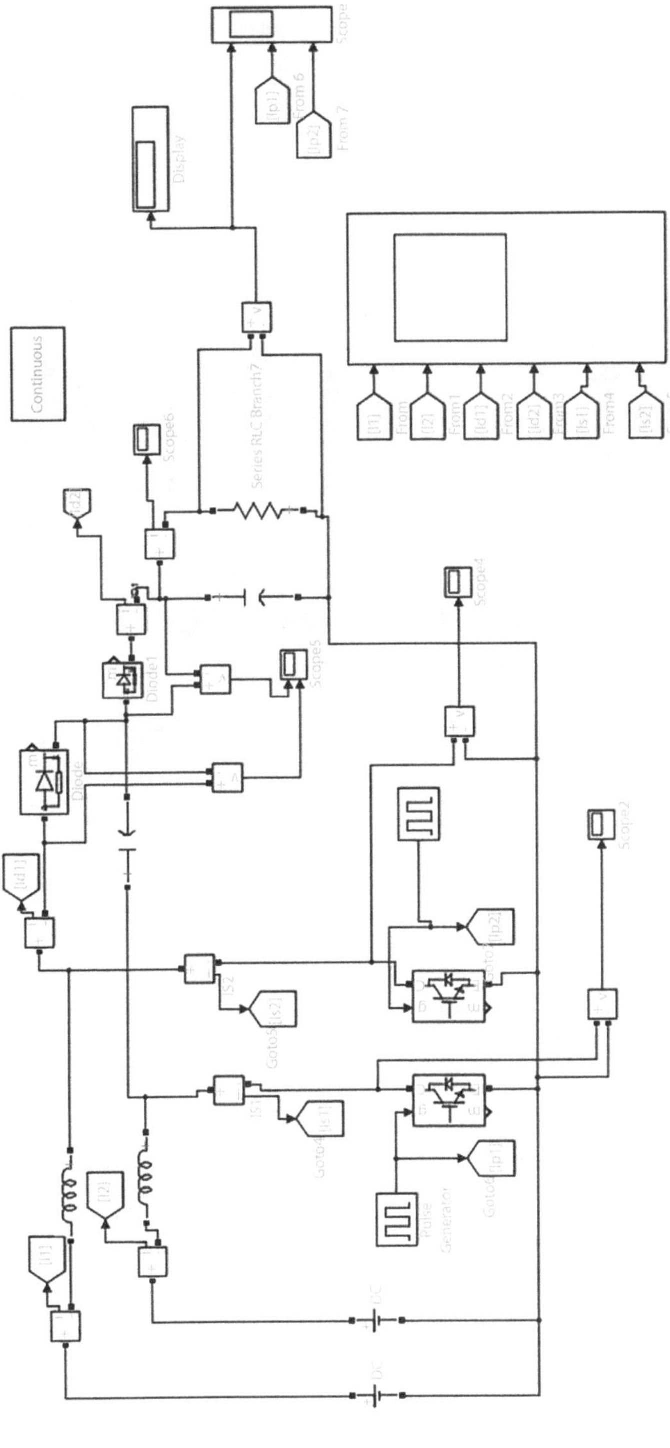
Continuous
Scope
Display
From 6
From 7
[Ip1]
[Ip2]
Scope1
From
From1
From2
From3
From4
From5
[I1]
[I2]
[Id1]
[Id2]
[Is1]
[Is2]
Series RLC Branch7
Scope6
Scope5
Scope4
Scope2
Diode
Diode1
IS1
IS2
Goto4
Goto5
Goto6
Pulse Generator
DC

(b)

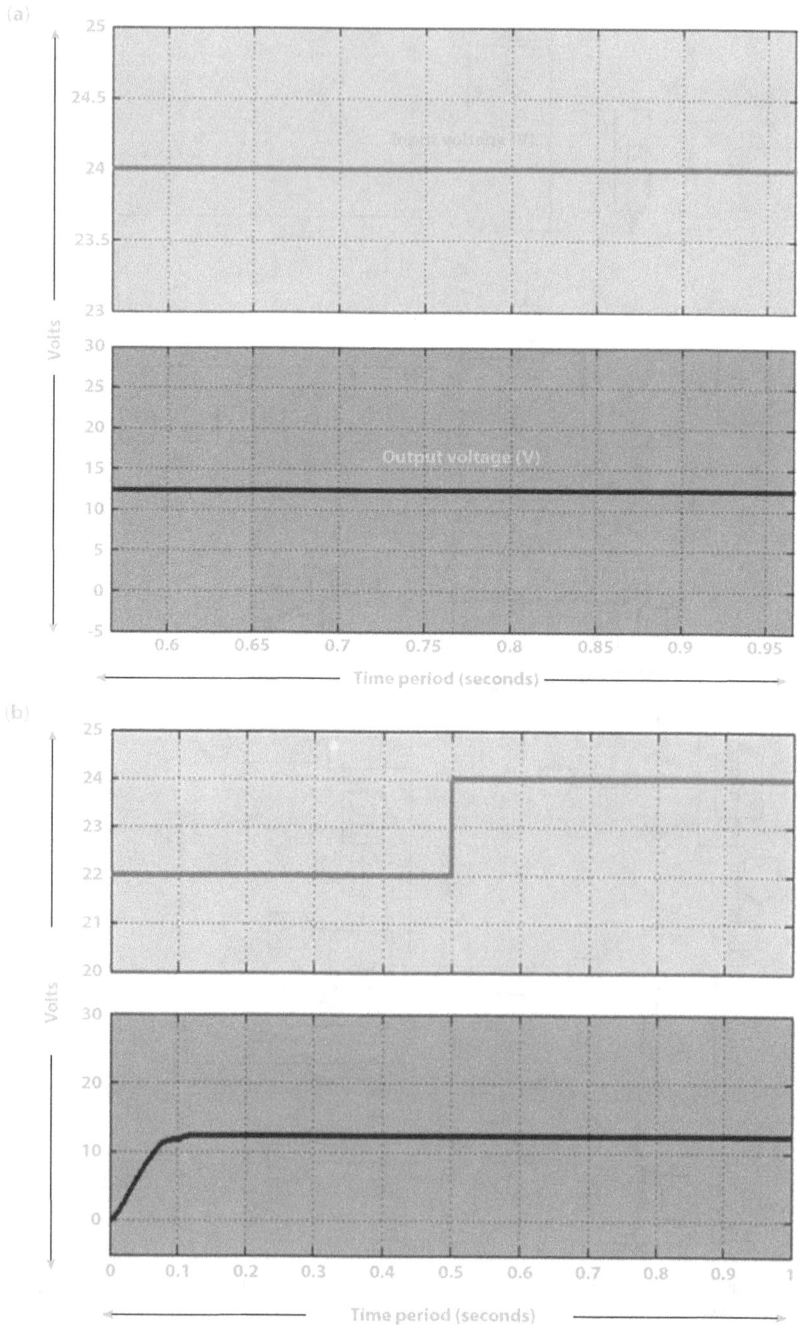
(a)
Volts
Output voltage (V)
Time period (seconds)
(b)
Volts
Time period (seconds)

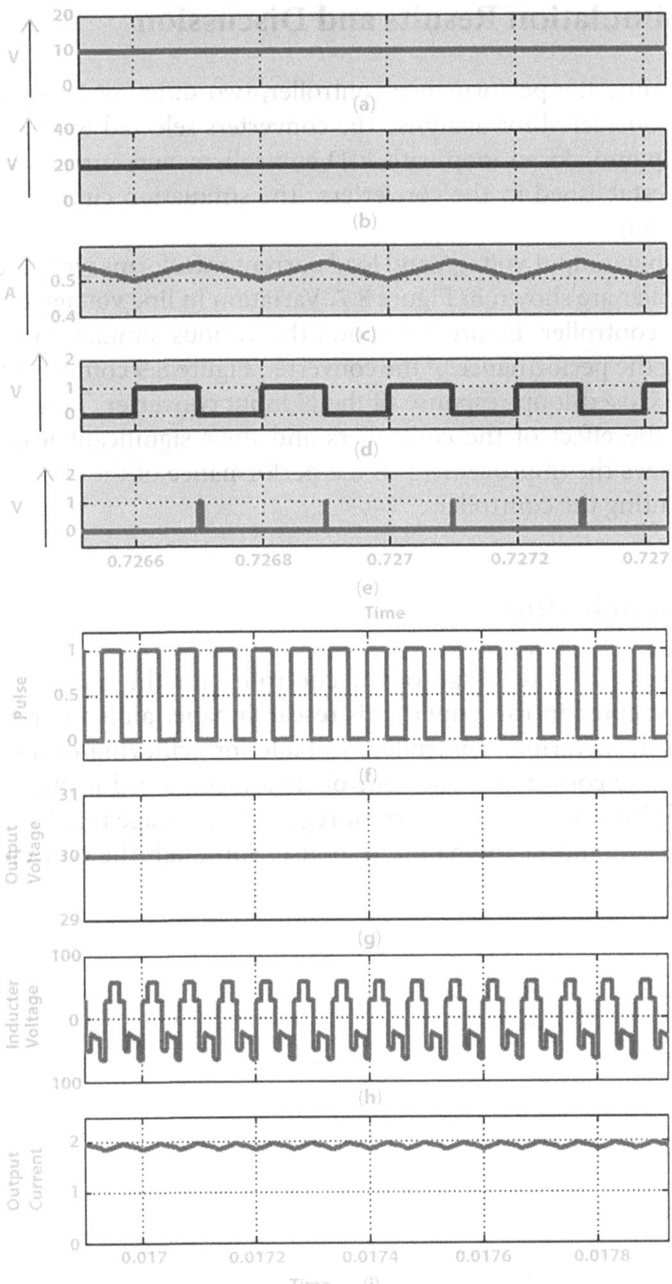
V
A
Time
Pulse
Output Voltage
Inducter Voltage
Output Current
0.7266
0.7268
0.727
0.7272
0.017
0.0172
0.0174
0.0176
0.0178
(a)
(b)
(c)
(d)
(e)
(f)
(g)
(h)
(i)

8.4 Simulation Results and Discussion

To determine the performance controller, two different N input topologies are considered for analysis. The converters selected were analyzed in open loop and closed loop with PID controllers, and current mode control was established in the converters. The simulation circuits are shown in Figure 8.6.

The input, output voltage, and load current waveforms with and without PI controller are shown in Figure 8.7. Variation in line voltage is neglected using PI controller. Figure 8.8 shows the various simulation waveforms depicting the performance of the converter. Figure 8.9 compares the open-loop and closed-loop response of the N input converter. Table 8.1 throws light on the effect of the controllers and their significant features. This also displays the improvement in the performance of the converters after implementing the controllers.

8.5 Conclusion

The importance and design of the controller for the multivariable system is presented in this paper. This research paper aims to bring out the comparison of various controllers suitable for achieving desired performance in the converters. The controllers are employed in these converters for multiple reasons like current regulation, voltage regulation, power management, and fast dynamic response. Although the conventional PI

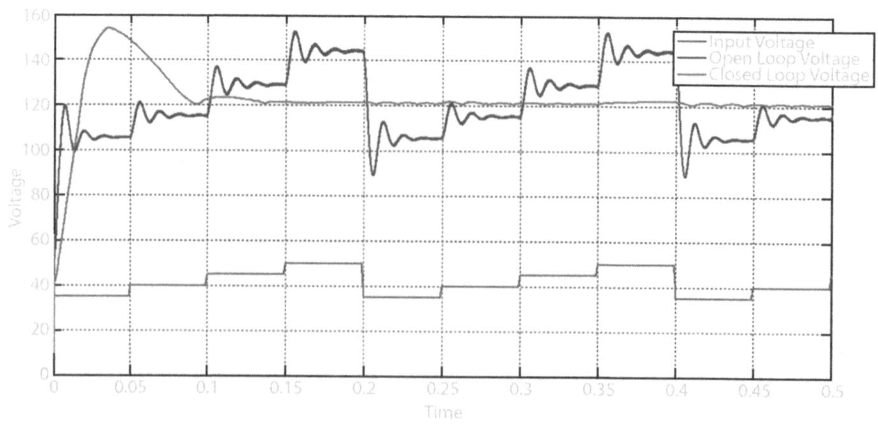

Table 8.1 Comparative analysis of state of art of controllers for multivariable systems.

Ref	Converter type	Controller	Objective	Features	Application
[14]	Double-Input Buck Converter (DIBC)	One-Cycle Control	To eliminate the interactions of the coupling loops in multi-input system	The control design is greatly simplified	Renewable energy application
[15]	Multiple Input Converter	Hybrid Fuzzy PI Controller	Power management control scheme	Robustness and high tracking capability	Electric vehicle
[16]	Two Input Fourth-Order Boost DC-DC Converter	Voltage-Mode Control-Based Controller	Bus voltage regulation and power management	Regulates the output bus voltage and concurrently supplies further load demand	Low voltage low power DC distribution system
[17]	DI-DO Converter	Constant Current Control and Direct Duty-Cycle Control	Deliver power to the loads simultaneously.	Flexible power control avoids cross-regulation problem and time-multiplexing control	Solar power supplies

(Continued)

Table 8.1 Comparative analysis of state of art of controllers for multivariable systems. (*Continued*)

Ref	Converter type	Controller	Objective	Features	Application
[18]	SIMO Bipolar Power Converters	Nonlinear Controller Based on Lyapunov Design	Output voltage regulation	A nonlinear Lyapunov control scheme gives improved performance during transient period than feedback conventional controller	Bipolar power supplies
[19]	DIDO Converter	One-Step Ahead Control	Utilization its operational modes based on the accessibility of the sources	Persistent current recovery from the input sources regardless of the load settings	Hybrid electric vehicle
[20]	Multi-Input Converter	Fuzzy Logic Controller	Multiplexed approach	Voltage regulation and time multiplexing	Smart DC microgrid

controllers are commonly used, tuning the parameters in these traditional controllers is very difficult from the above investigation. Therefore, controllers with the predictive model method are more suitable for the multivariable system. However, the conventional advanced controllers might result in ripples due to their variable switching frequency. After examination, it was realized that the hybrid fuzzy PI controller was chosen depending on the operation in steady-state and transient state. PI controller performs well in steady-state, and fuzzy PI outperforms during the transient period. Thus, this paper explores the significant features of the controllers and their effect in the performance of N input DC converters with the simulation results presented. The investigation carried out gives an idea for selecting suitable controllers depending on the objective function, simplicity, and accuracy.

References

1. Prabhala, V.A.K., Khazraei, M., Ferdowsi, M., A new control strategy for a class of multiple-input DC-DC converters, in: *2014 International Conference on Renewable Energy Research and Application (ICRERA)*, pp. 508–513, 2014, doi: 10.1109/ICRERA.2014.7016436.
2. Lavanya, A., Jayaseelan, N., Navamani, J.D., Kumar, K.V., Dual input DC-DC converter for renewable energy systems, in: *2017 International Conference on Inventive Systems and Control (ICISC)*, pp. 1–5, 2017, doi: 10.1109/ICISC.2017.8068598.
3. Affam, A., Buswig, Y.M., Othman, A.-K. B. H., Julai, N.B., Qays, O., A review of multiple input DC-DC converter topologies linked with hybrid electric vehicles and renewable energy systems. *Renew. Sust. Energ. Rev.*, 135, 110186, 2021. doi: 10.1016/j.rser.2020.110186.
4. Angalaeswari, S., Subbulekshmi, D., Deepa, T., Krithiga, S., Jha, A., Bidirectional DC-DC converter for regenerative fuel cell electric vehicle. *Mater. Today: Proc.*, Feb. 2021. doi: 10.1016/j.matpr.2021.01.243. (Article in press)
5. Pirasteh-Moghadam, M., Saryazdi, M.G., Loghman, E., Ali, K.E., Bakhtiari-Nejad, F., Development of neural fractional order PID controller with emulator. *ISA Trans.*, 106, 293–302, 2020. doi: 10.1016/j.isatra.2020.06.014.
6. Wang, H., Luo, Y., An, W., Sun, Q., Xu, J., Zhang, L., PID Controller-based stochastic optimization acceleration for deep neural networks. *IEEE Trans. Neural Netw. Learn. Syst.*, 31, 12, 5079–5091, 2020. doi: 10.1109/TNNLS.2019.2963066.
7. Sushita, K. and Shanmugasundaram, N., Performance and comparative analysis of Bldc motor with Pi and Pid controllers. *Ann. Rom. Soc.Cell Biol.*, 5, 3, 219–228, 2021. [Online]. Available: https://www.annalsofrscb.ro/index.php/journal/article/view/1500.

8. Nobile, G. *et al.*, Performance assessment of large photovoltaic (PV) plants using an integrated state-space average modeling approach. *Energies*, 13, 18, 4777, 2020. doi: 10.3390/en13184777.
9. Saadatmand, M., Mozafari, B., Gharehpetian, G.B., Soleymani, S., Optimal PID controller of large-scale PV farms for power systems LFO damping. *Int. Trans. Electr. Energy Syst.*, 30, 6, e12372, 1–14, 2020. doi: 10.1002/2050-7038.12372.
10. Kalyan, C.N.S. and Rao, G.S., Frequency and voltage stabilisation in combined load frequency control and automatic voltage regulation of multiarea system with hybrid generation utilities by AC/DC links. *Int. J. Sustain. Energy*, 39, 10, 1009–1029, 2020. doi: 10.1080/14786451.2020.1797740.
11. Ding, S., Park, J.H., Chen, C.-C., Second-order sliding mode controller design with output constraint. *Automatica*, 112, 108704, 2020. doi: 10.1016/j.automatica.108704.
12. Utkin, V., Poznyak, A., Orlov, Y., Polyakov, A., Conventional and high order sliding mode control. *J.Franklin Inst.*, 357, 15, 10244–10261, 2020. doi: 10.1016/j.jfranklin.2020.06.018.
13. Guo, V.P. *et al.*, Hybrid model predictive control for modified modular multilevel switch-mode power amplifier. *IEEE Trans. Power Electron.*, 36, 5, 5302–5322, 2021. doi: 10.1109/TPEL.2020.3031271.
14. Yang, D., Yang, M., Ruan, X., One-cycle control for a double-input DC/DC converter. *IEEE Trans. Power Electron.*, 27, 11, 4646–4655, 2012. doi: 10.1109/TPEL.2011.2164582.
15. Vidhya, S.D. and Balaji, M., Hybrid fuzzy PI controlled multi-input DC/DC converter for electric vehicle application. *Automatika*, 61, 1, 79–91, 2020. Available from: https: //dx.doi.org/10.1080/00051144.2019.1684038.
16. Kulshreshtha, A., Saxena, A.R., Veerachary, M., Non-isolated fourth-order boost DC-DC converter for power management in low voltage low power DC grids: Design and interaction analysis. *IEEE Access*, 8, 196500–196514, 2020. doi: 10.1109/ACCESS.2020.3034181.
17. Li, X.L., Dong, Z., Tse, C.K., Lu, D.D.-C., Single-Inductor Multi-Input Multi-Output DC–DC converter with high flexibility and simple control. *IEEE Trans. Power Electron.*, 35, 12, 13104–13114, 2020. doi: 10.1109/TPEL.2020.2991353.
18. Markkassery, S., Saradagi, A., Mahindrakar, A.D., Lakshminarasamma, N., Pasumarthy, R., Modeling, design and control of non-isolated single-input multi-output Zeta–Buck–Boost converter. *IEEE Trans. Ind. Appl.*, 56, 4, 3904–3918, 2020. doi: 10.1109/TIA.2020.2984190.
19. Santhosh, T.K. and Govindaraju, C., Dual input dual output power converter with one-step-ahead control for hybrid electric vehicle applications. *IET Electr. Syst. Transp.*, 7, 3, 190–200, 2017. DOI: 10.1049/iet-est.2016.0017.
20. Hanif, N.-U.-A., Kashif, S.A.R., Saqib, M.A.,., Multiplexed control strategy for a multi-input converter using fuzzy logic algorithm. *IET Electron. Lett.*, 52, 15, 1327–1329, 2016. DOI: 10.1049/el.2016.0940.

9

Fuzzy Logic Controlled Dual-Input DC-DC Converter for PV Applications

Nivas Jayaseelan*, A. Lavanya and J. Divya Navamani

Department of Electrical and Electronics Engineering, SRM Institute of Science and Technology, Kattankulathur, Tamil Nadu, India

Abstract

This paper recommends a dual-input DC-DC converter with fuzzy logic (FL) controller to interface renewable sources and to upgrade the reliability of the system. The ability of the proposed converter to work in distinct modes of operation is investigated and its performance is analyzed in open-loop and in closed-loop systems. For the dual-input converter, closed-loop control is carried out with PI control and FL control. A FL controller is proposed to regulate the flow of power, to and from the battery, and to provide fast regulation during fluctuations in the line and in the load side. MATLAB simulation tool is used for open and closed-loop control with PI and fuzzy controller, the results are compared and the effectiveness of dual-input converter to supply the load with FL control is proved. A 50-W experimental setup of the dual-input converter is realized and the outputs are discussed.

Keywords: Fuzzy, power management, closed loop, control, PI controller, dual input

9.1 Introduction

Primary energy sources are getting depleted rapidly; hence, there is a need for alternative renewable energy sources. Since renewable energy is cleaner and has less carbon count, it has attracted many researchers. Recently, there is an increase in the commission of more efficient solar power generation

**Corresponding author*: lavanyaa@srmist.edu.in

A. Chitra, V. Indragandhi, and W. Razia Sultana (eds.) Smart Grids and Green Energy Systems, (119–136)

systems in India as stated by Zhou *et al.* [1]. The newly launched 2,000-MW solar power plant in Karnataka will be the largest in the world in the future concerning land occupancy and generation capacity. In foreseeable future, it is anticipated that the power converters play an important role with renewable source as discussed by Lavanya *et al.* [2] and Prabhala *et al.* [3].

Earth's primary energy sources are coal and crude oil. These conventional energy sources are gradually depleting and also cause serious environmental problems. Hence, the most important measure in modernization is to harness renewable and clean energy sources. Multiple-input converters can use either one energy source or use more than one simultaneously or independently and hence can integrate the sources. Thus, the reliability of the system is improved. Multi-input converters can combine voltages and currents from various sources such as solar PV, batteries, fuel cells, supercapacitors, and further alternate sources, thereby using the energy optimally and economically as shown in Figure 9.1. Applications of multi-input converters in various fields are discussed by Rehman *et al.* [4] and Tseng *et al.* [8]. Wu *et al.* [5], Chen [6], Affam *et al.* [7] and Tseng *et al.* [8] proposes different structures for multi-input DC-DC converters. Wu H. *et al* [10] also proposes a family of multi-input DC-DC converters with buck-boost operation. PWM based Double input converter is discussed in Chen *et al.* [11]. G. Nobile *et al.* [12] analyses the performance of PV plant using state space average technique.

Since the voltages from renewable sources may fluctuate continuously, it is necessary to maintain the voltage constant. In addition, there is a need to convert the voltage at different levels, since every source may generate voltages of various magnitudes. High step-up conversion ratios can also be obtained using a tapped inductor, but there might be high voltage stress and reverse recovery problems due to current and voltage shoots caused by leakage These abovementioned problems might reduce the conversion efficiency and increase electro-magnetic interference (EMI) with more

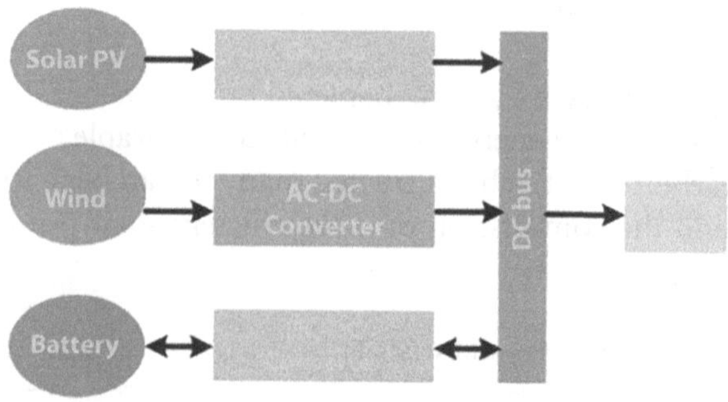

components in the work of Lee *et al.* [9]. Consequently, in the literature, various dual-input topologies and power management strategies were discussed Boutouba *et al.* [13]. Lavanya *et al.* [14] recommends an energy harvesting circuit for IoT applications. Jigang *et al.* [15] and Vidhya *et al.* [16] proposes fuzzy logic-based technique for electric drive applications.

9.2 D³ Converter Topology

The dual-input (DI)–dual-output (DO) DC (D^3) converter proposed is a multi-port converter with unique modes of operation. The structure facilitates DI and DO with an extra power flow path. The converter behaves as dual-input when both the sources concurrently fed the load. Similarly, when PV source simultaneously supplies both load and battery the converter acts as DO converter. The structure of the system is shown in Figure 9.2. Depending on the load demand, the converter can act in DI, DO, and SISO modes. The DI mode is activated when load demand is high and the battery will be in discharging mode. Then,

$$P_i < P_{out} \tag{9.1}$$

During DO mode, load demand is low, and the battery is in charging mode, and when the battery is fully charged, either battery or PV can supply the load (SISO) as given in Equations (9.1) and (9.2). Thus,

$$P_i \geq P_{out} \tag{9.2}$$

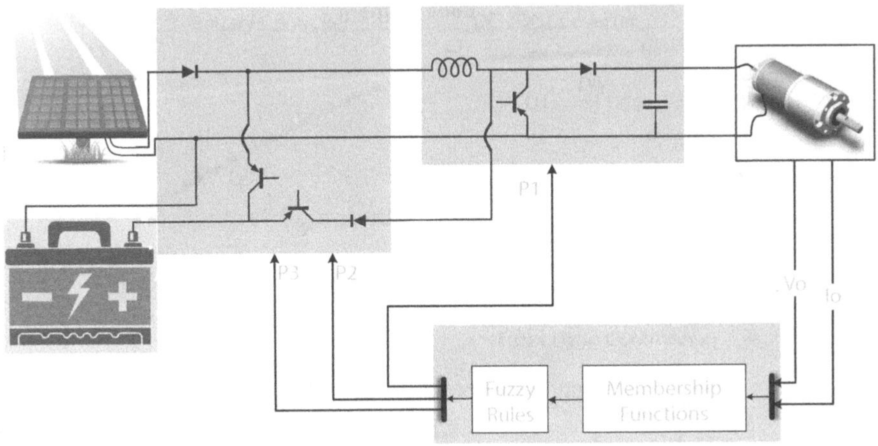

9.2.1 State-Space Model of the Converter

In the state-space model of the proposed converter and the system, input and output matrices are given as shown in Equation (9.3).

$$A=\begin{bmatrix} 0 & \frac{D-1}{L} \\ \frac{1-D}{C} & \frac{-1}{RC} \end{bmatrix} \quad B=\begin{bmatrix} \frac{1}{L} \\ 0 \end{bmatrix} \quad C=[0 \quad 1] \tag{9.3}$$

The transfer function is derived from the above state model of the converter. The general transfer function of the proposed converter is shown in Equation (9.4). The designed values are substituted in the general transfer function and the ratio is shown in Equation (9.5)

$$\frac{V_o}{V_g}=\frac{R(1-D)}{s^2RLC+sL+R(1-D)^2} \tag{9.4}$$

$$\frac{V_o}{V_g}=\frac{1.06e^5}{s^2+213s+5.32e^3} \tag{9.5}$$

Bode plot of the proposed converter is shown in Figure 9.3. The impulse response of the proposed converter is shown in Figure 9.4. With impulse

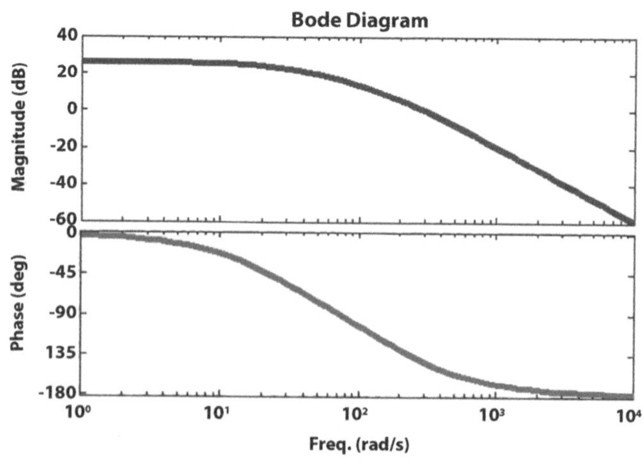

Figure 9.3 Bode plot of the proposed converter.

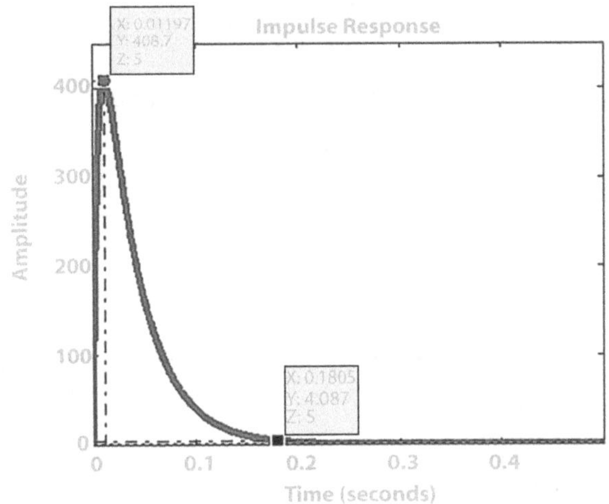

input, the settling time of the open-loop system is 0.18 s. Bode plot and impulse responses are shown to substantiate the behavior of the converter.

The battery controller circuit shown gives better control of the output voltage. Since the two switches S_2 and S_3 are responsible for the battery charging and discharging, it is necessary to control these switches. The optimum method of power flow management is achieved through the control battery. In this method, the battery is charged while the output current demand is low and the battery is discharged when the output current demand is high.

Thus, the output current is measured, and this current is used to set a threshold value to supply the switching pulses to the two switches. The controller circuit is shown in Figure 9.5. When the output current is below the threshold current, pulse 3 is supplied to the switch S_3, thereby discharging the battery and supplying current to the load [9].

During the first sub-mode in the dual input, battery or PV supplies power. Hence,

$$V_L = V_{pv} = V_b \tag{9.6}$$

The load voltage is the difference between the output voltage and the input voltage. Hence,

$$V_L = (V_{pv} - V_o) \tag{9.7}$$

Combining these equations and multiplying with the respective switch turned ON, inductor voltage is obtained as in Equations (9.8) from Equations (9.6) and (9.7)

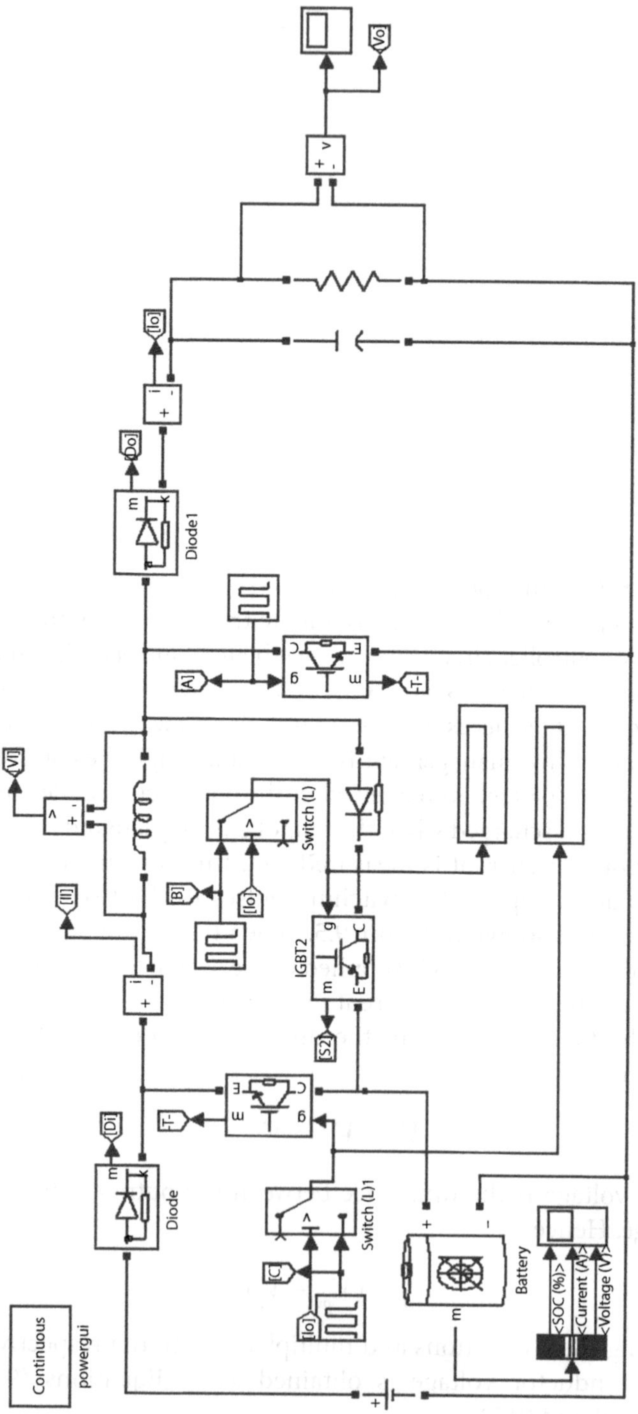

Figure 9.5 Battery controller circuit.

$$V_L = V_b D_3 T_s + V_{pv} (D_1 - D_3) T_s + (V_{pv} - V_o) (1 - D_1) T_s = 0 \tag{9.8}$$

From Equation (9.8), the inductor voltage second balance equation is written, and the equation for the output voltage is shown in Equation (9.9).

$$V_o = \frac{V_{pv}(1-D_3)+V_b D_3}{(1-D_1)} \tag{9.9}$$

During the first sub-mode in DO, the PV source supplies power only to the load as in Equation (9.10). Hence,

$$V_L = V_{pv} \tag{9.10}$$

In the second sub-mode, the PV source supplies to the battery alone as in Equation (9.11). Hence,

$$V_L = V_{pv} - V_b \tag{9.11}$$

In the third sub-mode, the load voltage is the difference between the output voltage and the input voltage as in Equation (9.12). Hence,

$$V_L = (V_{pv} - V_o) \tag{9.12}$$

From above equations, the output voltage is calculated and the equation for the output voltage is shown in Equations (9.13) and (9.14):

$$V_o = \frac{V_{pv} - V_b D_2}{(1-D_1-D_2)} \tag{9.13}$$

$$V_o = V_1 D_1 + D_2 (1 - D_1) V_2 \tag{9.14}$$

The output voltage ratio depends on the input voltages and the duty cycle. Since there are two voltages, the ratio between these two input voltages is also an important factor. The duty cycle for switch 3 is varied with fixed input voltages. The input voltage ratio is also varied with a fixed duty cycle.

The voltage gain for the existing converter in literature [10], whereas the voltage gain formula for the converter is as shown in Equation (9.13). The voltage gain for the two converters has been drawn in a graph as shown in

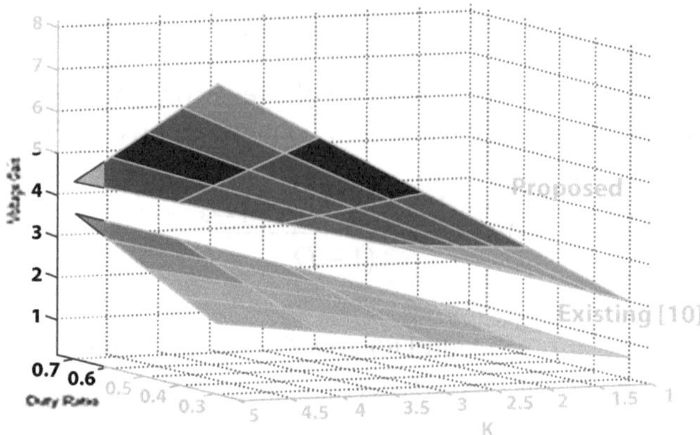

Figure 9.6 Voltage gain comparison.

Figure 9.6. It has been drawn considering the duty cycle and voltage ratio. It can be seen that the proposed converter has a higher voltage gain.

9.3 Closed-Loop Controller

A PI controller is used to control the pulse of the switch using the output voltage as feedback. The P value and I value are tuned using the Ziegler-Nichols method. The first boosting switch S1 is controlled by PI, whereas the other switches use pulse generator. The input is varied step by step to check the change in output. This is known as line regulation. Since PV is unreliable, the input voltage may change depending upon the time of the day and depending upon insolation. The input voltage is varied from 10 to 18 V, and it can be seen from Figure 9.7 that the output voltage remains constant.

The fuzzy logic controller (FLC) outdoes the performance of PI controller. It is simpler and offers better control compared to a PI controller. Both output current and output voltage are used as a feedback to control the switches. Contrary to PI where only the boost switch was controlled, in this study, all three switches using the FLC.

The membership functions are discussed below. Pulse 1 depends upon the output voltage. Hence, V_o is given as an input to the fuzzy controller. The rules for this are shown in Tables 9.1 and 9.2. The membership functions for V_o and Pulse 1 are shown in Table 9.1. Both the V_o and Pulse 1 have three membership functions each.

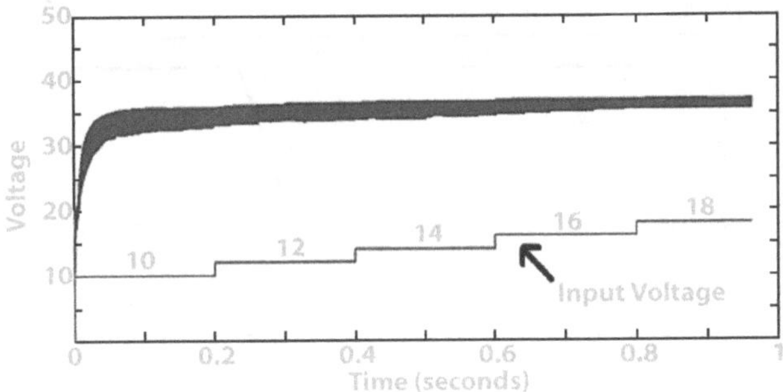

Pulse 2 and Pulse 3 depend upon the output current and the state of charge of the battery. The membership functions are prepared according to the rules shown in Tables 9.1 and 9.2. The input and output membership functions are shown in Figure 9.8(a) and (b) respectively. The rules for this are shown in Table 9.2. The output current and state of charge have three

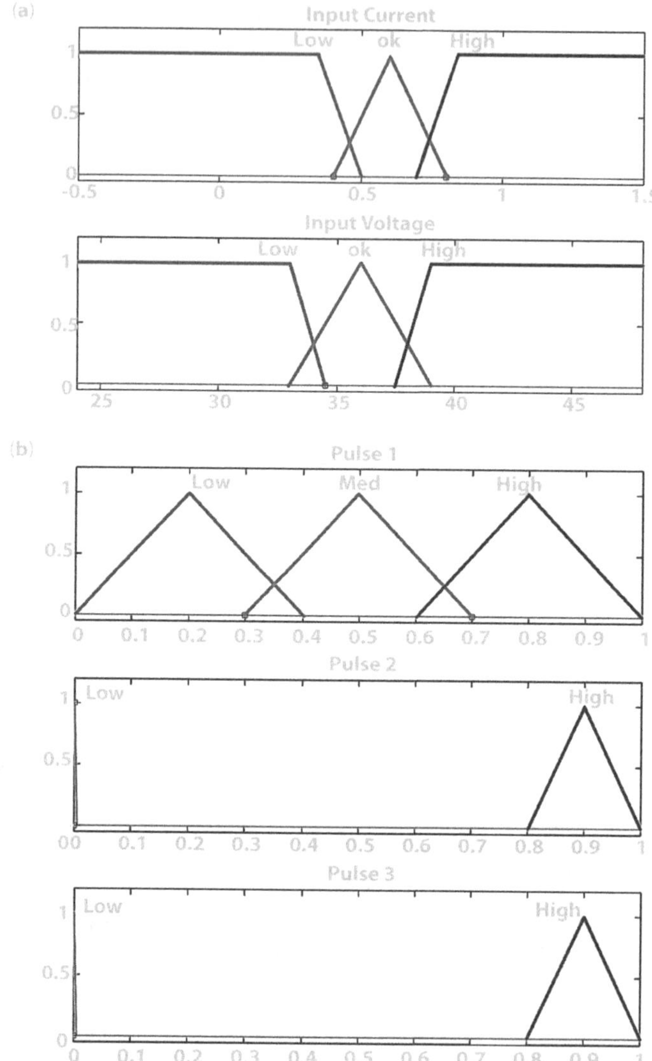

membership functions each, while the Pulses 2 and 3 have only two membership functions. The pulses are controlled as shown in Table 9.2. The battery charging switch and discharging switch are controlled depending upon the output current. The switching can be seen in Figure 9.9.

The PI controller and fuzzy controller outputs are compared and shown in Figure 9.10. It can be seen that, even though fuzzy output has a slightly

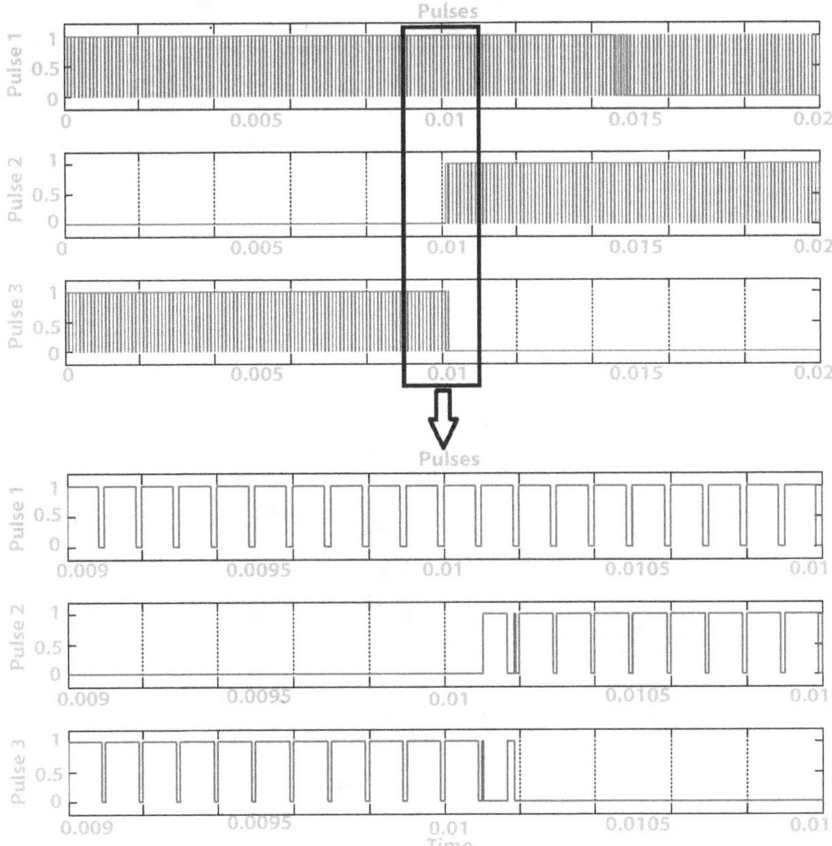

higher offshoot voltage, the settling time is significantly lower than the PI controlled output.

9.4 Experimental Verification

A 50-W experimental setup is built as established in Figure 9.11, and a clear view of converter is portrayed in Figure 9.12, validating the simulation results. The ratings of the various components used are as follows: Switch IRF840B; Diode IN4007; capacitor input, 1,000 μF, 25 V; Output, 1,000 μF, 50 V; Inductor, 47 μH; transformer 230/15 V, 1 A.

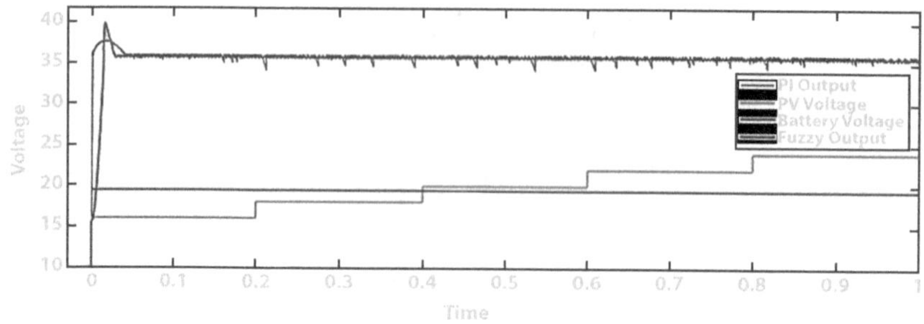

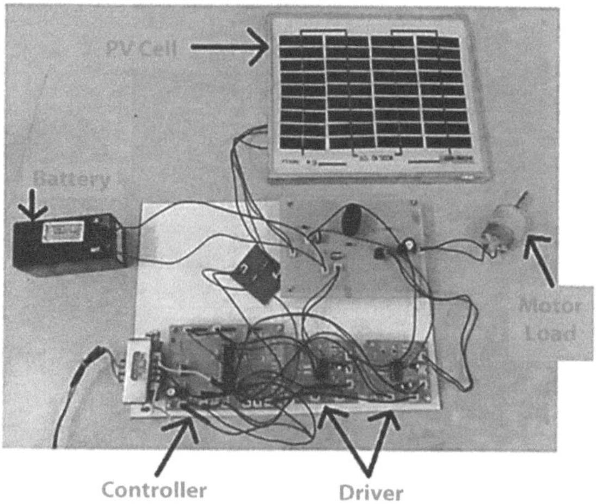

9.4.1 Result Discussion

Simulation and hardware outputs are presented in the waveforms below. Pulse 1 is kept constant at 50% in open-loop system, whereas it is varied according to the output voltage in the closed-loop system. Pulses P_2 and P_3 are 180° phase-shifted from each other, i.e., they are complementary switches and are never switched on together. The pulse given to the switch 1 in the simulation is depicted in Figure 9.13a, and the hardware results are shown in Figure 9.13b.

From the output voltage revealed in Figure 9.4, it is noticed that the required output value has been obtained simulation, and Figure 9.14b also shows the hardware output voltage.

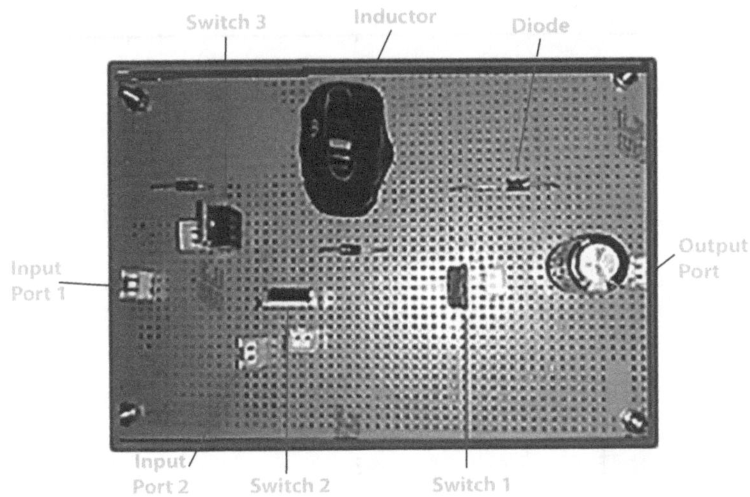
Switch 3
Inductor
Diode
Output Port
Input Port 1
Input Port 2
Switch 2
Switch 1

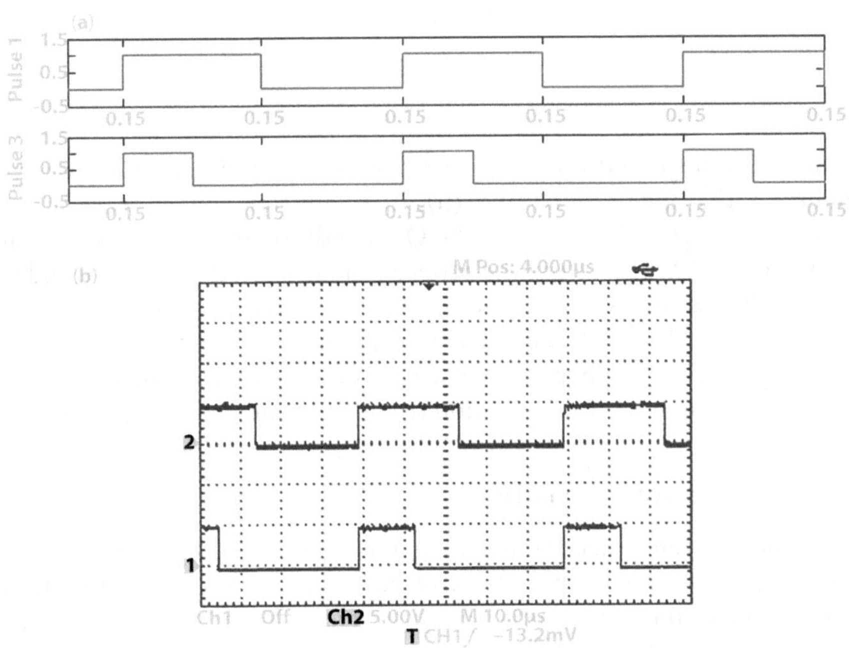
(a)
Pulse 1
Pulse 3
(b)
M Pos: 4.000μs
Ch1 Off Ch2 5.00V M 10.0μs
T CH1 / -13.2mV

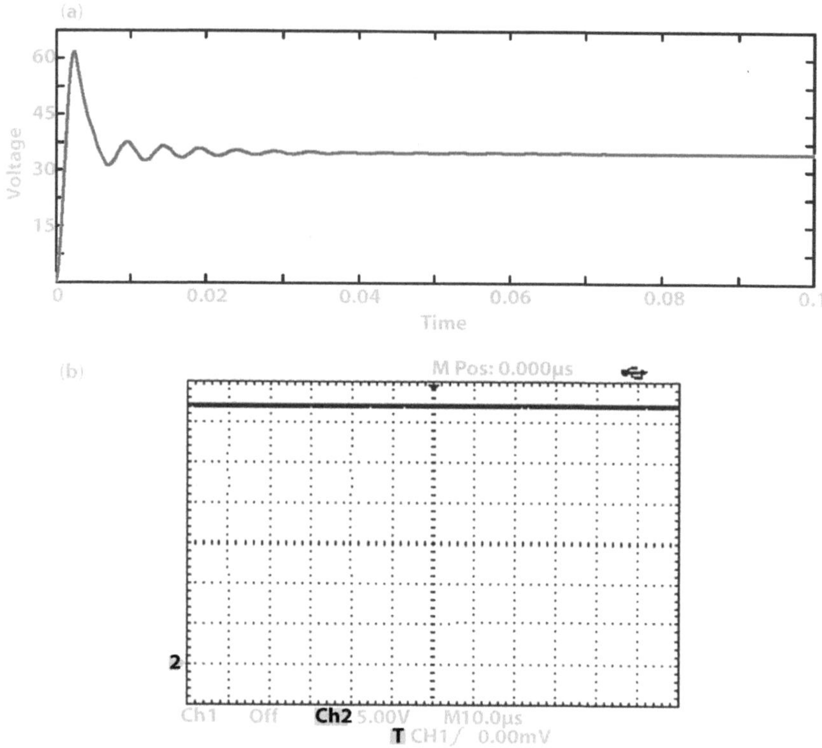

Switch voltage and diode voltage from the experiment are shown in Figures 9.15a and 9.15b(i) and (ii). The inductor voltage waveform is depicted in Figure 9.15. During the ON condition of the switch, the current from the PV source and the battery validates that it charges the inductor, and hence, we get a positive voltage across the inductor is obtained. When the switch is OFF, the polarity of the inductor changes, and it starts discharging. Hence, a negative voltage is measured across the inductor as exhibited in Figure 9.16a simulation result and validated in Figure 9.16b.

9.4.2 Comparative Analysis

Since the PV voltage keeps fluctuating, it is necessary to have a line regulation. The fluctuation in the input voltage is addressed and, as a result, the output voltage remains constant as illustrated in Figure 9.17a for constant PV voltage and Figure 9.17b illustrates for the constant battery voltage in the proposed system.

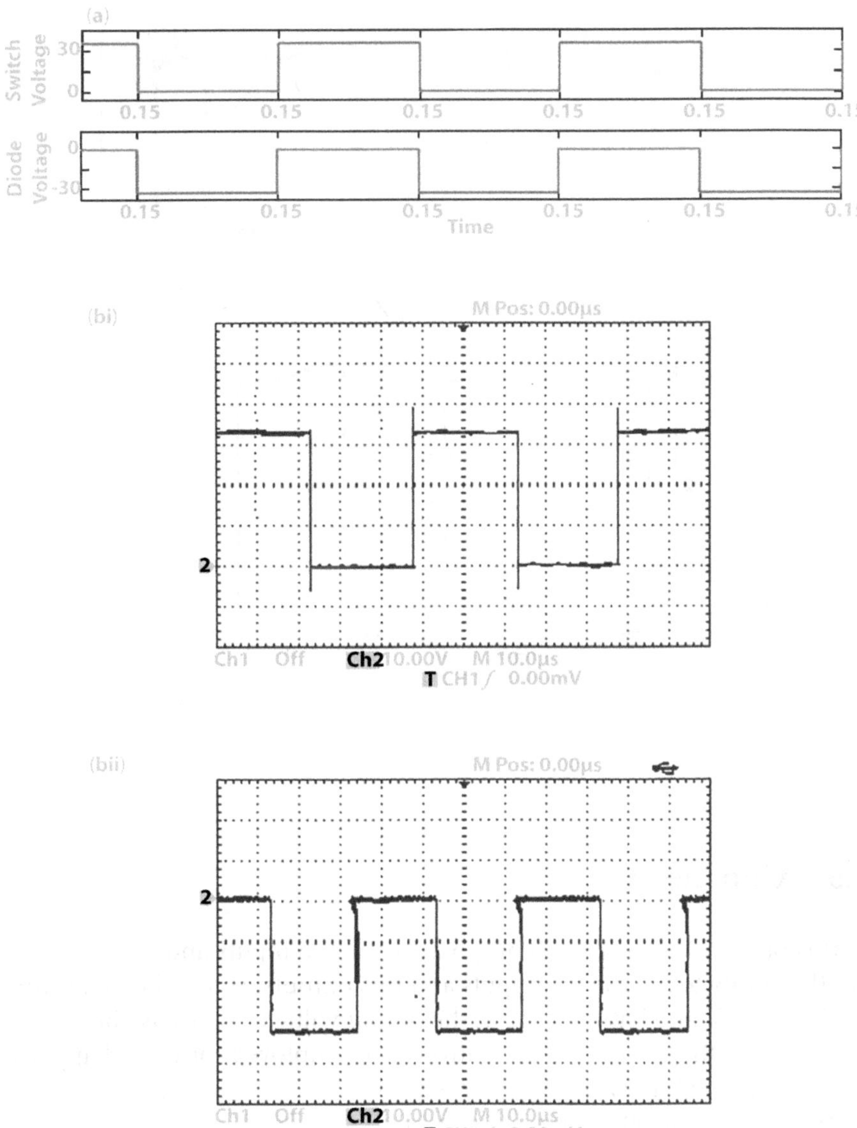
(a)
Switch Voltage
30
0
0.15 0.15 0.15 0.15 0.15 0.15
Diode Voltage
0
-30
0.15 0.15 0.15 0.15 0.15 0.15
Time
(bi)
M Pos: 0.00µs
2
Ch1 Off Ch2 10.00V M 10.0µs
T CH1 / 0.00mV
(bii)
M Pos: 0.00µs
2
Ch1 Off Ch2 10.00V M 10.0µs
T CH1 / 0.00mV

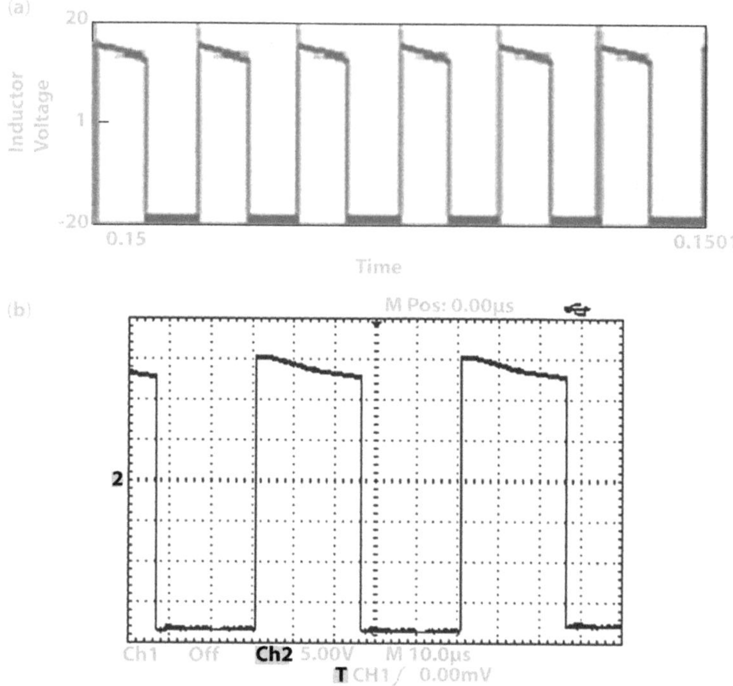

9.5 Conclusions

In this paper, D³ converter is proposed with two inputs and the converter has the ability to handle two inputs and two outputs. When the input from the PV is 18 V and battery is 12 V, the output voltage of 35 V is obtained as presented in this paper. The D³ converter is employed for hybrid applications where the high step-up is not required.

The fuzzy controller strategy is implemented to efficiently manage the required power and control the charging and discharging of battery based on the SOC of the battery. The performance of the converter PI and FLC is compared to prove the superiority of the proposed control strategy. The capability of D³ converter when subjected line and load regulation is proved with the help of the graphical analysis shown. The proposed D³ converter has been simulated using MATLAB and the results are proved with experimental results.

(a)

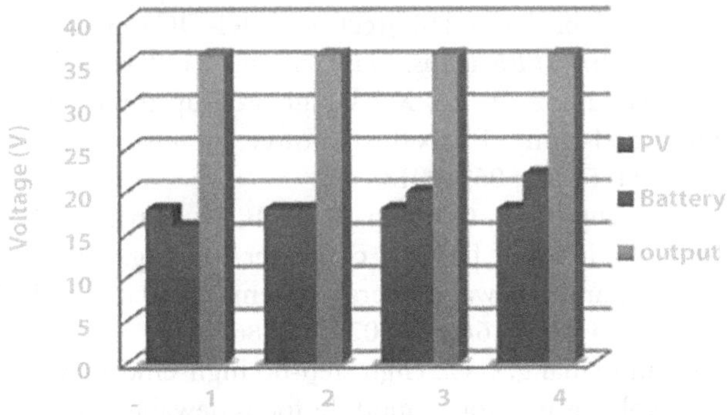

(b)

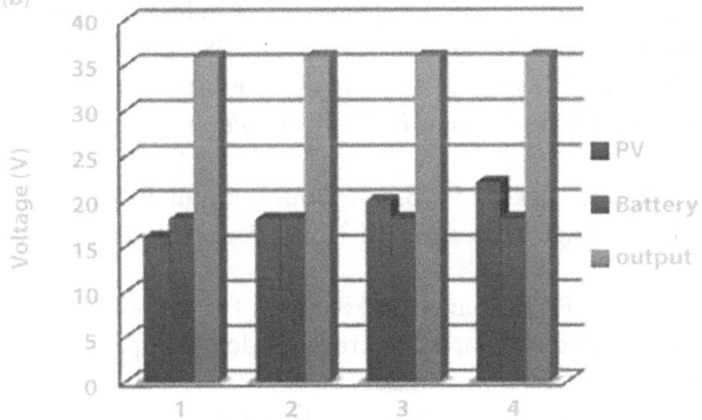

References

1. Zhou, L.W., Zhu, B.X., Luo, Q.M., High step-up converter with capacity of multiple input. *IET Power Electron.*, 5, 5, 524–531, 2012.
2. Lavanya, A., Jayaseelan, N., Navamani, J.D., Kumar, K.V., Dual input DC-DC converter for renewable energy systems, in: *2017 International Conference on Inventive Systems and Control (ICISC)*, pp. 1–5, 2017, doi: 10.1109/ICISC.2017.8068598Venkata.
3. Prabhala, A.K., Fajri, P., Gouribhatla, V.S.P., Baddipadiga, B.P., Ferdowsi, M., A DC–DC converter with high voltage gain and two input boost stages. *IEEE Trans. Power Electron.*, 31, 6, 4206–4215, 2016.
4. Rehman, Z., Al-Bahadly, I., Mukhopadhyay, S., Multi-input DC–DC converters in renewable energy applications – An overview. *Renew. Sust. Energ. Rev.*, 41, 521–539, 2014.

5. Wu, H., Xu, P., Hu, H., Zhou, Z., Xing, Y., Multiport converters based on integration of full-bridge and bidirectional DC–DC topologies for renewable generation systems. *IEEE Trans. Ind. Electron.*, 61, 2, 856–869, 2014.
6. Chen, Y.M., Huang, A.Q., Yu, X., A high step-up three-port DC–DC converter for stand-alone PV/battery power systems. *IEEE Trans. Power Electron.*, 28, 11, 5049–5062, 2013.
7. Affam, A., Buswig, Y.M., Othman, A.-K. B. H., Bin Julai, N., Qays, O., A review of multiple input DC-DC converter topologies linked with hybrid electric vehicles and renewable energy systems. *Renew. Sust. Energ. Rev.*, 135, 110186, 2021. doi: 10.1016/j.rser.2020.110186.
8. Tseng, K.C. and Huang, C.C., High step-up high-efficiency interleaved converter with voltage multiplier module for renewable energy system. *IEEE Trans. Ind. Electron.*, 61, 3, 1311–1319, 2014.
9. Lee, S., Kim, P., Choi, S., High step-up soft-switched converters using voltage multiplier cells. *IEEE Trans. Power Electron.*, 28, 7, 3379–3387, 2013.
10. Wu, H., Zhang, J., Xing, Y., A family of multiport buck–boost converters based on DC-link-inductors. *IEEE Trans. Power Electron.*, 30, 2, 735–746, 2015.
11. Chen, Y.-M., Liu, Y.-C., Lin, S.-H., Double-input PWM DC/DC converter for high/low voltage sources. *IEEE Trans. Power Electron.*, 27, 4, 1538–1545, 2012.
12. Nobile, G. *et al.*, Performance assessment of large photovoltaic (PV) plants using an integrated state-apace average modeling approach. *Energies*, 13, 18, 4777, 2020. doi: 10.3390/en13184777.
13. Boutouba, M., EI Ougli, A., Miqoi, S., Tifhaf, B., Assymetric fuzzy logic controlled DC-DC converter for solar energy system. *RESD*, 2, 1, 52–59, 2016.
14. Lavanya, A., Jegatheesan, R., Vijayakumar, K., Design of novel dual input DC–DC converter for energy harvesting system in IoT sensor nodes. *Wireless Pers. Commun.*, 117, 4, 2793–2808, 2021.
15. Jigang, H., Jie, W., Hui, F., An anti-windup self-tuning fuzzy PID controller for speed control of brushless DC motor. *Automatika*, 58, 3, 321–335, 2017.
16. Vidhya, S.D. and Balaji, M., Hybrid fuzzy PI controlled multi-input DC/DC converter for electric vehicle application. *J. Control. Meas. Electron. Comput. Commun.*, 61, 79–91, 2020.

10

A Smart IoT-Based Solar Power Monitoring System

O. Sobhana*, G.C. Prabhakar, N. Amarnadh Reddy and Rashmi Kapoor

EEE Department, VNR VJIET, Hyderabad, India

Abstract

One of the prominent renewable energy sources for electricity generation is photovoltaic (PV)–based solar system, which is an emerging alternative clean energy source to meet the demand of power. However, the inconsistent climatic conditions like variation in temperature and irradiance became barriers for the effective operation and performance of the PV system. Therefore, a real-time data assessment and smart monitoring technology is essential to intently evaluate the performance of the system. To make the system smart, an IoT-based power monitoring system was proposed for real-time remote monitoring. The acquired data was successfully transmitted as graphical representations with the smartphone application. The results manifest that the presented IoT-based system is a promising solution to the smart real-time monitoring of solar PV system.

Keywords: MPPT, sensors, ThinkSpeak, Aurdino

10.1 Introduction

Solar energy is at the forefront of clean energy because it is the simplest and most commercially effective method of generating electricity utilizing the sun's power. Power outages are fairly common in India. It is critical to use clean energy and monitor it as a result of this problem. Solar-powered families and communities may make the most of their energy production and consumption by keeping an eye on the weather forecast. The unique features of the photovoltaic (PV) are most simple installation, low running

**Corresponding author*: sobhana_o@vnrvjiet.in; oshreddy@gmail.com

A. Chitra, V. Indragandhi, and W. Razia Sultana (eds.) Smart Grids and Green Energy Systems, (137–146)

cost, and zero fuel requirement strived as future energy system [1, 6]. However, the system is instable and unpredictable for the amount of energy produced by the panels being the nature of solar energy. Distinct factors caused by PV are solar intensity, time duration of sunlight, and placement of panels, associated with temperature and prominent environmental situations such as wind, dust, and weather [2].

To improve the produced energy from a PV system, adequate measures should be employed during its installation. Since most of the PVs are located at remote or high-altitude places, elevations attribute for various maintenance and consecutive failures during operating state of the panels. Hence, a feasible system is required to monitor the amount of power delivered. When it comes to solar panels, the output power of the panels must be checked in order to receive the best power production from them. This necessitates the use of a real-time monitoring system. It can also be used to monitor the output power of each panel in a large solar power plant, which helps in recognition of build-up of dust. It also helps in averting any faults from occurring during operation.

The traditional monitoring methods such as reviewing the manual and remote wired monitoring that are time consuming and complex wiring setup necessitate new technology. In addition, the wiring complexity made this type of system as inefficient [3]. The emerging technology is Internet of Things (IoT), which provides approach of communication system. The IoT is embedded with network of systems enclosed with sensors, network connectivity, electronics, and software connectivity. This makes the object more accurate, sensed, and controllable monitoring with existing system results in the high efficiency, accuracy [7].

The cloud enables the IoT for quick and fast interaction with daily objects such as computer, laptop, smart devices, sensors, and measuring devices. The necessary components are actuators and network protocols [4, 5]. Hence, the IoT effectively implement with communication system for remarkable control of PV system in large fields and remote locations, which avoids the human involvement in terms of personnel inspection on time basis.

10.2 Phases of System Implementation Process

The proposed system is able to measure the electrical parameters, i.e., voltage, current, and the temperature along with intensity of the solar radiation. The recorded data was processed through the microcontroller-based ATMega-328P Arduino, and it has been sent through the wireless

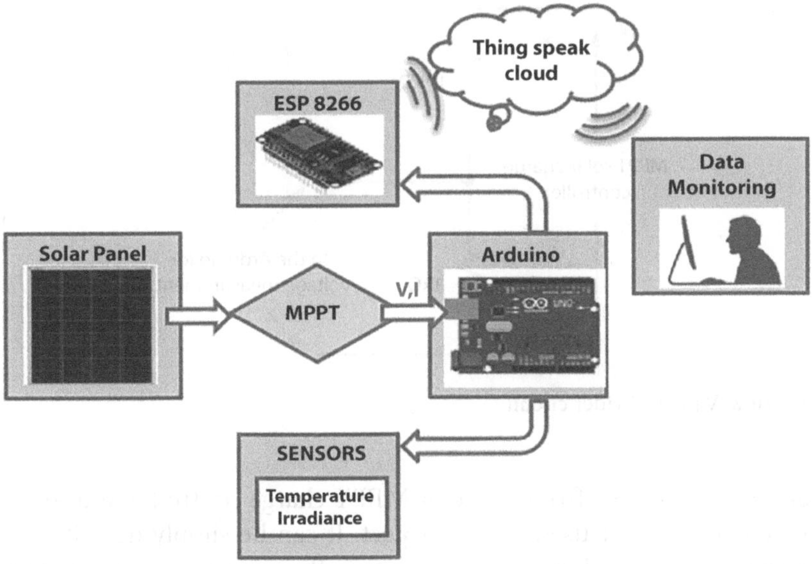

Figure 10.1 PV Panel monitoring system with IoT.

transmitter with the help of internet. ThingSpeak cloud platform is considered to foresee the collected data from the sensors and to visualize it in the graphical form. The cloud service is accessible only with the internet facility; thereby, ThingSpeak provides access to the website for the condition monitoring of the PV system. The system block diagram is shown in Figure 10.1.

10.2.1 Data Acquisition

A PV panel of monocrystalline type is used as solar PV system, which has maximum power output of 10 W and voltage at maximum power is 18 V, with maximum allowable current of 0.55 A is used to produce the energy needed for the system. A lead acid battery was used for storage with a capacity of 12 V and 5 Ah. In the present work, the power produced by the panels is continuously measured with the help of sensors and the data is further stored and transferred by the Arduino to the computer for the visualization of parameters in the graphical form. In this work, the voltage sensor is implemented with a voltage divider circuit to sense the solar PV voltage as illustrated in Figure 10.2.

$$V_o = V_m * {}^{1k\Omega}\!/_{1k\Omega + 10k\Omega}$$

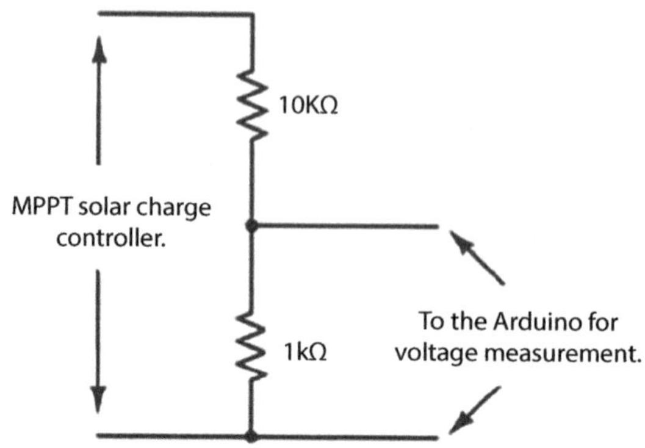

Figure 10.2 Voltage divider circuit.

where *Vm* is obtained from the solar MPPT charge controller and ACS712 is the current sensor used in the project. It can be simply described as a current sensor that calculates and measures the amount of current applied through its conductor. A temperature sensor is a temperature-sensing device that senses the temperature of the space surrounding it and converts the sensed data into electronic data so that it could be recorded, monitored, or alert us with changes in temperature. Contact-less temperature sensors commonly use infrared (IR) sensing devices. They detect an object's IR energy and send a signal to the electronically calibrated circuit. LM35 is the temperature sensor used in the project [10]. The LM35 is a temperature sensor that gives voltage proportional to the temperature. The output voltage is given in degrees celsius (°C) by the temperature sensor. The sensor has three terminals that measure ambient temperature between −54°C and 151°C. The temperature measured in LM35 is more exact when compared to a thermistor.

The Arduino Mega 2560 microcontroller connects all the sensors, components, and required statistics. Here, four sensors are connected with Arduino, which senses the parameters that have effect on the entire system. The data is transferred to the WIFI module at assigned Node MCU ESP8266 via serial communication protocol.

10.2.2 Data Interface

An open access IoT cloud-based communication system especially ThingSpeak is considered for this project, which retrieves the stored sensor data from Arduino through internet protocol (HTTP). The **WIFI** module pin in the Arduino enables and updates the data logs with received sensor

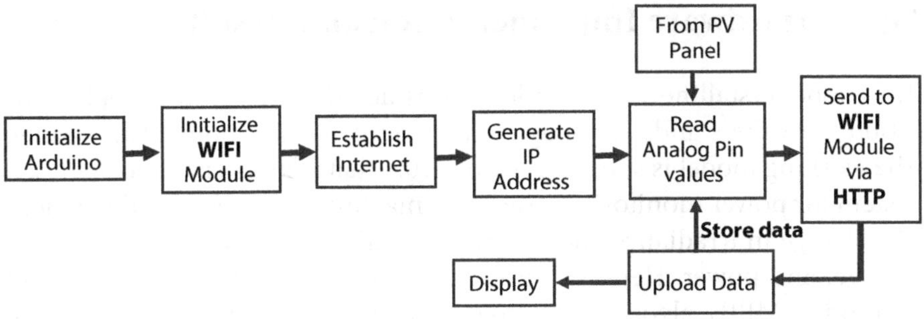

Figure 10.3 Implementation of data interface.

data and also updates the end application user with the present recordings. To access this feature, the end application user requires to register an account or need a specific IP address for the purpose, which has the facility of accessing different channels to monitor various parameters in the system [8, 9]. The implementation of data accessibility is given in Figure 10.3, which allow the user to visualize the graphical representations This kind of monitoring allows the data to access through smart phones and computers. The important observation of such system is that the panel information can be precisely monitored at anywhere with the accessibility of internet facility.

10.2.3 ThingSpeak Analytics

ThingSpeak is a cloud-based IoT analytics software that aggregate, visualize, and analyze live data streams. It allows submitting of data from our devices to ThingSpeak, which could generate real-time visualizations of live data and issue alarms.

ThingSpeak allows to save and analyze time-stamped numeric or alphanumeric data for free. Users can sign up for ThingSpeak by going to http://thingspeak.com and creating an account. Data is saved in channels in ThingSpeak. There is no limit to the amount of time stamped observations that can be made on a channel (think of these as rows in a spreadsheet). There are up to eight fields per channel (think of these as columns in a spreadsheet). Channels can be public, allowing anybody to view the material, or private, allowing only the owner and a few select individuals to view it. Each channel has a Write API Key connected with it, which is used to control who can write to it. Private channels also have one or more Read API Keys that control who can read from the channel. To read from public channels, you do not need an API Key. There are up to eight fields per channel. By default, only one field is produced.

10.3 Hardware Implementation and Results

The monocrystalline PV module is considered for hardware implementation with the standard voltage and current specifications of the panel. The working model is shown in Figure 10.4 that implements a solar-based system for power monitoring to achieve maximum power from the panels. The change in irradiance and temperature affects the PV module current, voltage, and power. Hence, to track the maximum power, classical P&O method of MPPT algorithm is programmed in the microcontroller. Once after converting the sun's photon energy into electricity, current and voltage parameters are recorded are shown in Figures 10.5 to 10.9 with the help of sensors. The cloud receives sensed signal from controller, and electrical characteristics are monitored and displayed to the smart system. The hardware setup is implemented to monitor the application for certain time duration to transmit through the data gateway and examined the visual display on ThingSpeak channel.

Based on the observations made from the graphical representations on ThingSpeak of Figures 10.5 to 10.9, the light intensity, i.e., irradiation and associated temperature, will affect operating voltage and current of PV system. The change in current is directly proportional to the variation in the light intensity and temperature, whereas the change in voltage is logarithmically related to temperature; therefore, rise in temperature

Figure 10.4 Hardware setup.

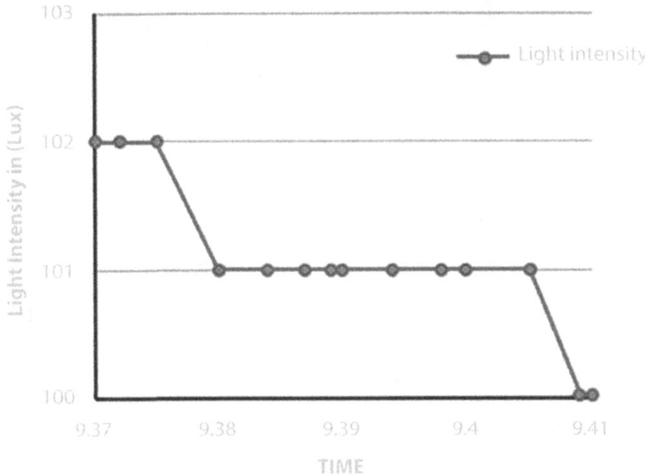
103
102
101
100
Light intensity
Light Intensity in (Lux)
9.37
9.38
9.39
9.4
9.41
TIME

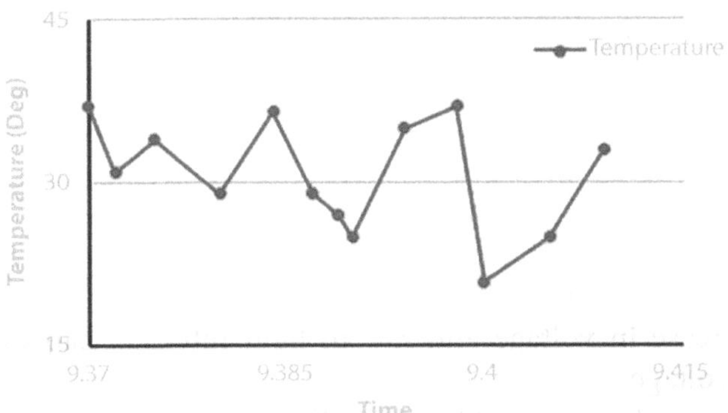
45
30
15
Temperature
Temperature (Deg)
9.37
9.385
9.4
9.415
Time

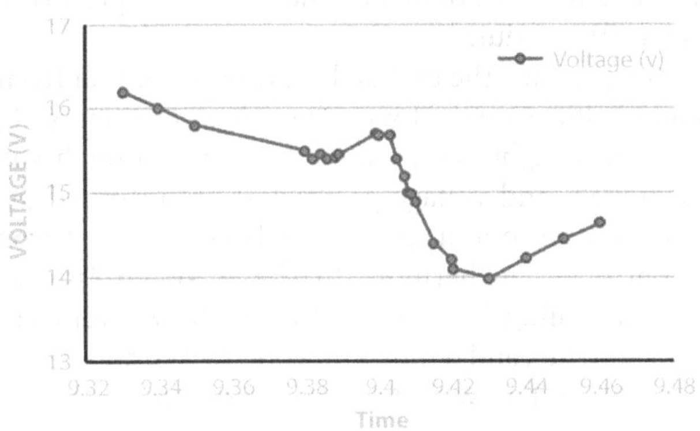
17
16
15
14
13
Voltage (v)
VOLTAGE (V)
9.32
9.34
9.36
9.38
9.4
9.42
9.44
9.46
9.48
Time

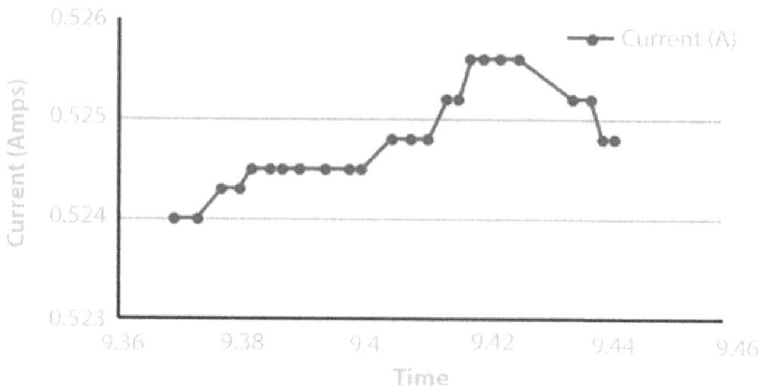

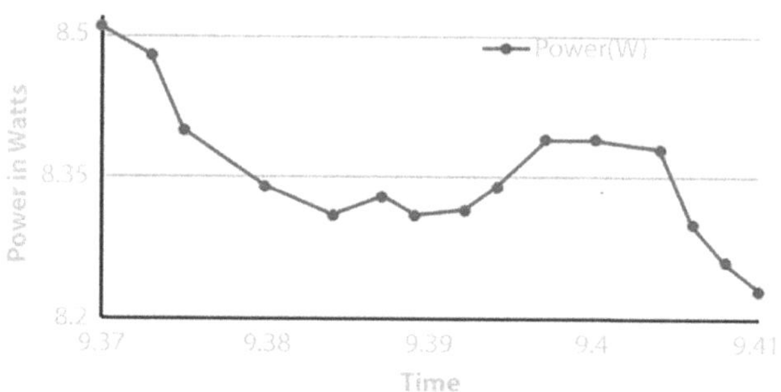

causes decrease in voltage and rise in light intensity causes a positive change in voltage.

The interpretation of output power is that the rise in temperature reduces the generated PV voltage and that, in contrast, the rise in irradiance fairly increases the current. Thus, these parameters are the performance indicators for the solar PV module.

Figure 10.10 represents the digital display of the output from the device. The current and voltage are the two influencing parameters of power.

The conversion of light energy into electricity through solar panels is achieved; the current and voltage parameters are recorded using sensors. Accordingly, maximum output power has been extracted from the panel with MPPT controller. Furthermore, IoT-based system is meant to alert if the solar panels are failing. Using the IoT technology, several characteristics like voltage, current, lux, and temperature are tracked and can be monitored on the website. This system with data acquisition, smart phone, and accessing provides information remotely with graphical interpretation of data.

Figure 10.10 Output on LCD display.

10.4 Conclusions

The proposed system implements power monitoring of PV system continuously using IoT. The readings of voltage, current, light, temperature, and power from their respective sensors are sent to the Arduino and successfully transmitted from Arduino to ESP8266 to the ThingSpeak cloud and, from thereon, data is monitored in real time. The continuous monitoring enables the information of uncertainty of solar radiations, and errors can be located easily. This system can be extended to large operating solar systems for preventive measures by daily monitoring the plant. It is highly suitable for the industrial and commercial applications.

References

1. Mellit, C.A., Pavan, A.M., Kalogirou, S.A., Fault detection method for grid-connected photovoltaic plants. *Renew. Energy*, 66, 99, 2014.
2. Arjyadhara, P., Ali, S.M., Chitralekha, J., Analysis of solar PV cell performance with changing irradiance and temperature. *Int. J. Eng. Comput. Sci.*, 2, 214, 2013.
3. Raza, A., Ikram, A.A., Amin, A., Ikram, A.J., A review of low cost and power efficient development boards for IoT applications. *Future Tech. Conf.*, IEEE, p. 786, 2016.
4. Atzori, L., Iera, A., Morabito, G., The internet of things. *Comput. Netw.*, 54, 2787, 2010.
5. Zanella, A., Bui, N., Castellani, A., Vangelista, L., Zorzi, M., Internet of things for smart cities. *IEEE Internet Things J.*, 1, 22, 2014.
6. Khadka, N., Adhikari, B., Bista, A., Shrestha, A., Solar Panel Cleaner Technology: A Review. *Int. Conf. Developments in Renewable Energy Technol. Proc.*, 2018.

7. Babu, R.L.R., Rambabu, D., Naidu, A.R., Prasad, R.D., Krishna, G.P., IoT enabled solar power monitoring system. *Int. J. Eng. Technol.*, 7, 526, 2018.
8. Li, Y.F., Lin, P.J., Zhou, H.F., Chen, Z.C., Wu, L.J., Cheng, S.Y., Su, F.P., On-line monitoring system of PV array based on internet of things technology. *IOP Conf. Ser.: Earth Environ. Sci.*, 93, 012078, 2017.
9. Khadka, N., Adhikari, B., Bista, A., Shrestha, A., Solar Panel Cleaner Technology: A Review. *Int. Conf. Developments in Renewable Energy Technol. Proc.*, 2018.
10. Arjyadhara, P., Ali, S.M., Chitralekha, J., Analysis of solar PV cell performance with changing irradiance and temperature. *Int. J. Eng. Comput. Sci.*, 2, 214, 2013.

11

Control of Multi-Input Interleaved DC-DC Boost Converter for Electric Vehicle and Renewable Energy

M. Bharathidasan and V. Indragandhi*

School of Electrical Engineering, Vellore Institute of Technology, Vellore, India

Abstract

The design and implementation of a new interleaved non-isolated DC-DC converter are proposed in this paper. Using a multi-input interleaved coupled inductor-based converter, the necessary with a constant level maximum output voltage may be obtained. Since it compensates for mismatches in photovoltaic devices and allows for separate and continuous power flow from these sources. The Proportional-Integral (PI) controller, which includes outer-loop voltage control and multi-inner loop current control, is used to regulate the multi-input boost converter. The proposed converter has the benefits of high gain, a low ripple in the output voltage, a low ripple in inductor current, greater power density, and efficiency. In the absence of one or two resources, the converter can provide the needed power to the load. Simulation results under various process conditions show the efficacy of the proposed controller in regulating the multi-input DC-DC step-up converter's output voltage.

Keywords: Multiple-input converter, coupled inductor, high gain, low ripple, PI controller

11.1 Introduction

Photovoltaic (PV) modules, wind turbines, and fuel cells (FCs) are all examples of renewable energy sources (RESs) that may be used to generate power. Conventional energy sources are gradually diminishing and

**Corresponding author*: indragandhi.v@vit.ac.in

A. Chitra, V. Indragandhi, and W. Razia Sultana (eds.) Smart Grids and Green Energy Systems, (147–158)

releasing greenhouse gases into the atmosphere. To overcome this problem, society must focus on alternative energy research and development. In grid-connected micro-grids [1, 2], hybridizing energy has gained popularity. The FCs stack in electric vehicles (EVs) usually used a perfect energy source. FCs is energy sources that transform chemical energy directly into electrical energy. To attain this, a multi-input power converter structure must be developed to provide the power of input supplied by RES to the load. These days, PV frameworks have the biggest development and establishment between other renewable energy systems. In comparison to the off-grid PV industry, the majority of the improvements within the introduced PV models have been proposed for grid-connected PV. Such a grid-connected PV framework is outlined in [3]. For plug-in EVs, a multi-functional power electronic connector capable of utilizing dual sources during charge preparation has also been suggested [4]. Multi-input converters have been classified into two methods: isolated multi-input DC-DC converter and non-isolated multi-input DC-DC converter [4]. Between the two sides of an isolated multi-input DC-DC converter, a large transformer offers electric isolation and impedance coordination. Isolated DC-DC converters are large and heavy due to the need of a transformer. A few switches are used in the input and output terminals of these converters, resulting in a more in losses and costs. A non-isolated multi-input DC-DC converter, by using various winding transformers, may share a common ground through the isolated one that is necessary to separate the ground. These issues continue in a non-isolated converter; it has a basic construction and is less costly than an isolated converter. The various multiport DC-DC converters topologies are described in [5]. A Y-source DC-DC converter was displayed in [6]; high voltage gain, constant input current, and minimal inrush current are all features of this device. For high-current applications, with a wide load range, authors in [7] presented a non-multiphase interleaved high step-up DC-DC resonant converter.

A block diagram of the new interleaved boost converter will be located in Figure 11.1. It can be seen that there are three input sources, comparative to a solar PV, fuel cell, and battery. Moreover, in comparison to traditional converters, the DC gain has been enhanced and utilized inductive components to exchange energy between sources and load. Under various working conditions, the appropriate controller needs to reduce the output voltage provided to the load. Among the most noticeable features of the proposed converter is that the battery bases are grounded, preventing noise from damaging the battery and therefore extending its life span. This converter has three working states, which are determined by the battery's charging or discharging status.

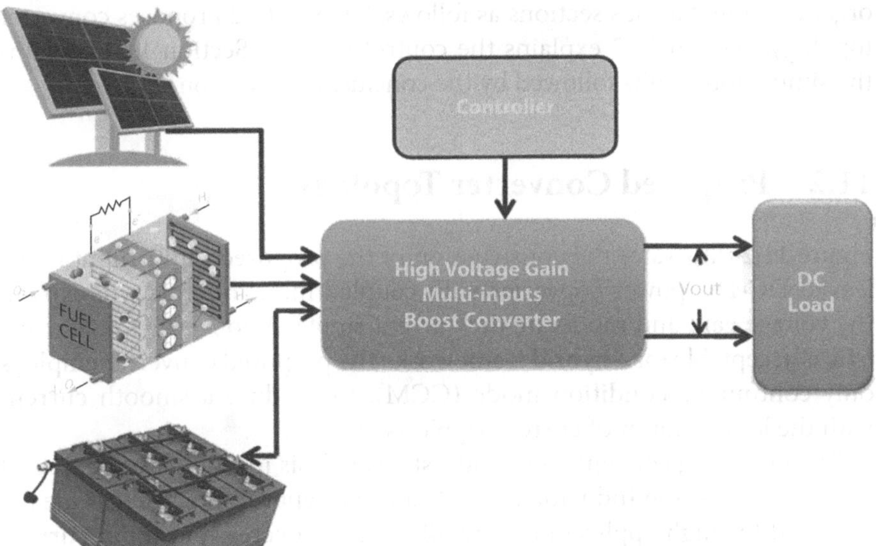

The grid is linked to a new step up multi-input DC-DC converter in [8]. Furthermore, the battery's DC gain is increased, and it may be charging and discharging in order to obtain power control. By simply increasing the turn ratio, the expansive DC gain may be obtained. In [9], a multiwinding transformer-based converter is shown. In the literature, a number of multi-input DC-DC converter topologies have been presented [10, 11]. Despite the fact that numerous multi-input DC-DC converters have been published for RES applications, only a few articles have discussed the control strategies for those converters under various working conditions [12]. In [13] investigates and classifies a variety of step-up DC-DC converters based on their properties and voltage-boosting methodologies. For a distributed generating system, a three-winding-coupled-inductor that blends a coupled-inductor and voltage-boost techniques is presented [14–17]. A detailed literature evaluation of PV device simulation models and determination methodologies was carried out [18]. In [19], state-of-the-art in the energy-storage topologies for HEVs and plug-in HEVs (PHEVs) are explored. The proposes a new multi-input multi-output (MIMO) DC-DC converter with excellent step-up capability over a wide power range in article [20]. The multi-input coupled inductor-based converter's control technique is described in this paper in order to stay the output voltage at the required level. The paper is organized into various sections as follows. Section 11.2 proposes converter topology. Section 11.3 explains the control strategy. Section 11.4 explains the simulation results followed by the conclusion in Section 11.5.

11.2 Proposed Converter Topology

Figure 11.2 illustrates the construction of the proposed multi-input interleaved DC-DC power converter with coupled inductor. This will expand the voltage gain and moderate the current ripple. A converter's characteristic is acceptable for a hybrid framework. The proposed converter employs only continuous condition mode (CCM), to produce a smooth current with the least amount of current ripple as feasible.

The operating concept and steady-state analysis in CCM are predicated on the fact that the inductor current does not approach zero, it is ideally intended for high applications and allows for a decrease in input current ripple and conduction mode. Then, V_{Pv} and V_{Fc} are two autonomous power sources that output is based on characteristics of them. PV and FC modules are changed by L1 and Lm in accordance with input sources. R_L is the load equivalent resistance associated with the DC bus. Switches S1, S2, S3, and S4 are control switches. Diodes D1, D2, D3, D4, D5, and D6 are

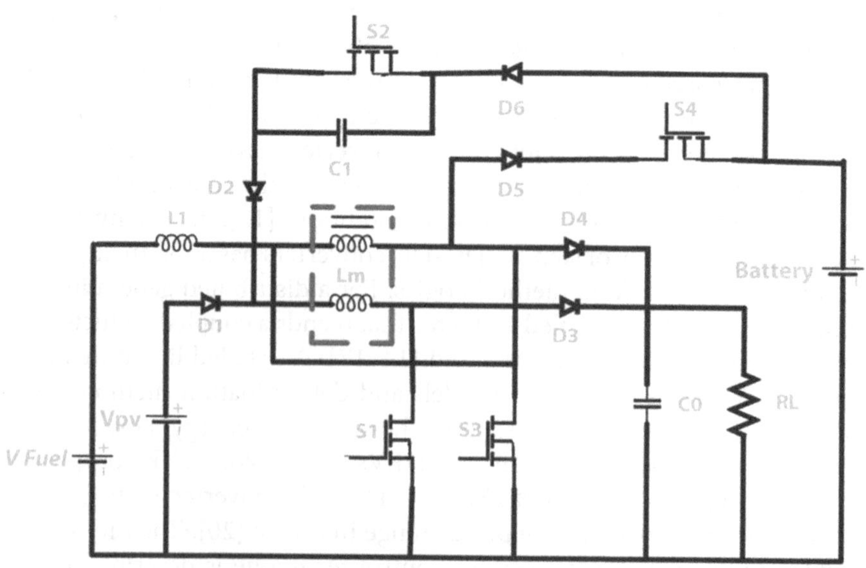

utilized to set up modes, which can be described. Capacitor C_1 is utilized to decrease the stress and output capacitor C_0 is as a voltage filter at the output. The leakage inductance of linked inductors might help to moderate diode reverse recovery. By expanding the stage of the interleaved converter, the ripple on the input current wills decrease. The magnetizing inductor and leakage are ideal models for the coupled inductor. The suggested converter achieves the specified output voltage V_0 with duty cycles due to its large voltage gain. This increases the efficiency of the proposed converter.

The proposed converter is operated in three various modes. In mode 1, PV and FC charge the battery and provide load energy. S1, S3, and S4 are ON. Lm stores energy. Besides, the magnetizing inductor Lm and the leakage inductance L1 are charged by DC input voltage. At last, the load energy is provided by the output capacitor C_0. In mode 2, PV and FC release the battery and provide load energy. Switch S4 is OFF, and switches S1, S2, and S3 are ON. The inductor L1 is charged by Vpv, and magnetic inductor Lm discharges the energy to the output capacitor. In mode 3, the load is provided by PV and FC, whereas the battery is not utilized. During this state, the system is working without battery charging or discharging. Switches S1 and S3 are ON, and the remaining switches are OFF. The inductor L1 and magnetic inductor Lm are charged through control sources Vpv and Vfc, individually.

To minimize the ripples in the input current and output voltage, the inductors and capacitors in this proposed converter must be appropriately designed. The input source inductor may be configured separately based on the amount of power provided by each renewable energy source. Equations (11.1) and (11.2) demonstrate the general formulas for designing the converter inductor and capacitor, respectively.

$$L = \frac{DV_{in}}{f\Delta I} = \frac{V_{in}(V_0 - V_{in})}{fV_0\Delta I} \tag{11.1}$$

$$C = \frac{DI_0}{f\Delta V_c} = \frac{I_0(V_0 - V_{in})}{fV_0\Delta V_c} \tag{11.2}$$

where D is the duty cycle, f is the switching frequency, Vin is the input voltage, V_0 is the output voltage, I_0 is the output current, ΔI is ripple current passing through the input inductor that is desirable, and ΔV_c is the output capacitor's desired ripple voltage. The voltage and current stress across the power devices are mostly determined by the turns ratio of the coupled inductors.

The magnetizing inductor currents on average may be calculated as

$$I_{Lm(avg)} = \frac{(N+1)I_{out}}{(1-D)} \tag{11.3}$$

The ripple in the magnetizing current may be written as

$$\Delta I_{Lm} = 0.15 \times I_{Lm(avg)} \tag{11.4}$$

The magnetizing inductance is estimated depending on the input current.

$$L_m = \frac{DV_{IN}}{f_{sw}\Delta I_{Lm}} \tag{11.5}$$

11.3 Control Strategy

V_{PV} and V_{FC} are two RESs that are being investigated. As a result, power switches must be correctly switched to maintain output voltage from the converter. Energy has been stored and transmitted using lead-acid

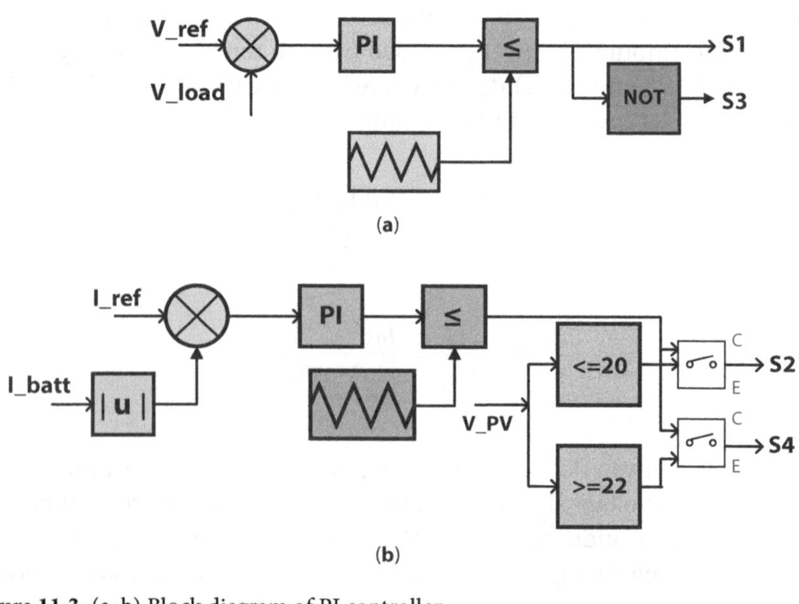

Figure 11.3 (a, b) Block diagram of PI controller.

batteries; the remaining energy from the PV panels will be used to charge it and, if PV power is inadequate, to assist in providing the load. The battery is associated to a bidirectional converter, allowing energy to flow in both directions. The stored and conveyed battery's power reference (Pdc_ref) is carefully associated with the voltage, and the current reference (Ibat_ref) of battery is evacuated from Pdc-ref by jumping the voltage of the battery. The PI controller is utilized to coordinate the battery in charger and discharger action as appeared in Figure 11.3.

11.4 Simulation Results

The proposed converter is simulated using the MATLAB/Simulink software package to ensure its validity. Both open-loop and closed-loop simulations are performed on the converter. As said prior, the proposed converter can be used for a wide range of domestic and industrial applications, such as smart homes, HEV, power sources, and DGs interface that are mostly PV array, fuel cell, etc. The input voltage is set to 20 V for both sources. As an energy storage component, a Li-Ion 24-V 3-Ah battery is utilized. The simulation parameters are given in Table 11.1.

For a better understanding of the results, a comparative study was carried out; the simulation results corroborate the newly implemented converter

Table 11.1 Simulation parameters for the proposed converter.

Symbol	Parameters	Range
Vpv, Vfc	Input voltages	20 V, 20 V
V_0	Output voltage	200 V
Po	Output power	400 W
fs	Switching frequency	15.5 KHz
D	Duty cycle	0.81
Lm	Magnetizing inductance	100 μH
L1	Leakage inductance	200 μF
C1, Co	Capacitors	22 nF, 1,200 μF
R_L	Load resistance	100 Ω

performance (Figure 11.4 and Figure 11.5), as well as the transient behavior. The converter is evaluated in open-loop and closed-loop mode under three separate working conditions, as previously stated. A gain of 10 V is achieved by tuning the converter; the output voltage is to be rise to 200 V. Figure 11.6 appears the output current, and voltage waveform has gotten from the simulation of the proposed converter. The system's output voltage maintains a good level of regulation without overshoots at the required level. In the steady state, the input inductor currents are non-variable. The same ripple content in the output voltage is obtained in the proposed converter. In any case, the magnitude of the output voltage is greater than what is observed from the proposed one for the same duty cycle. This will

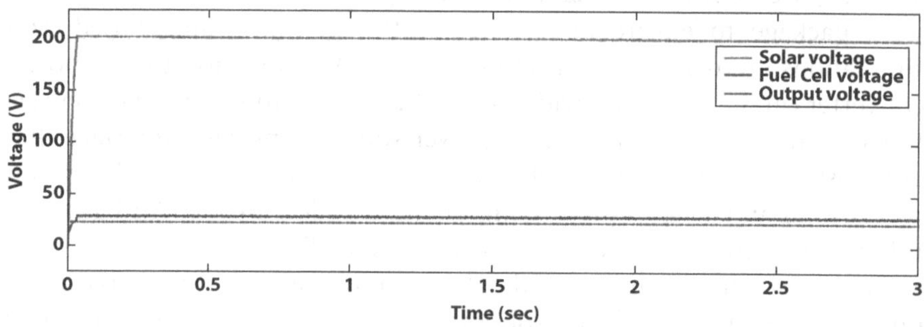

Figure 11.4 Simulation results of input voltages and output voltage.

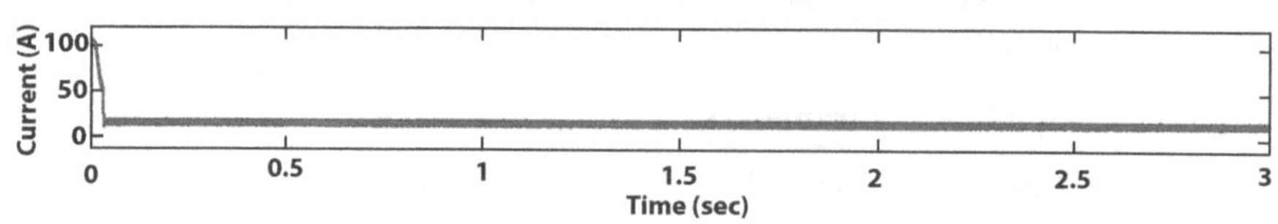

Figure 11.5 Simulation result of inductor current.

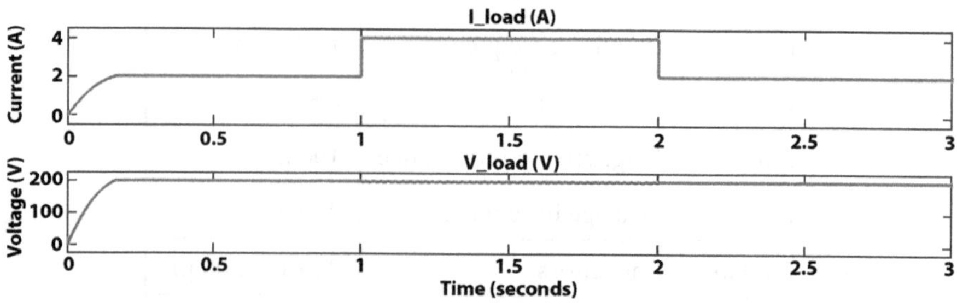

Figure 11.6 Simulation results of load current and voltage.

Table 11.2 Controller parameters.

	GV0	GL1	GLm
State 1	Kp = 0.5466	Kp = 7.000	Kp = -----
	KI = 48.3721	KI = 3.1439	KI =----
State 2	Kp = 0.4983	Kp = −0.0811	KP = 0.3303
	KI = 65.7330	KI = −9.7503	KI = −7.6737
State 3	Kp = 0.2983	Kp = 0.0001	KP = 0.4303
	KI = 65.7330	KI = 0.0150	KI = −8.0737

diminish the turn-OFF switching losses in each switch, leading to soft switching of a switch. This will further help in improving the efficiency of the system. For the closed-loop control, the PI controller is essential to attain the system stability, and it supports the converter in maintaining sustained output voltage. Initially, the load voltage is equated with a reference value and fed to the PI controller. The gain of the PI controller is provided in Table 11.2. The PI controller has the merits of less stable errors and faster response times. All of the simulation results show that the suggested multi-input interleaved coupled inductor-based boost converter control approach for renewable energy system and EV applications may deliver adequate performance even when converter operations vary.

11.5 Conclusion

In this study, a new three-input interleaved DC-DC step-up converter with coupled inductor has been proposed for EV and renewable energy system application. The proposed converter was being developed to provide a high gain while maintaining a low turn ratio. When one or more of the two supplies are unavailable, the converter can donate the needed power to a load. Three separate operational stages are modeled in the converter and then used to create a controller design plan. To accomplish minimal ripples in output voltage and input current, the converter parameters are along with simple-to-construct input inductors and output capacitors. The use of coupled inductors has grown, which decreases the volume and size of the magnetic core. The PI controller is used to implement the control strategy. Moreover, the design consideration and simulation results were displayed all together, which needed high gain and efficiency.

References

1. Moury, S., Lam, J., Srivastava, V., Church, R., New soft-switched multi-input converters with integrated active power factor correction for hybrid renewable energy applications, in: *Proc. 2016 IEEE Energy Convers. Cong. and Expo*, Milwaukee, WL, USA, 2016.
2. Sedaghati, R. and Shakarami, M.R., A novel control strategy and power management of hybrid PV/FC/SC/battery renewable power system-based grid-connected microgrid. *Sustain. Cities Soc*, 44, 830, 2019.
3. Forouzesh, M., Shen, Y., Yari, K., Siwakoti, Y.P., Blaabjerg, F., High-efficiency high step-up DC-DC converter with dual coupled inductors for grid-connected photovoltaic systems. *IEEE Trans. Power Electron.*, 33, 7, 5967, 2018.
4. Singh, A.K., Chinmaya, K.A., Badoni, M., Solar PV and grid based isolated converter for plug-in electric vehicles. *IET Power Electron.*, 12, 14, 3707, 2019.
5. Lavanya, A., Navamani, J.D., Vijayakumar, K.R., Multi-input DC-DC converter topologies - A Review, in: *Proc. 2016 International Conf. on Electric. Electron. and Optimiz. Tech*, Chennai, India, 2016.
6. Fang, X., Ding, X., Zhong, S., Tian, Y., Improved Quasi-Y-Source DC-DC converter for renewable energy. *CPSS Trans. Power Electron., Appl.*, 4, 2, 163, 2019.
7. Huang, Y., Tan, S.C., Hui, S.Y., Multiphase-interleaved high step-up DC/DC resonant converter for wide load range. *IEEE Trans. Power Electron.*, 34, 8, 7703, 2019.
8. Ahrabi, R.R., Ardi, H., Elmi, M., Ajami, A., A novel step-up multiinput DC-DC converter for hybrid electric vehicles application. *IEEE Trans. Power Electron.*, 32, 5, 3549, 2017.
9. Chen, Y.M., Liu, Y.C., Wu, F.Y., Multi-input DC/DC converter based on the multiwinding transformer for renewable energy applications. *IEEE Trans. Ind. and Appl.*, 38, 4, 1096, 2002.
10. Azizi, M., Mohamadian, M., Beiranvand, R., A new family of multi-input converters based on three switches leg. *IEEE Trans. Ind. Electron.*, 63, 11, 6812, 2016.
11. Hosseini, S.H., Haghighian, S.K., Danyali, S., Aghazadeh, S., Multi-input DC boost converter supplied by a hybrid PV/Wind turbine power systems for street lighting application connected to the grid, in: *Proc. 47th International Univ. Power Eng. Conf.*, Uxbridge, UK, 2012.
12. Mummadi, V. and Sawant, K.K., Control of multi-input integrated buck-boost converteri, in: *Proc. IEEE Reg. 10 and 3rd Int. Conf. Ind. Inf. Syst.*, Kharagpur, India, p. 1, 2008.
13. Forouzesh, M., Siwakoti, Y.P., Gorji, S.A., Blaabjerg, F., Lehman, B., Step-up DC-DC converters: A comprehensive review of voltage-boosting techniques, topologies, and applications. *IEEE Trans. Power Electron.*, 32, 12, 9143, 2017.

14. Liu, H., Li, F., Ai, J., A novel high step-up dual switches converter with coupled inductor and voltage multiplier cell for a renewable energy system. *IEEE Trans. Power Electron.*, 31, 7, 4974, 2016.
15. Kumar, A. and Sensarma, P., Ripple-free input current high voltage gain DC-DC converters with coupled inductors. *IEEE Trans. Power Electron.*, 34, 4, 3418, 2019.
16. Samuel, V.J., Keerthi, G., Mahalingam, P., Coupled inductor-based DC-DC converter with high voltage conversion ratio and smooth input current. *IET Power Electron.*, 13, 4, 733, 2020.
17. Hassan, W., Lu, D.D.C., Xiao, W., Analysis and experimental verification of a single-switch high-voltage gain ZCS DC–DC converter. *IET Power Electron.*, 12, 8, 2146, 2019.
18. Ma, T., Yang, H., Lu, L., Solar photovoltaic system modeling and performance prediction. *Renew. Sust. Energ. Rev.*, 36, 304, 2014.
19. Khaligh, A. and Li, Z., Battery, ultracapacitor, fuel cell, and hybrid energy storage systems for electric, hybrid electric, fuel cell, and plug-in hybrid electric vehicles: State of the art. *IEEE T. Veh. Technol.*, 59, 6, 2806, 2010.
20. Mohseni, P., Hosseini, S.H., Sabahi, M., Jalilzadeh, T., Maalandish, M., A new high step-up multi-input multi-output. *IEEE Trans. Ind. Electron.*, 66, 7, 5197, 2019.

12

Maximum Power Point Tracking Techniques for Photovoltaic Systems—A Comprehensive Review From Real-Time Implementation Perspective

Sudarshan B.S., Chitra A.*, Razia Sultana W., P.R. Chandrasekhar, Tanisha Ganguli and Ishita Sahu

School of Electrical Engineering, Vellore Institute of Technology, Vellore, India

Abstract

Solar energy is a limitless natural renewable energy source. It is distinguished by its adaptability and long-term viability. Due to its intermittent nature, high cost, and low efficiency, integrating and applying solar panels as clean energy resources in electricity and distribution remain a difficulty. As a result, a power converter that allows the photovoltaic (PV) system to operate at maximum power point tracking (MPPT) becomes increasingly important in order to maximize the utility of solar PV. This can be difficult because the voltage and current in the circuit fluctuate depending on the weather. When this occurs, a new maximum power point (MPP) must be determined in order to optimize system power output. For PV systems, there are numerous MPPT approaches available. Under rapidly changing environmental conditions, traditional MPPT approaches fail to follow the MPP. As a result, evolutionary algorithm–based MPPT and AI-based MPPT algorithms have been developed to increase the system's dynamic performance. As a result, evolutionary and AI-based MPPT techniques are becoming more popular. This study gives an in-depth examination of various MPPT algorithms, including their implementation and dynamic performance. An extensive review of the conventional and AI-based MPPT techniques is presented.

***Keywords*:** MPPT, implementation issues of MPPT, solar photovoltaics, partial shading, evolutionary algorithm

**Corresponding author*: chitra.a@vit.ac.in

A. Chitra, V. Indragandhi, and W. Razia Sultana (eds.) Smart Grids and Green Energy Systems, (159–196)

12.1 Introduction

Energy is critical for a country's economy and society's economic progress. The long-term supply of energy from sources that are inexpensive, non-polluting, non-destructive, secure, and accessible is critical for future development and economic progress [1]. At present, fossil fuels are the major energy supplier worldwide. Non-renewable energy resources are fossil fuels. Under the pressure of the earth's crust, fossil fuels take millions of years to produce, yet they are depleted at a far quicker pace than new ones are created [2, 3]. Researchers all around the world are working on green and clean renewable energy to fulfil growing energy demand and preserve the earth from pollution.

Out of all the renewable energy sources like wind, geothermal, solar, and biomass, solar energy has emerged as one of the best sources. Solar energy can be converted to electricity using a solar photovoltaic (PV) system.

A PV system's internal resistance is nonlinear, making it a nonlinear power source. Only when these systems are made to work at their maximum power point can they be used effectively. With changes in temperature and irradiation, the point at which maximum power may be harvested shifts. As a result, a methodology/system must be devised to track the greatest power point at any given moment, regardless of these fluctuations.

For enhanced system efficiency, a PV system with MPPT is used. Mechanical MPPT and electrical MPPT are the two main techniques of MPPT. The panel location is changed to track the MPP in mechanical MPPT. The panel is stationary in electrical MPPT, and the PV is operated at its MPP using a DC-to-DC power converter. It is possible to change the impedance as viewed by the PV panel by altering the duty ratio of the DC-to-DC converter. The maximum power point is attained when the PV panel's impedance equals the impedance connected to it.

The V-I relationship of a PV panel is exponential, and the maximum power point is the sole place on the V-I curve that delivers the highest power (MPP). To track the MPP, an analog or digital algorithm is created. The MPPT technology allows the converter circuit to harvest the greatest amount of power feasible from the PV system. Because the traditional methods such as open-circuit voltage (OCV) method, short-circuit current (SCC) method, constant voltage controller (CVC) method, perturb and observe (P&O) algorithm, incremental conductance (IncCond), hill climbing method, curve fitting–based method, look-up table method, parasitic capacitance method, extremum seeking (ES) algorithm, and Fibonacci series–based method have a poor dynamic responsiveness, evolutionary algorithm–based methods and AI-based strategies are being actively researched.

Modified P&O, particle swarm optimization (PSO) algorithm, Ant Colony Optimization (ACO) algorithm, Genetic Algorithm (GA), and Cuckoo Search (CS) algorithm are examples of evolutionary algorithm–based MPPT techniques, whereas AI-based MPPT techniques include Fuzzy Logic Controller (FLC)–based method, Artificial Neural Network (ANN)–based method, and Adaptive Neuro-Fuzzy Inference System (ANFIS).

Many research articles have been published that present the current state of MPPT methods. This study was written to describe and compare the most common MPPT methods in detail. The following is a breakdown of the structure of this publication. The mechanical MPPT is discussed in Section 12.2. The traditional MPPT techniques are discussed in Section 12.3. The evolutionary and AI-based MPP methods are discussed in the Section 12.4. Section 12.5 compares and analyzes all of the MPPT approaches' implementation difficulties, performance levels, complexity involved, and feasibility issues. Section 12.6 is a list of some of the most widely used commercial products that use various MPP tracking methods.

12.2 Conventional Electrical MPP Tracking Methods

According to the maximum power transfer theorem, the load's resistance must equal the source's resistance as seen from its output terminals to receive maximum power from a source with a finite internal resistance. Consider the situation where the solar panel is directly linked to a load. When this happens, the panel's operating point is seldom at maximum power. The impedance seen by the panel determines the panel's operating point. As a result, by varying the impedance perceived by the panel, the operating point may be shifted closer to the peak power point. Because panels are DC devices, DC-to-DC converters are required to match the source and load impedances. The operational point will be at the maximum power transfer point at a single amount of impedance (i.e., at the corresponding duty ratio of the DC-DC converter).

It is also worth noting that the duty ratio of the converter cannot be set to a constant number. This is because the panel's irradiation and temperature conditions change over time, causing the panel's output parameters to fluctuate. As a result, a system that changes the DC-to-DC converter's duty cycle must be built to track the peak power point. MPPT implementations use algorithms that sample panel voltages and currents often and change the converter duty ratio accordingly. The next subsections go through the most popular of these approaches.

12.2.1 Open-Circuit Voltage Method

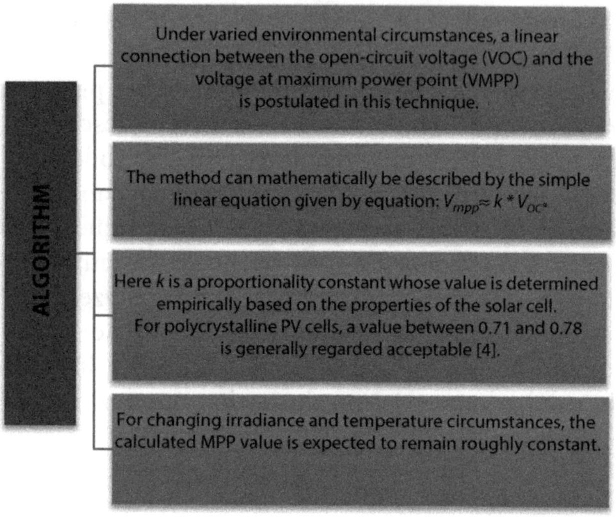

Implementation:
A voltage sensor and a microcontroller (for digital implementation) are required to implement the OCV technique. The processor's computations for digital implementation are likewise simplified. The main disadvantage is that the MPP may not be correctly tracked since the linear connection assumption is not particularly precise. In addition, measuring VOC necessitates maintaining the panel open circuit, i.e., the associated load is slashed. As a result, the consumer loses power. When pilot cells are taken into account [9], their selection becomes essential, since pilot cells must be carefully selected and positioned in order to replicate the behavior of the actual array and get the system's genuine OCV. This approach does not reliably measure MPP, particularly when PV panels are shaded. The control technique is indirect, and the implementation complexity is categorized as medium. There will only be one sensor required, and the implementation will be low-cost. The tracking efficiency, on the other hand, is low. The tracking speed is excellent.

12.2.2 Short-Circuit Current Method

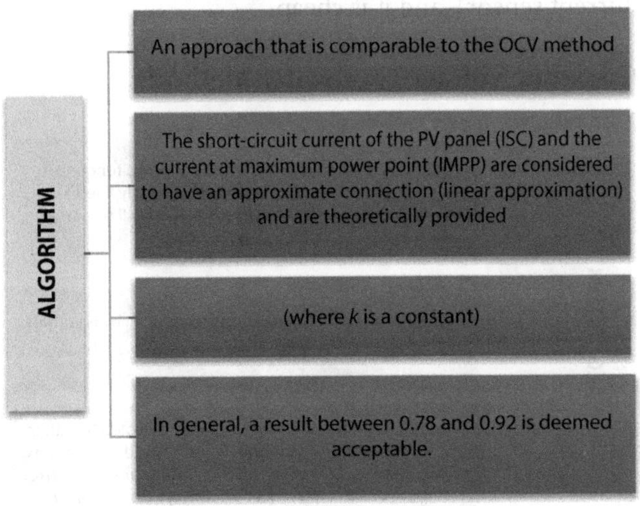

in [11] proposed online tracking and tuning of the proportionality constant, whereas the authors in [13] proposed a periodic sweep of the PV voltage from VOC to zero volts to constantly update the proportionality constant, ensuring that the tacked MPP is as close to the actual MPP as possible. The authors in [9] recommended that the method be implemented using DSP.

Implementation:

Although this approach is more precise and efficient than OCV and requires less computing load on the CPU, periodic monitoring of the SCC is a significant concern. Short-circuiting the PV array to measure ISC at every instant is not achievable. To get the same results as recommended in [9], an extra boost converter must be employed. This raises the cost of installation since the converter must be controlled by a separate control system. Microcontrollers may be utilized to improve accuracy since more processing power is required to sweep the panel from zero to the SCC value and record the output voltage profile [16]. The control technique in this method is indirect, comparable to the OCV method, and the implementation complexity may be categorized as medium. Only one sensor is required (current sensor), and it is cheap.

12.2.3 Constant Voltage Controller Method

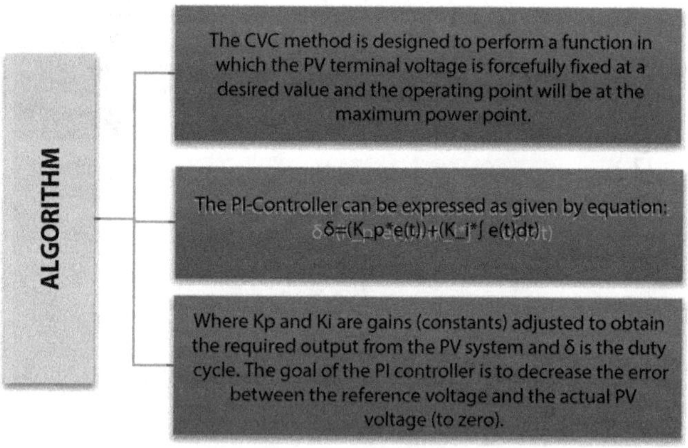

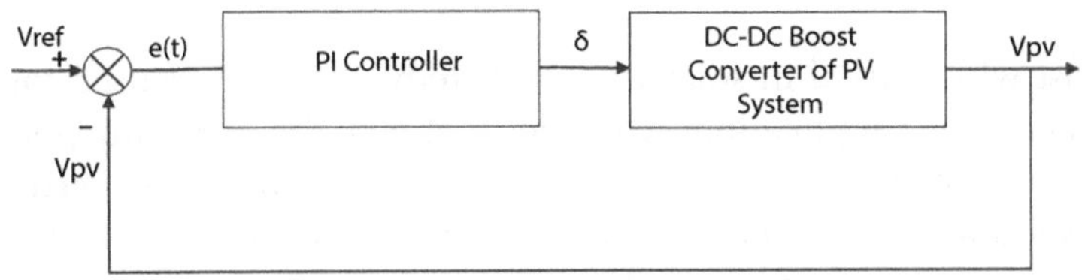

Figure 12.1 Constant voltage control method.

the system's nonlinear dynamics to track the MPP not only in steady state but also during dynamic situations. They argue that, despite the complexity of the dynamics, MPPT's resilience and accuracy surpass it. A constant voltage control method block diagram is given in Figure 12.1.

Implementation:

The most significant advantage of this approach is that it does not require a digital control to deploy. The controller may be implemented in analog utilizing operational-amplifier ICs with passive components. It is very straightforward to execute digitally. However, because the MPPT is dependent on the kind of solar panel, fresh designs must be created for each configuration (it is not a generic method) [15]. The authors conclude in [21] that the suggested CVC algorithm with online tracking replies in milliseconds based on their research.

12.2.4 Perturb and Observe Algorithm

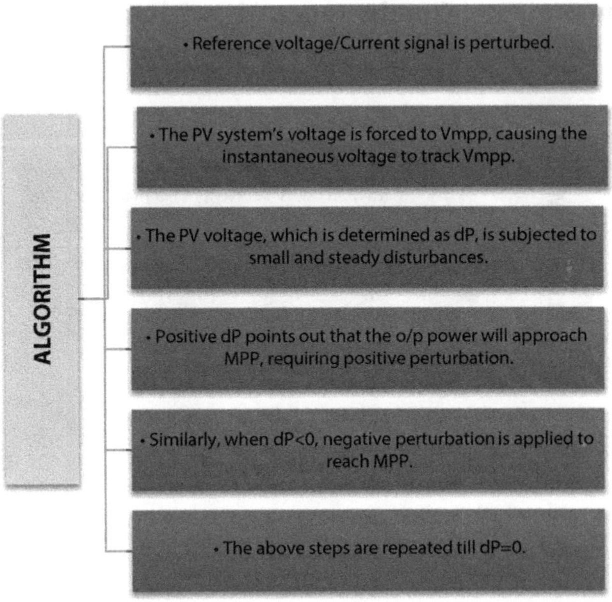

Brief:
It is most widely used in solar energy systems. It was first used for tracking for aerospace applications. Changes that take the operating point away from optimum point should be considered as oscillations. Perturbation period should be lower than the system settling time. Easy implementation of DSPs and has a simple structure. This algorithm can track almost the true MPP value. Speed control is also possible by changing the perturbation values. It has better accuracy than OCV and SCC methods.

Drawbacks:
It fails to track MPP during changing irradiance and temperature conditions. Determination of ideal perturbation value is difficult which depends on speed and accuracy of algorithm. Low perturbation amounts slow the procedure down, whereas large perturbation values cause higher oscillations around the MPP. The cost of implementation is high. To improve the efficiency of the traditional algorithm, a two-step P&O method can be utilized.

Implementation:
Basic implementation requires two sensors which are used to measure panel voltage and current respectively. One sensor can also be used; however, it depends on the type of system that is considered. A sampling control strategy is used which makes the implementation simple.

12.2.5 Incremental Conductance Algorithm

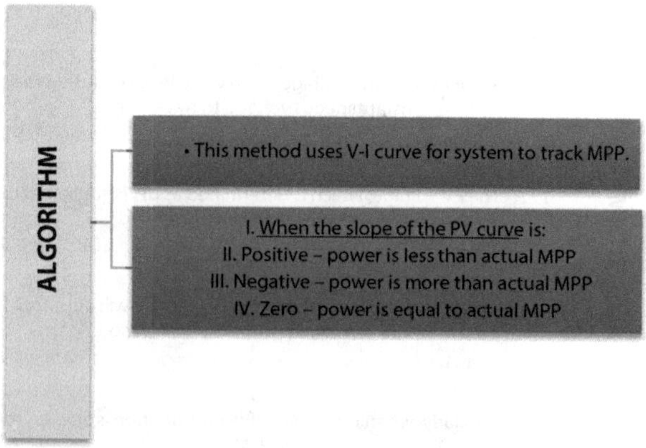

$$P_{MPP} = V_{MPP} * I_{MPP} \tag{12.1}$$

$$P = V * I \tag{12.2}$$

Differentiating with respect to the voltage, we get

$$\frac{dP}{dV} = I + \left(V * \frac{dI}{dV} \right) \tag{12.3}$$

For power to be maximum, the differential of the power should be equal to zero. In other words, $\frac{dP}{dV} = 0$

This implies that

$$I + \left(V * \frac{dI}{dV} \right) = 0 \tag{12.4}$$

which can be approximated as

$$\frac{\Delta I}{\Delta V} = \frac{-I_{MPP}}{V_{MPP}} \tag{12.5}$$

By evaluating Equation (12.6), the MPP can be tracked.

The conditions to track MPP are as explained by Equations (12.6) to (12.8):

$$\frac{dP}{dV} = 0 \rightarrow \frac{\Delta I}{\Delta V} = -\frac{I}{V} \text{ at MPP} \tag{12.6}$$

$$\frac{dP}{dV} > 0 \rightarrow \frac{\Delta I}{\Delta V} > -\frac{I}{V} \text{ left of MPP} \tag{12.7}$$

$$\frac{dP}{dV} < 0 \rightarrow \frac{\Delta I}{\Delta V} < -\frac{I}{V} \text{ right of MPP} \tag{12.8}$$

<u>Brief:</u>

It is better suited than P&O due to changing environmental conditions, since it can accurately find MPP during changing conditions. In [22], the authors described the principle of the algorithm. In this, the authors stated that, with designed controller, this system gives optimal solutions than P&O method. A variable self-dynamic method, also known as dynamic self-optimization is also used for better results.

Drawbacks:
It has complex implementation. System can oscillate around MPP due to large perturbation value. In [24], the authors pointed out that the system can only work when noise is negligible, and digital controller error tends to zero.

Implementation:
Two sensors are required, i.e., voltage and current sensors. It is implemented using microcontroller or DSP. PI controller increases effectivity. The error signal sent to the PI controller is as follows:

$$e = \frac{dI}{dV} + \frac{I}{V}$$

PI controller works to reduce error signal to zero. In [23], the authors stated that this method can work in changing environmental conditions.

12.2.6 Hill-Climbing (HC) Algorithm

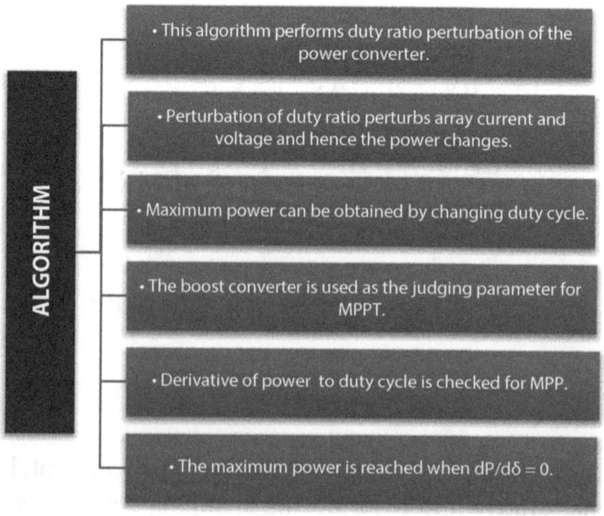

Brief:
Authors in [25] suggested a modified adaptive method which is better than the conventional method. On comparing hill climbing and P&O method, HC algorithm is considered to give better dynamic performance and regulated output. In [27], it is shown that the performance of this algorithm depends on the parameters of analog to digital converters and on digital filter parameters. It is easy to implement and has less cost.

Implementation:
In [26], authors have created a prototype adopting the conventional and modified HC algorithm and have got satisfactory results. It is helpful in fine tuning algorithms during practical implementation.

12.2.7 Other Conventional Methods

The authors described the curve fitting–based MPP and the mathematical model used to calculate the MPP in [28]. However, no instructions for calculating the equation's constants are provided. Authors in [29] described the mathematical method for determining the properties of solar cells. The constants can be determined using these. The authors in [29] state that for most solar cells, the values computed using the analytical method yield less than a 5% inaccuracy.

For MPP tracking of DC pumping systems, the authors in [30] adopted the look-up table technique. A look-up table with multiple MPP values associated with various situations is produced in this study. The values in the look-up table are the solar array's theoretical MPP locus. The authors in [31] described how they used the lookup table method to model and simulate a solar system.

The parasitic capacitance approach is referred to as a modified IncCond method by the authors in [32], because it is comparable to the latter but accounts for the solar cells' parasitic capacitance. The parasitic capacitance approach is described in [32], along with the equations that go along with it. The author in [33] gave a detailed description of the method and the circuitry needed to implement it. To achieve improved tracking, the authors recommended combining parasitic capacitance and IncCond approaches [34].

The stability of the MPPT scheme based on the ES algorithm was examined by the authors in [35]. This is achieved using Lyapunov's control approach. Also employed is the P&O algorithm. Analog-integrated circuits have been used to demonstrate the practical implementation. A new stochastic ES MPPT controller based on Newton is proposed in [36]. The new approach, according to the authors, addresses all stochastic perturbations and convergence issues. Instead of introducing sinusoidal disturbances, the authors in [37] advised that natural inverter ripple be used instead.

The control characteristics of a Fibonacci series–based MPP tracking system during shading of PV arrays are described in [38]. The authors also considered employing a wind turbine generator as a type of flywheel energy storage to compensate for the substantial output variability generated by the proposed MPPT. Authors in [39] proposed a modified method for tracking the global peak during shade conditions that takes into account power ripple and a large search range. The algorithm must be changed for real-time optimization. For slow irradiance change and rapid irradiance change, the modification is different [40].

12.3 Evolutionary Algorithm and Artificial Intelligence–Based MPP Tracking

12.3.1 Fuzzy Logic Controller–Based MPP Technique

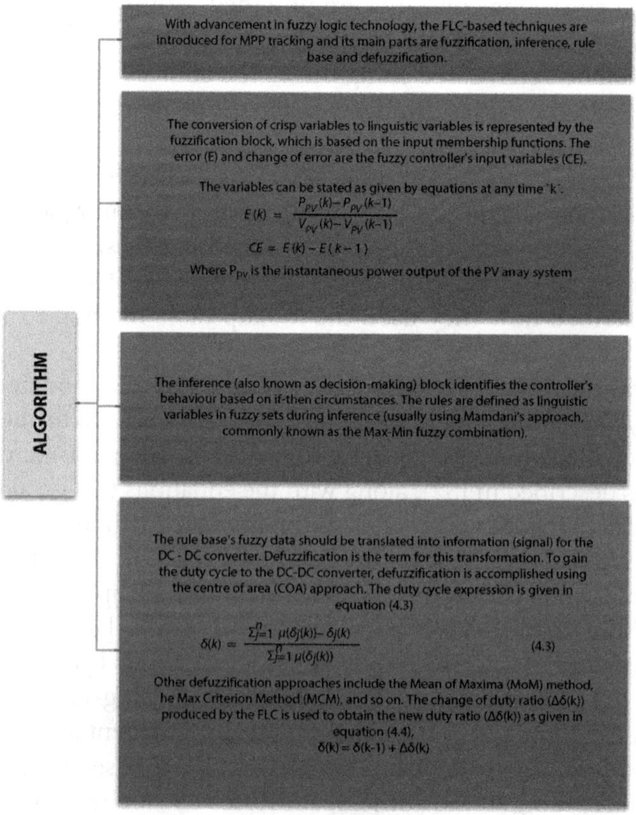

Brief:

Authors in [41] provided an overview of the fuzzy logic control-based MPP tracking approach. In this paper, a two-input FLC controller is proposed. The FLC method is faster than the traditional P&O method in tracking peak power point, according to the authors. The authors in [42] offered a fuzzy P&O technique for MPPT tracking. The authors determined that the fuzzy logic–based P&O approach had higher performance with smaller oscillations even under changeable irradiance and temperature situations, based on simulation results on the MATLAB/SIMULINK platform. The authors in [43] concluded that the FLC-based MPP is superior than hill-climbing approaches based on their experiments. The authors in [44] compared FLC and neural network-based MPPT and found that the FLC method is superior than the neural network method in terms of tracking. Authors in [45] proposed a fuzzy-based modified hill-climbing technique for adjusting irradiance and temperature. The proposed method, according to the authors, provides faster convergence, fewer MPPT oscillations in steady-state and accurate tracking in changing climates. The authors in [46] created a FLC system that can work in a variety of temperature and irradiance conditions. They find that, as compared to a system without a fuzzy controller, the fuzzy controller–based MPP can provide better tracking with less fluctuation. The authors in [47] suggested an FLC-based MPPT system and examined the controller performance using various membership functions in order to optimize the MPP. They compared the performance of FLC-based tracking to that of the traditional P&O method, concluding that the former outperforms the latter, even under varying climatic conditions. The authors in [48] proposed an asymmetrical FLC-based MPPT algorithm. To improve the effectiveness of the suggested algorithm, two approaches for creating the membership function have been developed, the second of which uses PSO to optimize the input membership function values. When compared to the P&O approach, the results of the testing suggest that the proposed algorithm performs better. The authors in [49] compared the P&O and FLC controllers and found that the FLC controller is faster in the transient state and produces a significantly smoother signal with fewer disruptions in the steady-state. The authors in [50] considered a dual MPPT controller based on an FLC. The first MPPT controller is a mechanical controller, while the second is an electrical controller that regulates the DC-DC converter. A FLC-based P&O mechanism has been implemented for the electrical controller. The FLC-based P&O method outperforms the conventional P&O algorithm, according to the results of a 150W prototype reported in [50]. The proposed FLC-based controller, according to the authors in [51], improves the hill-climbing search method by fuzzifying the rules of such techniques and eliminating their shortcomings. The tracking accuracy has been tested in a variety of meteorological conditions. The suggested controller is stated to have improved tracking precision, fewer oscillations around the MPP, no divergence during climatic changes, and faster convergence.

Implementation:

The system in [52] employs an FLC-based SEPIC converter. Rather of using a symmetrical distribution, a convergent distribution of membership functions was used. TMS320F28335 DSP was used to create the system. According to the scientists, the FLC-based system can properly follow the reference signal and transfer power at a rate of roughly 4.8% higher than the standard PI-based system. Authors in [53] suggested a controller that provides an ideal membership function for the FLC by combining FLC with Hopfield Neural Network. On the dSpace DS1104 controller, the controller has been installed and tested. When compared to conventional methods, the authors find that the proposed method is more reliable and accurate. FLC has been implemented in dSPACE and generated an <FLC.sdf> file using Matlab/Simulink. To control the converter, the file can be downloaded to the controller board [54]. The authors in [55] employed a 1N5822 schottky diode, an Opto-coupler TLP 250, and an ACS 712 current sensor to capture the controller's real-time input. The sensors detect input differences on a regular basis, and the FLC adjusts the switching pulses correspondingly. The authors in [56] employed the Virtex-II V2MB1000 Development Kit to implement FLC. It employs a probabilistic control method and is costly to deploy. The number of sensors needed varies according on the system. The tracking accuracy is very great, and the program tracks the global MPP very quickly.

The FLC approach has a number of advantages. The most significant benefit is that the system's accurate mathematical model is not necessary. In addition, the system can handle an infinite amount of inputs. Continuous integration causes it to develop a cumulative error. The PI-modified FLC (PI-FLC) approach is utilized to overcome this issue.

12.3.2 Artificial Neural Network–Based MPP Algorithm

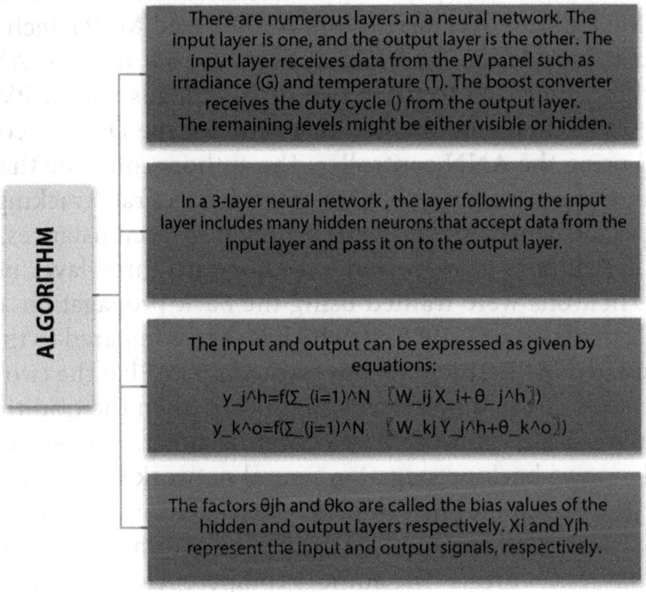

network control has been proposed. The algorithm, according to the authors, was adequate in aperiodic climate circumstances. The authors in [60] provided a brief overview of the ANN-based MPPT technique and discuss past work in the field. The authors in [61] used an ANN-based MPP tracking approach to drive two distinct loads with a PV-supplied separately excited DC motor. The duty ratio for the DC-DC converter is calculated using the ANN controller. The authors conclude that the suggested adaptive controller with ANN provides accurate tracking and outstanding performance under diverse irradiation circumstances. For MPP tracking, the authors in [62] used a feedforward three-layer neural network. The neurons were trained using the back-propagation algorithm. The ANN calculates an MPP first, which is then compared to the theoretically calculated MPP. The ANN has tracked the MPP if the two have very little difference. If the difference is significant, then the traditional P&O algorithm is used to find the best operating point. The authors in [63] suggested a novel back propagation neural network technique for tracking MPP. They compare their work's ANN-based method to traditional methods, stating that the proposed model is less difficult because the hidden layer has less neurons. The authors compare ANN-based MPP monitoring to the traditional IncCond tracking method in [64]. The authors in [65] used a modified IncCond technique based on a neural network for MPPT tracking. The results of the simulations were compared to those of a traditional P&O algorithm and a FLC-based algorithm, and the results demonstrate that the proposed approach is effective in tracking the MPP under partial shading situations.

Implementation:

The authors used an Altera Cyclone FPGA to implement the MPPT algorithm and generate PWM in [65]. The ANN program can be developed using a basic microcontroller like the PIC16F876. A simple neural network was employed by Petchjatuporn *et al.* to manage the power point on a solar charger [66]. For microcontrollers, neural networks are widely used. Any 8-bit microcontroller could be utilized, such as the Motorola 68HC711E9 [67]. The controller can be implemented using the TMS320C25 DSP. To track the MPP, the DSP-based controller senses the current and voltage of the solar array and calculates the solar array output power, power slope, and Vref. Depending on the implementation approach, the cost of implementation can be high, and the speed of tracking can also be high. With a probabilistic control method, the system also provides very accurate tracking.

12.3.3 Adaptive Neuro-Fuzzy Inference System MPP Tracking

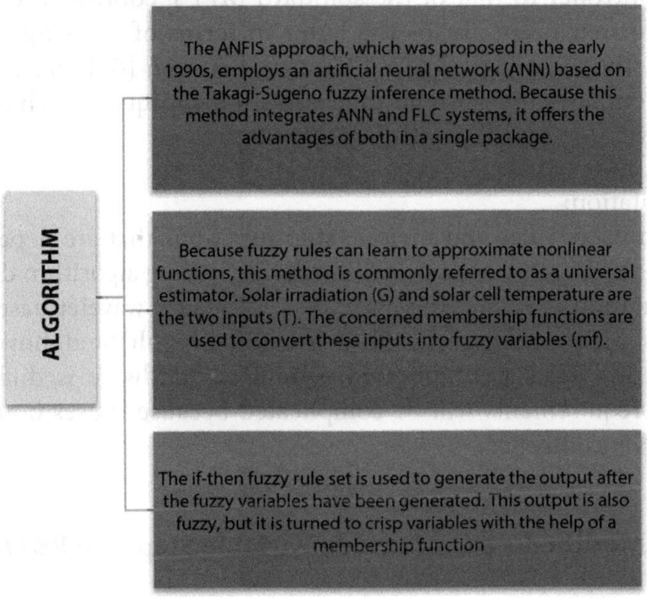

rules, which differs from [69]. They compare the performance of the ANFIS controller to that of the standard MPPT controller, concluding that the former outperforms the latter in terms of tracking speed and accuracy. The most significant difference between [69] and [72] is that the latter employs the sliding mode control technique, which eliminates the necessity for ANFIS controller training.

Implementation:
We search fuzzy rules and membership functions that are dependent on system behavior, which can be a difficult task during algorithm design and implementation [73, 74]. The authors developed a wavelet-based ANFIS system for MPP tracking in [75]. It should be highlighted, however, that implementing such a complicated system on hardware is difficult. The hardware requirements will be complicated because it uses both an FLC and an ANN controller.

12.3.4 Modified P&O Method (Variable Step Size P&O)

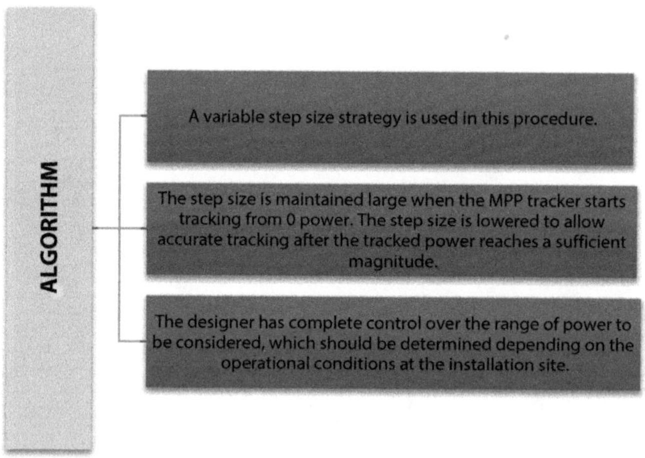

P&O MPPT with a zeta converter was used by the authors in [77]. For a step change in input voltage, the zeta converter is set to run in discontinuous conduction mode, and the algorithm's performance is evaluated. According to the authors in [77], the proposed strategy produces more accurate results.

Drawback:

The step size must be assessed and altered on a regular basis when the environment is rapidly changing.

Implementation:

This strategy is better suited to being implemented digitally. The key benefit of this method is that it can track MPP more precisely and effectively than the traditional P&O algorithm, and it is reasonably simple to implement. The authors presented a new adaptive P&O approach with variable step size based on fuzzy logic control in [78]. The converter used here is a DC-DC buck converter, and the proposed method's output was sufficient. The authors in [79] proposed using the FPGA XC2C384 to build a variable step size P&O algorithm. The most significant benefit is that the system does not require the addition of an external sensory unit. As a result, the system is cost-effective. The value of D, which represents duty perturbation, is variable in this approach based on the current operating point. Two approaches for determining the step size have been proposed in [80]. The variable step size is calculated using the derivative of the $[(1-D) * I_{PV}]$ vs. D curve in the first technique. Due to the step size, there may be some steady-state oscillations. The second technique uses two sequential samples of $[(1-D) * I_{PV}]$ to tune the step size. The second technique provides excellent steady-state and transient performance. For the implementation of variable step-size MPPT, the authors in [81] employed Intel's 80C196KC low-power consumption, CMOS microcontroller.

12.3.5 Particle Swarm Optimization Algorithm

ALGORITHM

The particle swarm optimization algorithm is an optimization technique based on the behaviour of flocks of birds and fish schools.

It is a stochastic method that is commonly used for global optimization.

The individuals of a flock interact and then the most successful individuals are imitated for optimization.
The particles movement depends on the velocity (v), acceleration coefficients (φ) and position (P). The movement is based on the following equation (4.7),

$$v_i^{k+1} = v_i^k + c_1R_1^k(Pbest_i - x_i^k) + c_2R_2^k(Gbest_i - x_i^k) \quad (4.7)$$

This can further be simplified into,

$$v_i^{k+1} = w \times v_v^k + c_1R_1\, Pbest_i + c_2R_2\, Pbest_i \quad (4.8)$$

$$x_i^{k+1} = x_i^k + v_i^{k+1} \quad (4.9)$$

The velocity updating is influenced by the acceleration coefficients. The user determines which weight factor (w) to apply.

A large value of w speeds up the process of reaching the global maximum and is less impacted by local locations, whereas a small value of w accelerates convergence.

The objective function of this algorithm with respect to MPP is the power delivered by the PV array at the present iteration must be greater than that in the previous iteration. Mathematically it can be expressed as given by equation (3.38),

$$P_{pv}^k > P_{pv}^{k-1} \quad (4.10)$$

The stopping criterion can be taken as,

$$\left(\frac{P_{pv}^{k-1} - P_{pv}^k}{P_{pv}^k}\right) > \Delta P \quad (4.11)$$

where P_{pv}^{k-1} is the power obtained in the previous iteration, P_{pvk} is the power obtained in the present iteration and ΔP is the minimum error in power.

The algorithm transmits three duty cycles δi (i=1, 2, 3) to the power converter. After each iteration, the duty cycle value approaches a constant, ensuring that the operational point is maintained.

The author in [63] utilized PSO to calculate the array voltage and then follows the MPP in a standalone PV system. Authors in [84] showed that tracking efficiency is more than 98% with a 14-ms convergence time. The authors in [85] employed PSO with direct duty cycle capabilities to track the MPP of the system in order to eliminate the PI control loop that was used to adjust the duty cycle. When compared to standard hill-climbing algorithms, the findings show that the suggested technique performs better.

Drawback:
The algorithm's computations are straightforward, and the algorithm's global MPP point is extremely near to the real (theoretical) MPP point. Convergence may not occur, though, if the coefficient values are chosen incorrectly. The technique is partly optimistic, and there is no precise rule for determining coefficient values.

Implementation:
To decrease steady-state oscillations following MPP tracking, the authors in [86] suggested a modified PSO technique. The suggested technique has been validated for a variety of irradiance and temperature scenarios, and its superiority has been demonstrated. The authors in [87] suggested a hybrid PSO-ANN method for global peak detection under partial shading circumstances. The Initialization and Repulsion PSO (IR-PSO) approach for MPP tracking is proposed by the authors in [88]. It has two phases: the first of which is the initialization PSO and the second of which is the repulsion PSO. The system was simulated and tested under a variety of shading circumstances, demonstrating the superiority of the suggested technique. When the range of the search space is broad, a novel approach combining PSO and P&O is presented in [89] to minimize the convergence time. After the P&O technique has found the local maxima, the PSO is employed to find the global peak. The authors suggest a hybrid P&O–PSO algorithm for global peak tracking in [90]. The hardware implementation of PSO with high calculating speed and high stability based on large-scale integration (LSI) is an essential topic to address, according to the scholar's dissertation [91]. The authors in [76] used the TMS320F240 DSP to build a modified PSO-based MPPT. In other words, low-cost microcontrollers can readily implement the PSO algorithm. The Atmega328P microcontroller was utilized by the authors in [92] to build a differential evolutionary algorithm-based PSO. The results of the implementation revealed good tracking and no steady-state oscillations.

12.3.6 Ant Colony Optimization–Based MPP Tracking

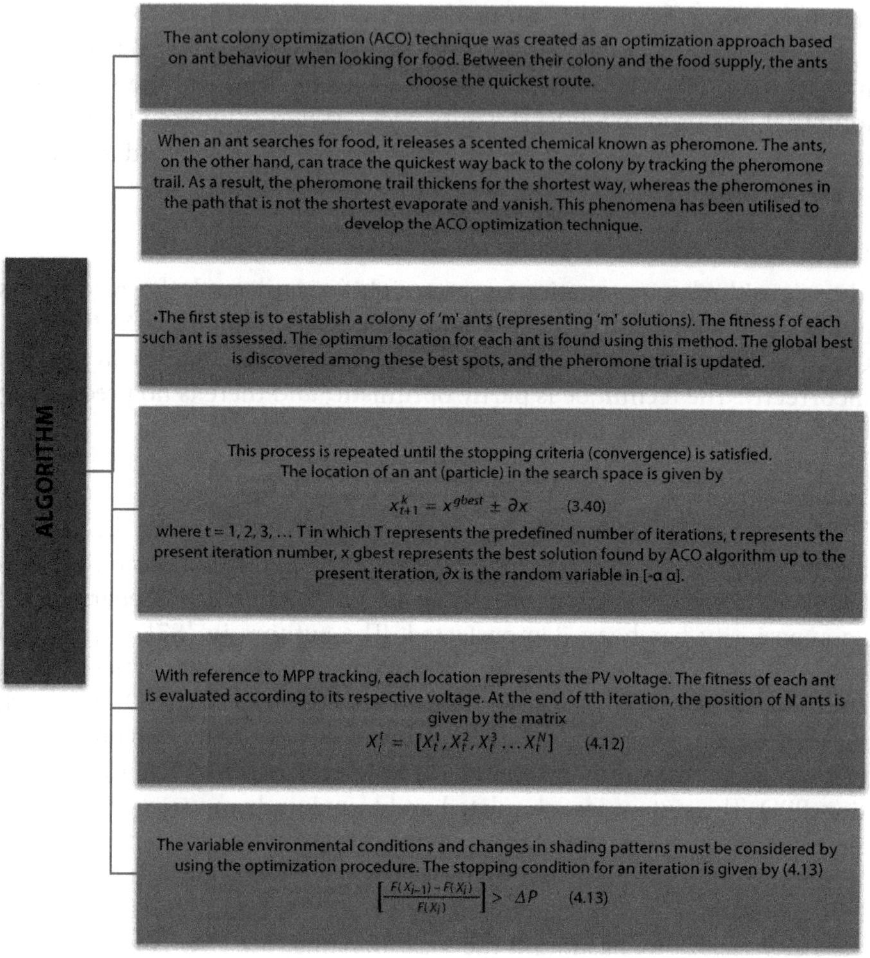

Implementation:

Authors in [93] proposed a unique ACO approach that is based on a heuristic algorithm and uses a simplified control system. As a control variable, the PV current is employed. During steady-state and transient irradiance circumstances, the suggested approach has been shown to provide superior tracking performance and efficiency. The authors in [94] suggested a fuzzy logic–based ACO technique for MPP tracking optimization. The ACO technique is used to fine-tune the fuzzy parameters. To find the local peak, a conventional technique such as the OCV method is utilized, and the suggested algorithm works from that local peak to discover the MPP. In the early phases of tracking, the authors in [95] recommended using a colon-based search followed by the P&O technique. The suggested technique integrates P&O's local search capability and ACO's global search capability. This results in faster convergence, improved performance, and increased efficiency. The MPPT ACO algorithm may be implemented using a PIC microcontroller, and it can also be tweaked to provide a better PI–MPPT control algorithm. The authors in [96] presented a novel pheromone-based ACO algorithm (ACO–NPU), claiming that it outperforms all other conventional, fuzzy, and neural network-based algorithms in terms of speed, accuracy, stability, and resilience.

12.3.7 Genetic Algorithm–Based Tracking

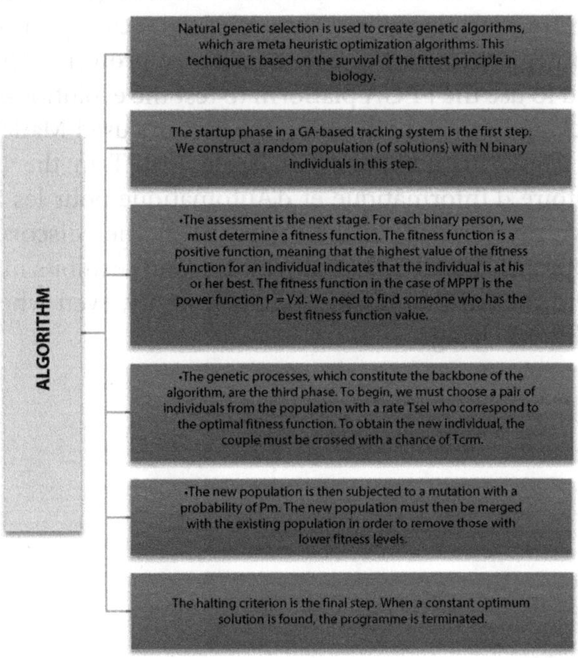

Brief:

The authors in [97] provided an overview of the GA and how to use it. During partial shading, they employ a GA to track the global maximum power point. Simulated investigations using MATLAB have revealed shorter tracking times and higher tracking accuracy. The optimal subsets of the membership functions are obtained using GAs. The suggested approach is more resilient, has a quicker reaction, and reduced steady-state oscillations. The performance of GA-based MPP tracking is compared to that of conventional P&O and IncCond techniques in [100]. When compared to traditional algorithms, the findings demonstrate that the suggested method has greater stability, quicker response, and superior performance. The authors discussed the GA-based MPPT and the test results in [101]. The suggested technique is supposed to be less noise-sensitive, with fewer oscillations around the MPP.

Implementation:

The authors in [98] used a GA in an ANN system to track global peak. The system load is a single-stage inverter-driven direct-coupled induction motor drive. Because of the variable step size, the suggested approach was shown to follow the peak faster, and the oscillations around the MPP were also reduced when compared to the traditional P&O method. Authors in [99] presented a hybrid Genetic k-Means Method (GKA) that incorporates the OLS training algorithm. The suggested technique uses a superior clustering strategy to solve the over-sensitivity problem. The authors in [102] planned to use the FPGA platform to test the evolutionary algorithm for MPP tracking in the lab. In [103], the authors used Matlab/Simulink® and Control Desk® to build the GA-based MPPT in the LIAS laboratory (Laboratoire d'Informatique et d'Automatique pour les Systémes) of Poitiers University (France) in the dSpace1005. They discovered that the GA-based approach with P&O (and with InCond) exhibits minimal oscillations around the MPP and low noise sensitivity, even when the atmospheric conditions change.

12.3.8 Cuckoo Search–Based MPPT

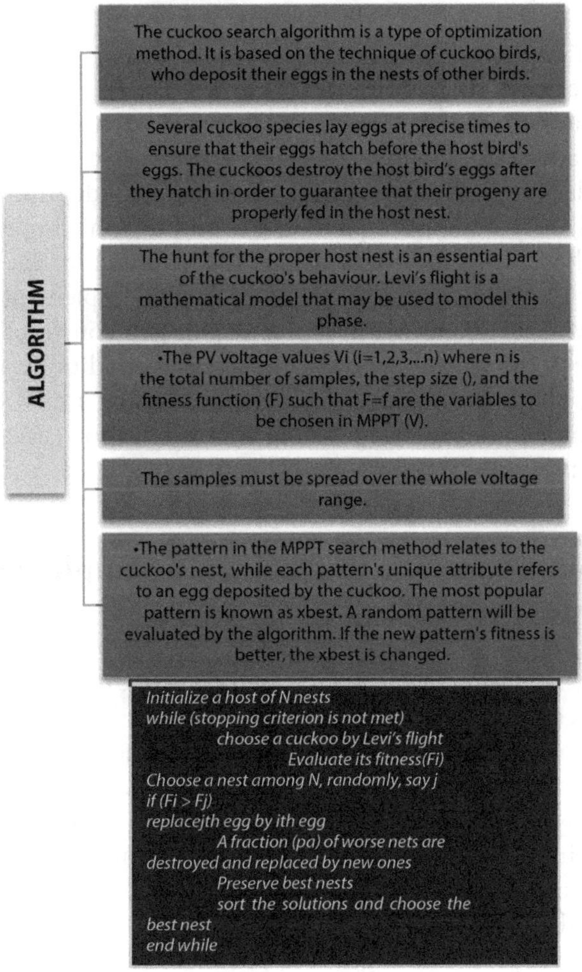

reference [105]. To validate the suggested approach, the parameters of two commercial PV generators were computed.

Implementation:
The developers of [106] utilized PID controllers and the CS method to efficiently track the MPP. Because it decreases transient overshoot and increases steady-state stability, the PID controller is preferred over the PI or P controller [107]. However, noise on the input side (due to the derivative control) may have an impact on the output, which may be reduced by filters. When compared to PSO, the programming for the algorithm is simpler. Any 16-bit microcontroller can be used to implement the method. The DSP TMS320F28335 was utilized in [108] to construct the method for firing the boost converter in partial shading conditions. Data such as array power, voltage, and current are obtained using a different DSP. For communication and data storage, serial communication (RS232) and a PC monitoring interface are utilized.

12.4 Comprehensive Review on the Implementation Issues of MPPT

Ease of implementation of any MPPT method is one of the most important factors to be considered for any solar PV system.

Table 12.1 describes the possible advantages and disadvantages in the implementation of the solar MPP tracking systems.

The various factors considered for comparison involve the number of sensors required for implementation, type of control strategy, mode of implementation, cost involved, tracking accuracy, ability to track global MPP during shading conditions, system efficiency, sensitivity to parameter changes, need for prior training, and application suitability. In probabilistic control, randomized algorithms with polynomial complexity are used to identify satisfactory controllers.

12.5 Commercial Products

From the implementation point of view, it is very much desirable to find a suitable inverter for the solar PV system. The type of inverter for solar PV systems is important to know in order to explore the commercial products that have been successfully implemented and tested which will be giving the researchers awareness in this regard. Hence, an attempt is made in this

Table 12.1 Comparison of various MPP tracking techniques.

Parameters considered	**MPP algorithms**												
	OCV	**SCC**	**CVC**	**P&O**	**InCond**	**FLC**	**ANN**	**ANFIS**	**Modified P&O**	**PI-FLC**	**N-FLC**	**PSO**	**ACO**
Variables to be sensed and/or controlled [28]	V	I	V	V and I	V and I	V and I	G and T	G and T	V and I	V and I	G, T, and V	G, T, V, and I	G, T, V, and I
Controlled Strategy [109]	Indirect	Indirect	Indirect	Sampling	Sampling	Probablistic	Probablistic	Probablistic	Sampling	Probablistic	Probablistic	Probablistic	Probablistic
Implementation [109]	Analog or Digital	Analog or Digital	Analog	Analog or Digital	Digital	Digital	Digital	Digital	Digital	Digital	Digital	Digital	Digital
Coast [28]	Less costly	Less costly	Less costly	Relatively expensive	Relatively expensive	Relatively expensive	Relatively expensive	Relatively expensive	Relatively expensive	Relatively expensive	Relatively expensive	Relatively expensive	Relatively expensive
Accuracy of Tracking [28]	Low	Low	Low	Relatively better	Good	Excellent	Excellent	Excellent	Very good	Excellent	Excellent	Excellent	Excellent
Global MPP tracking (during shading) [28, 109]	No	No	No	In some cases only	Yes	Yes	Yes	Yes	Yes	Yes	Yes	Yes	Yes
Efficiency	Low	Low	Low	Medium	High	Very High	Very High	Very High	Very High	Very High	Very High	Very High	Very High
Sensitivity to parameter changer [28]	Low	Low	Low	Medium	Medium	High	High	High	Medium	High	High	High	High

(*Continued*)

Table 12.1 Comparison of various MPP tracking techniques. (*Continued*)

Parameters considered	MPP algorithms												
	OCV	**SCC**	**CVC**	**P&O**	**InCond**	**FLC**	**ANN**	**ANFIS**	**Modified P&O**	**PI-FLC**	**N-FLC**	**PSO**	**ACO**
Complexity of Implementation	Low	Low	Low	Medium	Medium	Medium	Medium	High	Medium	High	High	High	High
Requirement of prior training of control parameters [28]	No	No	No	No	No	No	Yes	Yes	No	No	Yes	No	No
Application suitability [28]	Standalone	Standalone	Standalone	Standalone	Standalone	Standalone and Grid-connected	Standalone and Grid-connected	Standalone and Grid-connected	Grid-connected	Standalone and Grid-connected	Standalone and Grid-connected	Standalone and Grid-connected	Stand alone and Grid-connected

Table 12.2 Commercial inverters and MPPT methods used.

Sl. no.	Product	MPPT method
1	Outback MX-60, FlexMx60	Curve Fitting
2	Model Enphase, Enphase Energy	One-Cycle Control (OCC) Technique
3	Uni-Lyx PV Inverter, Danfoss	Differentiation Technique
4	Triple Lynx PV Inverter, Danfoss	Feedback Voltage or Current Technique
5	AERL COOLMAX SR Maximizer of AERL Company	Load Current/Load Voltage Maximization Technique
6	Morningstar-Trakstar MPPT Charge Controller, Solar Electric Supply	Intelligence MPPT Techniques
7	Parallux, eIQ Energy, Powerbox, Solar-Edge Technologies, Solar Magic, National Semiconductros, Maximizer-ES, Maximizer-EP, TIGO Energy, Sun Mizer, Xandex	Distributed Maximum Power Point Tracking
8	Genasun, USA GV Boost Charge Controller with MPPT, Blue Chip Energy Solutions, India - Solar MPPT Charge Controller	Look-Up Table Method
9	Conext™ CL 25000E, Conext™ Core XC Series Grid Tie Photovoltaic Inverter, Schneider Electric	P&O/P&O routine with prediction of the power change trend

paper to provide an insight into this issue. A list of some of such products has been given below with the MPPT method used [28, 110]. Table 12.2 consolidates the same.

12.6 Conclusion

Since the initial investment involved is high with the PV systems, it is vital to harvest them with maximum efficiency. The paper has explored from the conventional MPPT techniques till the recent evolutionary and AI-based MPPT techniques. The implementation issue is a major concern for real-time execution, so the various MPPT techniques have been discussed from their realization factor.

The described MPPT techniques are compared in terms of their tracking accuracy, cost, type of control strategy, mode and complexity of implementation, efficiency, and ability to track global maxima under shading and suitability of application. It can be observed from the review that the conventional MPPT methods are less expensive and, at the same time, their performance is moderate whereas, the AI-based and evolutionary algorithm–based MPPT methods offer better dynamic performance under partial shading condition. However, from the implementation point of view, these techniques are expensive and involve higher complexity. Hence, it is required to identify an MPPT technique with promising performance and at affordable cost in order to populate the solar PV system utility.

For interfacing with utility loads, the solar PV system is in need of an inverter that aids the conversion from DC to AC. It is quite appealing to review the commercial solar PV inverters that are employed in the industrial sector. This paper envisages revealing the solar PV inverter along with the type of MPPT technique employed in few practical systems. This exhaustive review paper will really be a promising source for the future researchers in the solar PV systems.

References

1. Oyedepo, S.O., Energy and sustainable development in Nigeria: The way forward. *Energy Sustain. Soc.*, 2, 1, 15, 2012.
2. Miller, G. and Spoolman, S., *Environmental Science: Problems, Connections and Solutions*, Cengage Learning, Chicago, 2007.
3. Ahuja, S. (Ed.), *Food, Energy, and Water: The Chemistry Connection*, Elsevier, Chicago, 2015.
4. Stamatescu, I. *et al.*, Design and implementation of a solar-tracking algorithm. *Proc. Eng.*, 69, 500–5075, 2014.
5. Montecucco, A. and Knox, A.R., Maximum power point tracking converter based on the open-circuit voltage method for thermoelectric generators. *IEEE Trans. Power Electron.*, 30, 2, 828–839, 2015.
6. Siddhant, K., *Implementation of fractional open circuit voltage mppt algorithm in a low cost microcontroller*, Diss. National Institute of Technology, Rourkela, 2014.
7. Kumari, J.S., Ch, S.B., Kullayappa, T.R., Design and analysis of open circuit voltage based maximum power point tracking for photovoltaic system.

8. Ovaska, S., *Maximum power point tracking algorithms for photovoltaic applications*, Diss. Aalto University, Vancouver, 2010.
9. Esram, T. and Chapman, P.L., Comparison of photovoltaic array maximum power point tracking techniques. *IEEE T. Energy Conver.*, 22, 2, 439–449, 2007.
10. Noguchi, T., Togashi, S., Nakamoto, R., Short-current pulse-based maximum-power-point tracking method for multiple photovoltaic-and-converter module system. *IEEE Trans. Ind. Electron.*, 49, 1, 217–223, 2002.
11. Xiong, Y., Qian, S., Xu, J., Research on constant voltage with incremental conductance MPPT method. *2012 Asia-Pacific Power and Energy Engineering Conference (APPEEC)*, IEEE, 2012.
12. Yuvarajan, S. and Xu, S., Photo-voltaic power converter with a simple maximum-power-point-tracker. *Proceedings of the 2003 International Symposium on Circuits and Systems, 2003. ISCAS'03*, vol. 3, IEEE, 2003.
13. Bekker, B. and Beukes, H.J., Finding an optimal PV panel maximum power point tracking method. *AFRICON, 2004. 7th AFRICON Conference in Africa*, vol. 2, IEEE, 2004.
14. Rosu-Hamzescu, M. and Oprea, S., *Practical guide to implementing solar panel MPPT algorithms*, Microchip Technology Inc, Application Note, Harvard, AN1521, 2013.
15. Gupta, A., Chauhan, Y.K., Pachauri, R.K., A comparative investigation of maximum power point tracking methods for solar PV system. *Sol. Energy*, 136, 236–253, 2016.
16. Xiong, Y., Qian, S., Xu, J., Research on constant voltage with incremental conductance MPPT method. *2012 Asia-Pacific Power and Energy Engineering Conference (APPEEC)*, IEEE, 2012.
17. Leedy, A.W., Guo, L., Aganah, K.A., A constant voltage MPPT method for a solar powered boost converter with DC motor load. *2012 Proceedings of IEEE Southeastcon*, IEEE, 2012.
18. Ye, Z. and Wu, X., Compensation loop design of a photovoltaic system based on constant voltage MPPT. *2009 APPEEC Power and Energy Engineering Conference*, IEEE, 2009.
19. De Cesare, G., Caputo, D., Nascetti, A., Maximum power point tracker for portable photovoltaic systems with resistive-like load. *Sol. Energy*, 80, 8, 982–988, 2006.
20. Tafticht, T. *et al.*, An improved maximum power point tracking method for photovoltaic systems. *Renew. Energy*, 33, 7, 1508–1516, 2008.
21. Aganah, K.A. and Leedy, A.W., A constant voltage maximum power point tracking method for solar powered systems. *2011 IEEE 43rd Southeastern Symposium on System Theory (SSST)*, IEEE, 2011.
22. Radhia, G. *et al.*, MPPT controller for a photovoltaic power system based on increment conductance approach. *2013 International Conference on Renewable Energy Research and Applications (ICRERA)*, IEEE, 2013.

23. Safari, A. and Mekhilef, S., Incremental conductance MPPT method for PV systems. *2011 24th Canadian Conference on Electrical and Computer Engineering (CCECE)*, IEEE, 2011.
24. Elgendy, M.A., Zahawi, B., Atkinson, D.J., Assessment of the incremental conductance maximum power point tracking algorithm. *IEEE T. Sustain. Energ.*, 4, 1, 108–117, 2013.
25. Yu, T.-C. and Lin, Y.-C., *A Study on Maximum Power Point Tracking Algorithms for Photovoltaic Systems*, Lunghwa University of Science and Technology, Dept. of Electrical Engineering, Chicago. 25 Apr. 2017.
26. Abuzed, S.A., Foster, M.P., Stone, D.A., *Variable PWM step-size for modified Hill climbing MPPT PV converter*, pp. 4–5, 2014.
27. Kumar, R., Dependence of hill climbing MPPT algorithm on ADC and Digital filter parameters. *2012 IEEE 5th India International Conference on Power Electronics (IICPE)*, IEEE, 2012.
28. Subudhi, B. and Pradhan, R., A comparative study on maximum power point tracking techniques for photovoltaic power systems. *IEEE T. Sustain. Energ.*, 4, 1, 89–98, 2013.
29. Phang, J.C.H., Chan, D.S.H., Phillips, J.R., Accurate analytical method for the extraction of solar cell model parameters. *Electron. Lett.*, 20, 10, 406–408, 1984.
30. Ibrahim, H. E.-S. A. *et al.*, Microcomputer controlled buck regulator for maximum power point tracker for DC pumping system operates from photovoltaic system. *FUZZ-IEEE'99. 1999 IEEE International Fuzzy Systems Conference Proceedings*, 1999, vol. 1, IEEE, 1999.
31. Ilyas, A., Khan, M.R., Ayyub, M., Lookup table based modeling and simulation of solar photovoltaic system. *2015 Annual IEEE India Conference (INDICON)*, IEEE, 2015.
32. Zainudin, H.N. and Mekhilef, S., *Comparison study of maximum power point tracker techniques for PV systems*, 2010.
33. Pongratananukul, N., *Analysis and simulation tools for solar array power systems*, University of Central Florida, Chicago, 2005.
34. Hohm, D.P. and Ropp, M.E., Comparative study of maximum power point tracking algorithms using an experimental, programmable, maximum power point tracking test bed. *2000 Conference Record of the Twenty-Eighth IEEE Photovoltaic Specialists Conference*, IEEE, 2000.
35. Leyva, R. *et al.*, MPPT of photovoltaic systems using extremum-seeking control. *IEEE Trans. Aerosp. Electron. Syst.*, 42, 1, 249–258, 2006.
36. Li, H. *et al.*, A Newton-based extremum seeking MPPT method for photovoltaic systems with stochastic perturbations. *Int. J. Photoenergy*, 2014, 2014.
37. Brunton, S.L. *et al.*, Maximum power point tracking for photovoltaic optimization using ripple-based extremum seeking control. *IEEE Trans. Power Electron.*, 25, 10, 2531–2540, 2010.
38. Miyatake, M. *et al.*, Control characteristics of a fibonacci-search-based maximum power point tracker when a photovoltaic array is partially shaded.

The 4th International Power Electronics and Motion Control Conference, 2004. IPEMC 2004, vol. 2, IEEE, 2004.

39. Ramaprabha, R., Balaji, M., Mathur, B.L., Maximum power point tracking of partially shaded solar PV system using modified Fibonacci search method with fuzzy controller. *Int. J. Electr. Power Energy Syst.*, 43, 1, 754–765, 2012.
40. Ahmed, N.A. and Miyatake, M., A novel maximum power point tracking for photovoltaic applications under partially shaded insolation conditions. *Electr. Power Syst. Res.*, 78, 5, 777–784, 2008.
41. Cheikh, M.S.A. *et al.*, Maximum power point tracking using a fuzzy logic control scheme. *Rev. Energ. Renouv.*, 10, 3, 387–395, 2007.
42. Chim, C.S. *et al.*, Fuzzy logic based MPPT for photovoltaic modules influenced by solar irradiation and cell temperature. *2011 UKSim 13th International Conference on Computer Modelling and Simulation (UKSim)*, IEEE, 2011.
43. Won, C.-Y. *et al.*, A new maximum power point tracker of photovoltaic arrays using fuzzy controller. *25th Annual IEEE Power Electronics Specialists Conference, PESC'94 Record*, vol. 1, IEEE, 1994.
44. Ben Salah, C. and Ouali, M., Comparison of fuzzy logic and neural network in maximum power point tracker for PV systems. *Electr. Power Syst. Res.*, 81, 1, 43–50, 2011, 2010. doi:10.1016/j.epsr.2010.07.005.
45. Alajmi, B.N. *et al.*, Fuzzy-logic-control approach of a modified hill-climbing method for maximum power point in microgrid standalone photovoltaic system. *IEEE Trans. Power Electron.*, 26, 4, 1022–1030, 2011.
46. Mahalakshmi, R., Aswin, K.A., Kumar, A., Design of Fuzzy logic based maximum power point tracking controller for solar array for cloudy weather conditions. *2014 Power and Energy Systems Conference: Towards Sustainable Energy*, IEEE, 2014.
47. Narendiran, S. *et al.*, Fuzzy logic controller based maximum power point tracking for PV system. *2016 3rd International Conference on Electrical Energy Systems (ICEES)*, IEEE, 2016.
48. Cheng, P.-C. *et al.*, Optimization of a fuzzy-logic-control-based MPPT algorithm using the particle swarm optimization technique. *Energies*, 8, 6, 5338–5360, 2015.
49. Cheikh, M.S.A. *et al.*, Maximum power point tracking using a fuzzy logic control scheme. *Rev. Energ. Renouv.*, 10, 3, 387–395, 2007.
50. Al Nabulsi, A. and Dhaouadi, R., Efficiency optimization of a DSP-based standalone PV system using fuzzy logic and dual-MPPT control. *IEEE Trans. Industr. Inform.*, 8, 3, 573–584, 2012.
51. Altas, I.H. and Sharaf, A.M., A novel maximum power fuzzy logic controller for photovoltaic solar energy systems. *Renew. Energy*, 33, 3, 388–399, 2008.
52. El Khateb, A. *et al.*, Fuzzy-logic-controller-based SEPIC converter for maximum power point tracking. *IEEE Trans. Ind. Appl.*, 50, 4, 2349–2358, 2014.

53. Subiyanto, S., Mohamed, A., Hannan, M.A., Intelligent maximum power point tracking for PV system using Hopfield neural network optimized fuzzy logic controller. *Energ. Buildings*, 51, 29–38, 2012.
54. Mohamed, A. and Hannan, M.A., Hardware implementation of fuzzy logic based maximum power point tracking controller for PV systems. *2010 4th International Power Engineering and Optimization Conference (PEOCO)*, IEEE, 2010.
55. Jasmin, E.R. and James, J., Implementation of fuzzy logic based maximum power point tracking in photovoltaic system. *Proc. of Int. Conf. on Control, Communication and Power Engineering, CCPE*, 2014.
56. Chekired, F. *et al.*, Implementation of a MPPT fuzzy controller for photovoltaic systems on FPGA circuit. *Energy Proc.*, 6, 541–549, 2011.
57. Elobaid, L.M., Abdelsalam, A.K., Zakzouk, E.E., Artificial neural network based maximum power point tracking technique for PV systems. *IECON 2012-38th Annual Conference on IEEE Industrial Electronics Society*, IEEE, 2012.
58. Giraud, F. and Salameh, Z.M., Analysis of the effects of a passing cloud on a grid-interactive photovoltaic system with battery storage using neural networks. *IEEE T. Energy Conver.*, 14, 4, 1572–1577, 1999.
59. Baek, J.W. *et al.*, Maximum power point tracking control of photovoltaic system using neural network. *2010 International Conference on Electrical Machines and Systems (ICEMS)*, IEEE, 2010.
60. Ishaque, K. and Salam, Z., A review of maximum power point tracking techniques of PV system for uniform insolation and partial shading condition. *Renew. Sust. Energ. Rev.*, 19, 475–488, 2013.
61. Veerachary, M. and Yadaiah, N., ANN based peak power tracking for PV supplied DC motors. *Sol. Energy*, 69, 4, 343–350, 2000.
62. Alabedin, A.M.Z., El-Saadany, E.F., Salama, M.M.A., Maximum power point tracking for photovoltaic systems using fuzzy logic and artificial neural networks. *2011 IEEE Power and Energy Society General Meeting*, IEEE, 2011.
63. Islam, M.A. and Kabir, M.A., Neural network based maximum power point tracking of photovoltaic arrays. *TENCON 2011-2011 IEEE Region 10 Conference*, IEEE, 2011.
64. Anzalchi, A. and Sarwat, A., Artificial neural network based duty cycle estimation for maximum power point tracking in photovoltaic systems. *SoutheastCon 2015*, IEEE, 2015.
65. Punitha, K., Devaraj, D., Sakthivel, S., Artificial neural network based modified incremental conductance algorithm for maximum power point tracking in photovoltaic system under partial shading conditions. *Energy*, 62, 330–340, 2013.
66. Petchjatuporn, P. *et al.*, A solar-powered battery charger with neural network maximum power point tracking implemented on a low-cost PIC-microcontroller. *TENCON 2005-2005 IEEE Region 10*, IEEE, 2005.

67. Binfet, J. and Wilamowski, B.M., Microprocessor implementation of fuzzy systems and neural networks. *IJCNN'01. International Joint Conference on Neural Networks, 2001. Proceedings*, vol. 1, IEEE, 2001.
68. Iqbal, A., Abu-Rub, H., Ahmed, S.M., Adaptive neuro-fuzzy inference system based maximum power point tracking of a solar PV module. *2010 IEEE International Energy Conference and Exhibition (EnergyCon)*, IEEE, 2010.
69. Aldobhani, A.M.S. and John, R., *Maximum power point tracking of PV system using ANFIS prediction and fuzzy logic tracking*, De Montfort University, Chicago, 2008.
70. Singh, R. and Pandit, M., Controlling output voltage of photovoltaic cells using ANFIS and interfacing it with closed loop boost converter. *Int. J. Curr. Eng. Technol.*, 3, 2, 417–423, 2013.
71. Abu-Rub, H. *et al.*, Quasi-Z-source inverter-based photovoltaic generation system with maximum power tracking control using ANFIS. *IEEE T. Sustain. Energ.*, 4, 1, 11–20, 2013.
72. Khaehintung, N., Sirisuk, P., Kurutach, W., A novel ANFIS controller for maximum power point tracking in photovoltaic systems. *The Fifth International Conference on Power Electronics and Drive Systems, 2003. PEDS 2003*, vol. 2, IEEE, 2003.
73. Karanjkar, D.S. *et al.*, Real time simulation and analysis of maximum power point tracking (MPPT) techniques for solar photovoltaic system. *2014 Recent Advances in Engineering and Computational Sciences (RAECS)*, IEEE, 2014.
74. Sharma, M. and Gupta, S., Adaptive neuro fuzzy inference system technique for photovoltaic array system. *Int. J. Ind. Electron. Electr. Eng.*, 4, 8, Aug. 2016.
75. Hassan, S.Z. *et al.*, Neuro-Fuzzy Wavelet Based Adaptive MPPT Algorithm for Photovoltaic Systems. *Energies*, 10, 3, 394, 2017.
76. Al-Diab, A. and Sourkounis, C., Variable step size P&O MPPT algorithm for PV systems. *2010 12th International Conference on Optimization of Electrical and Electronic Equipment (OPTIM)*, IEEE, 2010.
77. Satapathy, S., Dash, K.M., Babu, B.C., Variable step size MPPT algorithm for photo voltaic array using zeta converter-A comparative analysis. *2013 Students Conference on Engineering and Systems (SCES)*, IEEE, 2013.
78. Aashoor, F.A.O. and Robinson, F.V.P., A variable step size perturb and observe algorithm for photovoltaic maximum power point tracking. *2012 47th International Universities Power Engineering Conference (UPEC)*, IEEE, 2012.
79. Khaehintung, N., Wiangtong, T., Sirisuk, P., FPGA implementation of MPPT using variable step-size P&O algorithm for PV applications. *2006 ISCIT'06, International Symposium on Communications and Information Technologies*, IEEE, 2006.
80. Ahmed, E.M. and Shoyama, M., Variable step size maximum power point tracker using a single variable for stand-alone battery storage PV systems. *J. Power Electron.*, 11, 2, 218–227, 2011.

81. Koutroulis, E., Kalaitzakis, K., Voulgaris, N.C., Development of a microcontroller-based, photovoltaic maximum power point tracking control system. *IEEE Trans. Power Electron.*, 16, 1, 46–54, 2001.
82. Liu, Y.-H. *et al.*, A particle swarm optimization-based maximum power point tracking algorithm for PV systems operating under partially shaded conditions. *IEEE T. Energy Conver.*, 27, 4, 1027–1035, 2012.
83. Khare, A. and Rangnekar, S., A review of particle swarm optimization and its applications in solar photovoltaic system. *Appl. Soft Comput.*, 13, 5, 2997–3006, 2013.
84. Azab, M., Optimal power point tracking for stand-alone PV system using particle swarm optimization. *2010 IEEE International Symposium on Industrial Electronics (ISIE)*, IEEE, 2010.
85. Ishaque, K., Salam, Z., Shamsudin, A., Application of particle swarm optimization for maximum power point tracking of PV system with direct control method. *IECon 2011-37th Annual Conference on IEEE Industrial Electronics Society*, IEEE, 2011.
86. Ishaque, K. *et al.*, An improved particle swarm optimization (PSO)–based MPPT for PV with reduced steady-state oscillation. *IEEE Trans. Power Electron.*, 27, 8, 3627–3638, 2012.
87. Ngan, M.S. and Tan, C.W., Multiple peaks tracking algorithm using particle swarm optimization incorporated with artificial neural network. *Measurement*, 5, 10, 1325–1331, 1820.
88. Phimmasone, V. *et al.*, Verification of efficacy of the improved PSO-based MPPT controlling multiple photovoltaic arrays. *2011 IEEE Ninth International Conference on Power Electronics and Drive Systems (PEDS)*, IEEE, 2011.
89. Miyatake, M. *et al.*, Maximum power point tracking of multiple photovoltaic arrays: A PSO approach. *IEEE Trans. Aerosp. Electron. Syst.*, 47, 1, 367–380, 2011.
90. Lian, K.L., Jhang, J.H., Tian, I.S., A maximum power point tracking method based on perturb-and-observe combined with particle swarm optimization. *IEEE J. Photovolt.*, 4, 2, 626–633, 2014.
91. Jia, M., *Hardware Implementation of Particle Swarm Optimization and its Application for Adaptive Signal Processing*, PhD diss., Waseda University, Chicago, 2013.
92. Seyedmahmoudian, M. *et al.*, Simulation and hardware implementation of new maximum power point tracking technique for partially shaded PV system using hybrid DEPSO method. *IEEE T. Sustain. Energ.*, 6, 3, 850–862, 2015.
93. Jiang, L.L., Maskell, D.L., Patra, J.C., A novel ant colony optimization-based maximum power point tracking for photovoltaic systems under partially shaded conditions. *Energ. Buildings*, 58, 227–236, 2013.
94. Adly, M. and Besheer, A.H., An optimized fuzzy maximum power point tracker for stand alone photovoltaic systems: Ant colony approach. *2012 7th*

IEEE Conference on Industrial Electronics and Applications (ICIEA), IEEE, 2012.

95. Sundareswaran, K. *et al.*, Development of an improved P&O algorithm assisted through a colony of foraging ants for MPPT in PV system. *IEEE Trans. Industr. Inform.*, 12, 1, 187–200, 2016.
96. Titri, S. *et al.*, A new MPPT controller based on the ant colony optimization algorithm for photovoltaic systems under partial shading conditions. *Appl. Soft Comput.*, 58, 465–479, 2017.
97. Ramaprabha, R. and Mathur, B.L., Genetic algorithm based maximum power point tracking for partially shaded solar photovoltaic array. *IJRRIS*, 2, 2012.
98. Kulaksız, A.A. and Akkaya, R., A genetic algorithm optimized ANN-based MPPT algorithm for a stand-alone PV system with induction motor drive. *Sol. Energy*, 86, 9, 2366–2375, 2012.
99. Liao, C.-C., Genetic k-means algorithm based RBF network for photovoltaic MPP prediction. *Energy*, 35, 2, 529–536, 2010.
100. Hadji, S., Gaubert, J.-P., Krim, F., Experimental analysis of genetic algorithms based MPPT for PV systems. *2014 International Renewable and Sustainable Energy Conference (IRSEC)*, IEEE, 2014.
101. Hadji, S., Gaubert, J.-P., Krim, F., Genetic algorithms for maximum power point tracking in photovoltaic systems. *Proceedings of the 2011-14th European Conference on Power Electronics and Applications (EPE 2011)*, IEEE, 2011.
102. Shaiek, Y. *et al.*, Comparison between conventional methods and GA approach for maximum power point tracking of shaded solar PV generators. *Sol. Energy*, 90, 107–122, 2013.
103. Hadji, S., Gaubert, J.-P., Krim, F., Theoretical and experimental analysis of genetic algorithms based MPPT for PV systems. *Energy Proc.*, 74, 772–787, 2015.
104. Ahmed, J. and Salam, Z., A maximum power point tracking (MPPT) for PV system using Cuckoo Search with partial shading capability. *Appl. Energy*, 119, 118–130, 2014.
105. Ma, J. *et al.*, Parameter estimation of photovoltaic models via cuckoo search. *J. Appl. Math.*, 2013, 2013.
106. Kishnani, M., Pareek, S., Gupta, R., Optimal Tuning of PID controller by Cuckoo Search via Lévy flights. *2014 International Conference on Advances in Engineering and Technology Research (ICAETR)*, IEEE, 2014.
107. Vaigundamoorthi, M. and Ramesh, R., Experimental investigation of chaos in input regulated solar PV powered Cuk converter. *Int. J. Comput. Appl.*, 43, 10, 11–6, 2012.
108. Shi, J.-Y. *et al.*, Improved global maximum power point tracking for photovoltaic system via cuckoo search under partial shaded conditions. *J. Power Electron.*, 16, 1, 287–296, 2016.

109. Husain, M.A. *et al.*, Comparative assessment of maximum power point tracking procedures for photovoltaic systems. *Green Energy Environ.*, 2, 1, 5–17, 2016.
110. Verma, D. *et al.*, Maximum power point tracking (MPPT) techniques: Recapitulation in solar photovoltaic systems. *Renew. Sust. Energ. Rev.*, 54, 1018–1034, 2016.

13

Reliability Analysis Techniques of Grid-Connected PV Power Models

Raghavendra Rao N. S.*, Chitra A. and Daki Krishnachaitanya

School of Electrical Engineering, Vellore Institute of Technology, Vellore, India

Abstract

Solar PV (photovoltaic) and wind renewable energy sources have been widely used compared to other renewable energy sources. The PV systems offer many benefits over wind energy systems like simple technology, minimal influence on the local environment, cost, and maintenance. Thus, the PV system speeds up its energy production to fulfill energy demand. To meet this goal, the reliability of grid-connected PV systems should be considered. This paper compares possible failure rates of power model components of PV grid-connected systems among widely used reliability prediction standards using ALD (Advanced Logistics Development) software. The resultant values from this software for each standard, the reliability, and the MTBF (mean time between failures) of grid-connected PV power models components are estimated and examined respectively. Finally, some interesting conclusion concerning to vulnerable components of PVs power models is obtained.

***Keywords*:** Reliability, MTBF, PV system, failure rate

13.1 Introduction

In the present era, renewable energy-oriented systems gain extra interest due to the demand for clean energy. This demand can be compensated by different renewable energy sources, namely, wind, solar photovoltaic (PV), geothermal, and biomass. Among them, power generation using a solar PV system is an important method. Grid-tied PV renewable energy system is

**Corresponding author*: raghavendrarao.ns2019@vitstudent.ac.in

A. Chitra, V. Indragandhi, and W. Razia Sultana (eds.) Smart Grids and Green Energy Systems, (197–206)

a powerful model designed by semiconductor devices like diodes, transistors, printed circuit boards (PCBs), and integrated circuits (IC). During operation, such grid-tied PV systems may have intermittent faults or failure (refer to Figure 13.1), reducing the system's reliability. In terms of reliability, various studies provide extensive analysis of the effect of adding particular renewable energy resources to the utility grid as distributed generation.

The author in [1] used Monte Carlo simulation to analyze the reliability of off-grid standalone microgrid system using PV along with other renewable systems. Hybrid AC-DC power systems reliability indices and their consumer performance of each node are evaluated in [2] using a multi-state model. The findings demonstrated the importance of reliability assessment in determining AC-DC power system risks. The authors in [3] and [4] conducted a case study to minimize the power losses in grid inter-connected renewable energy sources by improving the reliability of the system.

The findings revealed that an unreasonable reliability assessment technique had a significant impact on reliability evaluation, and reliability evaluation accuracy had a critical impact on distribution system performance and future planning. Therefore, this paper will focus on reliability investigation techniques on grid-connected PV power models components. In Section 13.2, the empirical relations in the reliability analysis and reliability standard are explained. The comparative study of grid-connected PV power models components for several reliability standards is conducted in section 13.3. Finally, the paper will conclude in the last section.

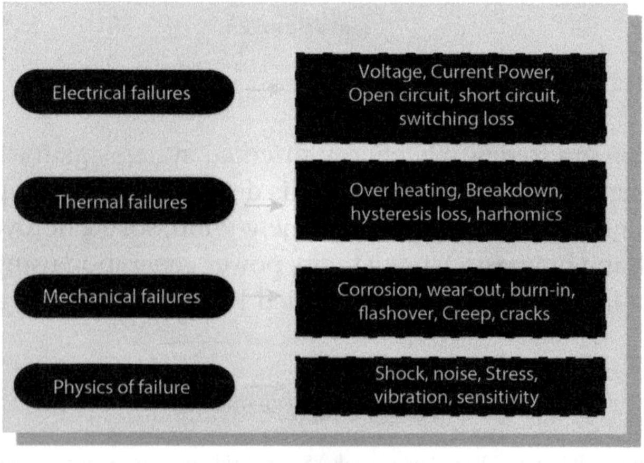

13.2 Reliability Empirical Relations and Standards

The failure is a stochastic process of a system. The probability of its analysis under desired working conditions at the time is called reliability [5]. Mathematically, the reliability function R (t) is a probability of successful operation of a failure-free system between the time intervals 0 to time t.

$$\text{i.e., } R(t) = P(T > t)\ t \geq 0 \tag{13.1}$$

Here, T is a random variable that represents failure time. The failure rate λ (t) function is the normalized number of failures per unit time of the system that is still operating [6]. The relation between R (t) and λ (t) is given by

$$\mathrm{R(t)} = \mathrm{e}^{-\lambda(\mathrm{t})} \tag{13.2}$$

The meantime operation of equipment intended between failures is called mean time between failures (MTBF) [7], and it is given by

$$MTBF = \int_0^\infty \mathrm{R(t)dt} = \frac{1}{\lambda} \tag{13.3}$$

The R (t), λ (t), and MTBF are closely related one and another (refer to Figure 13.2). If any one of them is estimated, then the remaining functions can be estimated by elementary calculus.

Several standards, models, and manuals have been used in reliability estimation; some of the field database reliability analysis standards are

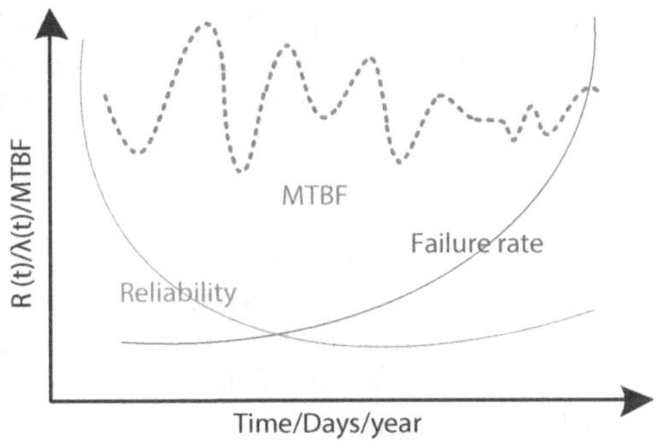

Figure 13.2 Typical reliability, failure rate, and MTBF plot.

Table 13.1 Reliability prediction models.

Standards	Advantages	Disadvantages
MIL-HDBK-217 (Military Handbook): It is first field based reliability prediction model.	It includes more the 14 environment classification. It support both parts count and stress analysis for an electronic device.	Field test, lab testing, burn-in data and wear out factors are not considered.
Telcordia (Bellcore): It is designed by Bell company, a telecommunication research company.	Field test, lab testing, burn-in data and wear out factors are considered.	Number of ground environments is limited. Compared to MIL-HDBK-217, it as support few model parts.
IEC62380 (International Electro-technical Commission): It is based on the UTEC80-810 (UTE-Union Technique de L'Electricite) data.	The temperature and power on/off cycling is taken into account.	Its mathematical analysis is cluttered.
Siemens SN29500: It is based on IEC61709 concept.	Used as a basis for modification of failure rate data at the given condition to the actual operating condition.	It is exclusively for Siemens devices.
299B/C: It is a Chinese Military model; GJB/z and 299B are its standards.	In reliability prediction, few data are required for part count analysis than that of part stress.	Not for the hybrid models. Primarily used in china.
FIDES: It is created under the DGA (Delegation Generale pour l'Armement) supervision by French ministry of defense.	It is based on physic of failure; considering existing models, test data, and field returns.	It applies to nonlinear systems only. It strongly needs critical data.

discussed in Table 13.1 [8, 9]. Such standards include stress-strength interference methods, field data, physics-of-failure, and basic parts count methods. Many of these approaches can be applied using commercial software that is freely available.

13.3 Reliability Estimation of Grid-Connected PV Power Models

The PV system integrated into the utility grid is shown in Figure 13.3 [10] incorporate majorly four power modules, namely, a converter, inverter, controller, and a DC-link capacitor. The converter has been used to regulate the DC bus voltage, an inverter converts the DC voltage to an AC voltage, and the entire operation is monitored by the controller. The components of such models are given in Figure 13.4. The failure of any one of such module or component might cause the entire system to fail. However, how much each components of respective models affect the reliability of the overall system and which components is the least reliable in the entire power model. These challenges are still an unknown. For further studies, the reliability analysis of Power models Components of grid-connected PV system is carried out using ALD MTBF calculator software [11].

ALD is a free reliability prediction tool that is exclusively applied irrespective of electrical and mechanical components. This tool is built with more than 26 standards to support them. Among them, the popularly used

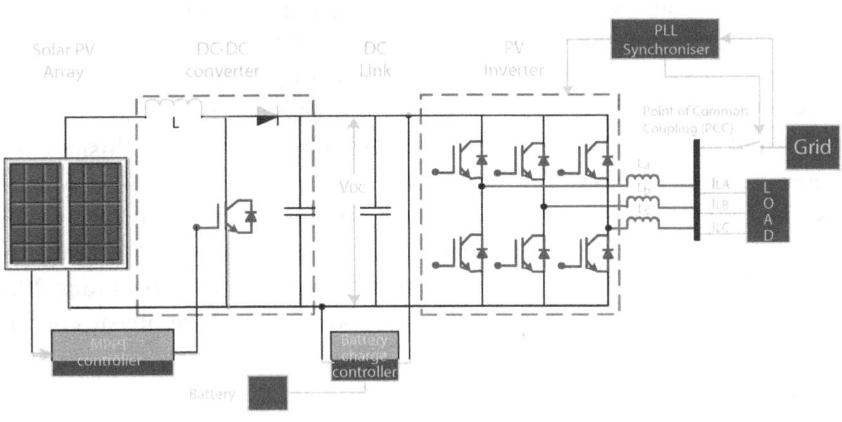

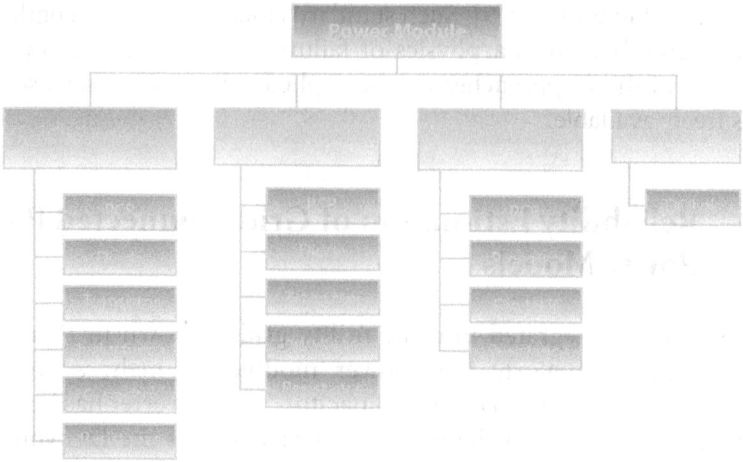

standards for electronics components are MIL-HDBK-217E/F issues [12, 13], Telcordia issues, Fides issues [14, 15], IEC issues [16], HDBK-217 Plus, and Siemens series. Thus, using such reliability standards, the failure rates of various components of grid-connected PV systems power models are estimated and tabulated in Table 13.2. Estimated values from the ALD tool for different standards in Table 13.2, reliability indices graphs that define the performance and behavior of power models components of the grid-connected PV system can be drawn by substituting the estimation in (1), (2), and (3); results are shown in Figure 13.5.

Figure 13.5a shows the total failure rate of grid-connected PV system power models at the time t for various reliability standards. The failure values of each component of a grid-connected PV system, particular to standards, are given in Figure 13.5b. Figure 13.5c illustrates the reliability of grid-connected PV system power models specific to the standards and time t. The grid-connected PV system power models indusial components MTBF is plotted in Figure 13.5d. The resultant reliability indices value of grid-connected PV power models slightly various within the standards. This is because most of these standards are based on the military handbook with additional or altered factors. Here, Fides-2009 values are slightly more compare to other standard, as the physics of failure factor incorporated.

Furthermore, the model-wise failure rate of the grid-connected PV system is illustrated in Figure 13.6 for respective reliability standards. These vulnerable models can be identified. From the plot, an

Table 13.2 Estimated failure rates of components using ALD software.

Components	Failure rates (λ) in FPMH					
	MIL217F-2 (PC)	**Telcordia-3**	**IEC62380**	**SN29500-1**	**299C (PC)**	**Fides-2009 (PC)**
DC link capacitor	1.3273	0.1147	0.0033	0.0166	0.804	1.16
Capacitor	0.1845	0.165	0.0065	0.019	0.98	0.205
Diode	0.0026	0.0118	0.0102	0.0125	0.002	0.065
Transistor	0.0046	0.0071	0.0101	0.0045	0.005	0.0078
Inductance	0.143	0.0115	0.0054	0.0044	0.08	0.18
Controller	0.285	0.0426	0.012	0.0048	0.189	0.35
Digital IC	0.0343	0.032	0.0287	0.045	0.014	0.0501
Resistance	0.0134	0.0042	0.0018	0.005	0.012	0.004
PCB	1.018	0.78	0.705	0.908	0.897	1.12

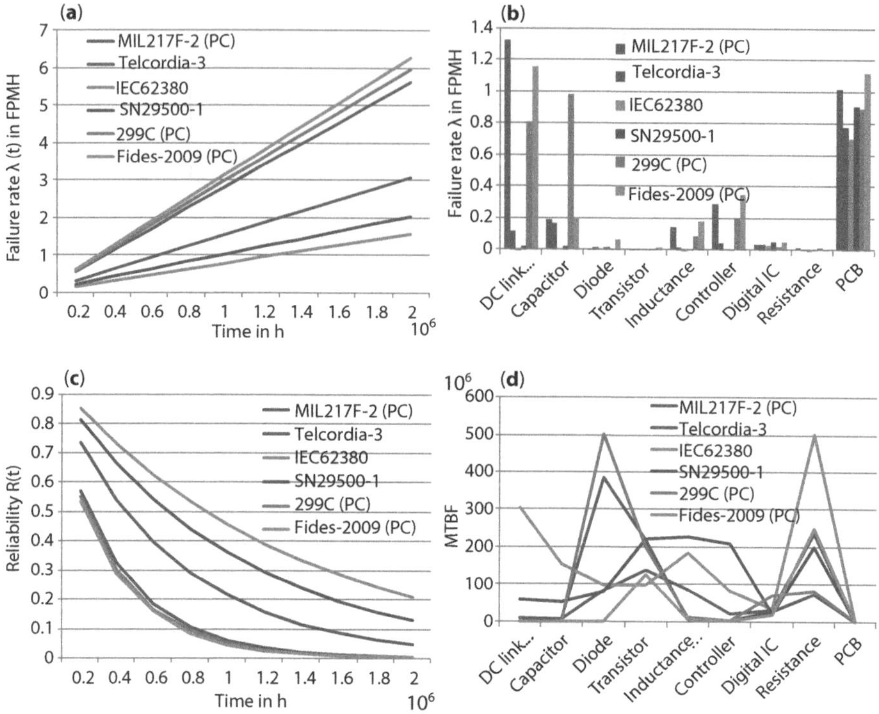

Figure 13.5 Reliability indices of power models components of grid-connected PV system (a) Total failure rate of grid-connected PV system. (b) Grid-connected PV systems components failure values. (c) Reliability of grid-connected PV systems. (d) MTBF of grid-connected PV systems.

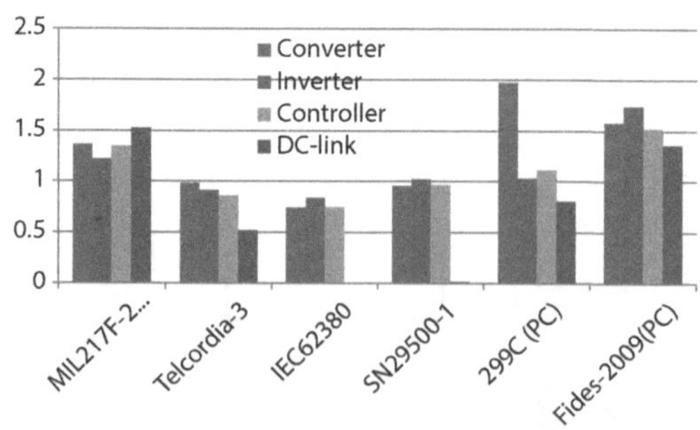

Figure 13.6 Estimated failure rates of power models of grid-connected PV system.

inverter is a highly vulnerable model in the case of IEC62380, SN29500-1, and Fides-2009 (PC) standards. Similarly, DC-link in MIL217F-2 (PC), converter in Telcordia-3, and 299C (PC) are highly vulnerable models, respectively.

13.4 Conclusion

To understand the reliability of grid-connected PV power models, a comparative reliability study is conducted using ALD software considering widely used reliability standards. Through the analysis, some important conclusions are discussed as follows.

- The type of reliability analysis depends on the data availability, nature of deployment of the PV system, environmental factors, etc.
- Identifying the most vulnerable component or model in the system varies from one reliability standard to another because of the factors considered during estimation.
- The accuracy of predicted grid-connected PV power models reliability values can be improved by updating the reliability standards periodically.

References

1. Rathore, A. and Patidar, N.P., Reliability assessment using probabilistic modelling of pumped storage hydro plant with PV-wind based standalone microgrid. *Int. J. Electr. Power Energy Syst.*, *106*, 17–32, 2019.
2. Guo, L., Ding, Y., Bao, M., Shao, C., Wang, P., Goel, L., Nodal reliability evaluation for a VSC-MTDC-based hybrid AC/DC power system. *IEEE T. Power Syst.*, *35*, 3, 2300–2312, 2019.
3. Reddy, S.S., Optimization of renewable energy resources in hybrid energy systems. *J. Green Eng.*, *7*, 1, 43–60, 2017.
4. Reddy, S.S., Optimal scheduling of wind-thermal power system using clustered adaptive teaching learning based optimization. *Electr. Eng.*, *99*, 2, 535–550, 2017.
5. Chung, H.S.H., Wang, H., Blaabjerg, F., Pecht, M., *Reliability of power electronic converter systems*, Institution of Engineering and Technology, 2016.
6. Pham, H. (Ed.), *Handbook of Reliability Engineering*, Springer, 2003, ISBN: 978-1-85233-841-1.
7. Bra, E., *Reliability engineering a life cycle approach*, Taylor & Francis, 392, 2016, https://doi.org/10.1201/9781315367422, eBook ISBN9781315367422.
8. European Power Supply Manufacturers Association, *Guidelines to understanding reliability prediction*, Revision October 13, 2004. https://www.energisa.com.br/Documents/Tocantins/Guidelines%20to%20Understanding%20Reliability%20Prediction.pdf (Accessed: July 4, 2021).
9. IEEE guide for selecting and using reliability predictions based on IEEE 1413, in: IEEE Std 1413.1-2002, pp.1-106, 2003.

10. Gupta, N., Garg, R., Kumar, P., Sensitivity and reliability models of a PV system connected to grid. *Renew. Sust. Energ. Rev.*, 69, 188–96, 2017 Mar 1.
11. Free MTBF calculator user guide (no date) Aldservice.com. Available at: https://aldservice.com/Free-MTBF-Calculator-User-Guide.html (Accessed: July 4, 2021).
12. Department of Defense, *Reliability Prediction of Electronic Equipment (Military Handbook)*, p. 205, Department of Defense, USA, 1991.
13. *Military Handbook Reliability Prediction of Electronic Equipment MIL-HDBH-217E*, Department of Defense Assistance Mil., Federal Agencies, Ind., Washington, DC, USA, 1990.
14. FIDES, *FIDES guide 2009 Edition A September 2010 Reliability Methodology for Electronic Systems*, pp. 1–347, French armament Industry Supervision agency FIDES, September 2010.
15. Airbus, F. *et al.*, FIDES guide 2004 issue A reliability methodology for electronic systems, Aldsoftware.com. [Online]. Available: https://www.aldsoftware.com/download/ramc/RAMC_KB/Reliability/Methods/FIDES.PDF. [Accessed: 07-July-2021].
16. Комиссия, М.Э., Reliability data handbook-universal model for reliability prediction of electronics components, PCBs and equipment, Iec.ch, 2004. Available at: https://webstore.iec.ch/preview/info_iec62380%7Bed1.0%7Den.pdf (Accessed: July 7, 2021).

14

DC Microgrid: A Review on Issues and Control

D. Anitha and K. Premkumar*

Department of Electrical and Electronics Engineering, Rajalakshmi Engineering College, Chennai, India

Abstract

Over the past two decades, DC microgrids have achieved a spectacular progression, owing to it is to simplicity, efficiency, and wide DC power utilization. However, the DC microgrid has encountered enormous power quality problems in its operation and control due to the incorporation of multiple distributed energy sources (DERs) in the common DC bus. This chapter presents an explicit review on the different types of technical challenges faced by the DC microgrid. The readers can understand the root cause of the technical challenges faced by the DC microgrid and identify the appropriate remedial actions to mitigate the problems aroused. In addition, this chapter analyzes the different types of control strategy implemented in the DC microgrid from the earlier stages of evolution of microgrids. The different types of control strategies are broadly classified into single-layer and multilayer control strategies. A comprehensive analysis of this control strategy and the selection of appropriate control technique according to the size and nature of DC microgrid is explained in this chapter.

Keywords: DC microgrids, distributed energy sources, DC circuit breakers, Master-Slave control, droop control

**Corresponding author*: premkumar.k@rajalakshmi.edu.in

A. Chitra, V. Indragandhi, and W. Razia Sultana (eds.) Smart Grids and Green Energy Systems, (207–230)

14.1 Introduction

Non-conventional sources like solar energy, wind energy, and tidal energy are getting under limelight for the past two decades, owing to its abundant availability [1]. The renewable energy sources are the key component of these microgrids. Although a system with solar panels, fuel cells, wind turbines, etc., cannot be called as microgrid. The term microgrid includes integration major components such as renewable energy sources, loads, a proper control strategy and energy storage system (ESS) [2, 3] in order to have a seamless flow of power during both grid connected mode and islanded mode. The environmental impact of microgrid is zero emission of greenhouse gases like carbon monoxide and carbon dioxide [4, 5]. This is an implicit advantage compared to power generation due to conventional sources. In addition, the depletion of fossil fuels in the upcoming decades gives way to microgrid technologies without any option [6, 7]. However, the probabilistic nature of these sources depending upon the seasons and climate changes reduces the overall reliability of these systems. To remove this difficulty distributed generation (DG) such as photovoltaic panels, micro-turbine, and fuel cell are incorporated along with these sources and enhancing the reliability. This chapter concentrates on DC microgrid system even though AC microgrid is also in play in the current scenario. In comparison with AC microgrids, the DC microgrid (DCMG) systems provide good efficiency, robust stability, and higher reliability with marginal loss. The minimal ejection of electromagnetic field and absence of technical effects such as skin effect, harmonic distortion, and proximity effect [8] in DC system makes it lot more compatible and safer on comparison with analogous AC microgrids. The DC currents produce lesser range of damage comparatively [9]. Owing to the absence of skin effect, the cables used for DC transmission can decrease their losses from 15% to 20% [10]. In addition, the DC microgrid has further more advantages such as lower transmission losses, zero reactive power, lesser stages of power conversion, and well controllability as the frequency of the system is zero. The outputs obtained from the renewable sources like solar and fuel cell can be incorporated with the power system readily by using a single converter with lesser loss and good efficiency. The counterpart AC sources like wind power generation, gas turbines are provided with a minimum of two back to back converters to convert the obtained AC supply to DC and convert it back to AC again. This surges the losses due to switching and downturns the efficiency of the overall system [11–13]. Another important aspect of DC system is that it completely eliminates the synchronization issues, which is a major problem in case of AC grids [14]. Apart from the pros enlisted above the DC MG also suffers from certain cons such as impedance mismatch on source side, identification of proper grounding, standardization issues, and zero probability of identifying the zero crossing in voltage and current magnitude. However, the invent of technology in power semiconductor devices are capable of minimizing the disadvantages faced by DC system [15–17]. The control strategy employed for DC microgrid became more efficient lately after the development of semiconductor and protection relays. In [18, 19], a detailed control of DC microgrid is proposed. This chapter is categorized as follows. In Section 14.2, all the power quality issues and technical difficulties faced by DCMG is enlisted. Section 14.3 depicts the control strategies adopted for efficient operation of microgrid system, and Section 14.5 concludes the chapter.

14.2 Challenges Incurred in DCMG

Although the DC microgrid is efficient and has less losses when compared to AC microgrid, the construction of DCMG is a formidable task when considering the following technical issues: 1) difficulty experienced in suppressing the arc due to absence of zero crossing DC current; 2) identification of proper grounding; 3) effect of short-circuit fault current and its sensitivity toward the inverter; and 4) electromagnetic interference and inrush currents.

14.2.1 Difficulties in Extinguishing Arc

Protection is the key component to maintain a healthy power system. The biggest challenge in DC microgrid is to design protective devices such as circuit breakers, relays, and fuses based on the fault current ratio. The working principle of these breakers purely depends upon zero crossing of the current component. Thus, AC breakers can be easily developed based on the zero crossing of the current and fault magnitude. The arc produced in AC system gets separated within other half-cycle. While moving to DC systems owing to non-availability of zero crossing [20], it becomes tedious to separate and extinguish the arc current. Thereby, the arc current develops in magnitude within short time and the overall systems become more prone to heavy fault current [21]. The breaker contacts face high erosion problems due to this developing fault current. The overall span of the DC breakers decreases in short time leading to danger in personal safety and also flow of fault current into the healthy system [22].

Later, MCCBs (Molded Case Circuit Breakers) were employed to decrease the time constant of breaker [23] and minimize the arc current with the help of extinguishing compartment and relevant devices. Later MCCB also went through many drawbacks, owing to quick discharge of filter capacitor in the network. Thus, zero crossing is a major challenge in DC microgrid to ensure power system protection. In the near future, the development of power electronic devices like IGBT and IGCT will provide quick solutions for these issues [24].

14.2.2 Lack of Adequate Grounding

Grounding is a crucial strategy in DC microgrid and is very difficult to employ based on the type of fault (either pole to pole or pole to ground fault). The protection of a power system is enhanced with proper grounding technique [25–27]. A DC system is said to be properly grounded if it minimizes the leakage current, identifies the location of fault, reduces the common mode touch voltage for human safety, reducing stray current magnitude, etc. Three major types of grounding configuration are employed in current scenario for high-voltage DC microgrids based on the type of fault characteristics, namely, TN, IT, and TT grounding configurations [28–30]. In the TN configuration, the system is solidly grounded with low impedance and location of fault becomes easier as high fault magnitude current flows in the system. Due to high fault current magnitude, this system is very dangerous to human safety and device safety. On the contrary, detection of fault is tedious in the IT configuration as this type of grounding has high impedance and low fault current magnitude. This system provides good human safety against touch voltage but fails to identify the fault location soon. In the TT configuration, the fault current reinstates within the system continually and, thereby, voltage stress increases in it. In addition, this configuration requires multiple grounding points. Therefore, selecting the type of grounding becomes arduous task in DC microgrids.

14.2.3 Effect of Short-Circuit Fault Current and Inverter Sensitivity

The occurrences of fault in a power system are undeniable and unpredictable in nature. In case of AC microgrid, when a short-circuit fault exists, it draws fault current (I_f) nearly 10 times the magnitude of rated current prevailing in the system [31]. Due to such a large magnitude of current, the relays that are associated nearer to the fault location senses the high current and provides the necessary protection scheme required to maintain safe operation of grid connected mode. Conversely, moving to DC microgrid, the characteristics of short-circuit fault current varies depending upon the mode of operation of the power system and the converter topologies. In isolated mode of operation, when a fault occurs, it draws five times greater current than that of threshold current. This fault current is contributed by the distributed generators present in the isolated grid [32, 33]. In case of grid tied mode, the fault current is 20–50 times more than the threshold magnitude, which is very pernicious to the power system. This fault current is jointly contributed by the DG present in the microgrid together with the main utility grid. This high amount of short-circuit fault current requires protection relays operating at high speed in order to break the circuit in case of fault. This is a big challenging task in DC microgrid as breaking the circuit in DC takes several complications as mentioned prior. The inverter circuits present in the DC microgrid have low current and high sensitivity characteristics. When the fault magnitude is tremendously high, it causes detrimental issues in the inverter and current based relays. Owing to the high sensitive nature of the inverters and current based relays, they are provided with restricted operation. This, in turn, delays the process of tripping the circuit and the fault current magnitude rises continuously. This complexity needs high focus to establish inverters with greater range of flexibility to control the fault [34].

14.2.4 Electromagnetic Interference and Inrush Currents

In a DC microgrid, the voltage conversion, stability issues, faults, and other maintenance are controlled through PWM-based converter controller circuits. These converter circuits are connected to DC capacitors on both front and rear end of it which also acts as a filter for safe operation. These converters work at a very high switching frequency for quick response and control. These DC-DC converters with DC side link capacitors connected to DC bus cables generate multiple resonant frequencies when operating at very high speed and cause strong Electromagnetic Interference in the microgrid. When any of these resonating frequency matches or gets tuned to the harmonics emanated from the converter, it instantly affects the power quality, destabilizes the DC bus, affects the voltage and current magnitude, and thereby reduces the overall system efficiency. To minimize the root of cause (the presence of electromagnetic interference due to multiple resonating frequency) an electromagnetic interference filter is placed in parallel at the output end of the power electronic converters. The objective of this filter is to maintain the EMI within electromagnetic interference compatibility limits. Another issue arises even EMI filters are connected at the output end of converters due to inrush currents. In an AC system, inrush currents arise when energizing an induction motor, transformers, etc. In the DC counterpart, when power electronics DC-DC converters are coupled to DC bus, inrush current passes through EMI filter that, in turn, causes power quality issues and voltage fluctuation and subsequently affects the performance of all other equipment connected to the DC bus [35, 36].

14.3 Control Strategies Adopted in DC Micro-Grid

In anticipation of obtaining an efficient microgrid with greater stability, an effective and optimal control strategy needs to be implemented in a system. The control system plays a major role when the microgrid is in both grid-tied mode and isolated mode. The control strategy should be implemented based on the type of control needed in the particular geographical area. Many types of controls exist in DC microgrid with single-layer and multi-layer complex hierarchy [37–41]. However, the commonly employed controls in DC microgrids are classified based on the scheme of control, namely, centralized control, decentralized control, and distributed control. Apart from these control, a hierarchical control approach is also implemented in current scenario in microgrids, which works in a multi-layered fashion. Before analyzing the individual methods of control, the necessary objectives of a control strategy should be well understood [37–46].

- The transient stability and dynamic stability of the system should not be affected during fault conditions.
- The small signal stability of the microgrid is to be maintained when connected to non-linear and constant power loads [47, 48].
- An efficient and quick transition should be achieved when the grid switches from grid-tied mode to autonomous mode and vice-versa.
- When the microgrid is connected to the utility grid through the point of common coupling, all the synchronizing parameters should be effectively monitored by the control strategy, thus enabling a proper synchronization.
- In both the mode of operation the microgrid should be capable of controlling the voltage and current effectively.
- Decreasing the transmission and distribution losses within the grid.

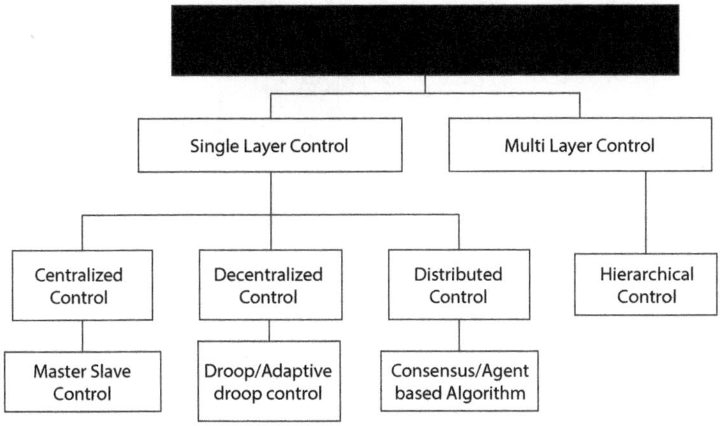

Figure 14.1 Classification of DC micro-grid control strategies.

- Maintaining the power quality of the grid.
- Unit commitment and economic dispatch.

The type of controller which has to be adopted is governed by many factors like number of DGs present, voltage rating of the grid, type and scheme of connection of the grid, and area of coverage of the grid. Among all these factors the geographical area of extension is the most important parameter because based on it, the type of control which has to be adopted in that particular zone will be decided. The different classification of DC microgrid control strategies are represented pictorially in Figure 14.1.

14.3.1 Centralized Control

This type of control is an effective automated controller which is even more efficient in a localized small sized grid. The operation of this control can be interpreted with a well-known example named, Master-Slave control strategy. It has a single Central Master Controller unit (CMC) common for all the microgrids connected in the particular zone. These individual microgrids have their own individual Slave Control (SC) that, in turn, gets all the prevailing information from respective Distributed Energy Resources (DERs) [49–51]. Thus, the master unit gets all the information from their slave units and continuous monitoring is done in the CMC. Figure 14.2a shows a centralized control strategy with four MGs connected to it. Each MG has its own slave unit which passes the information to the single master controller. The Digital Communication Links (DCLs)

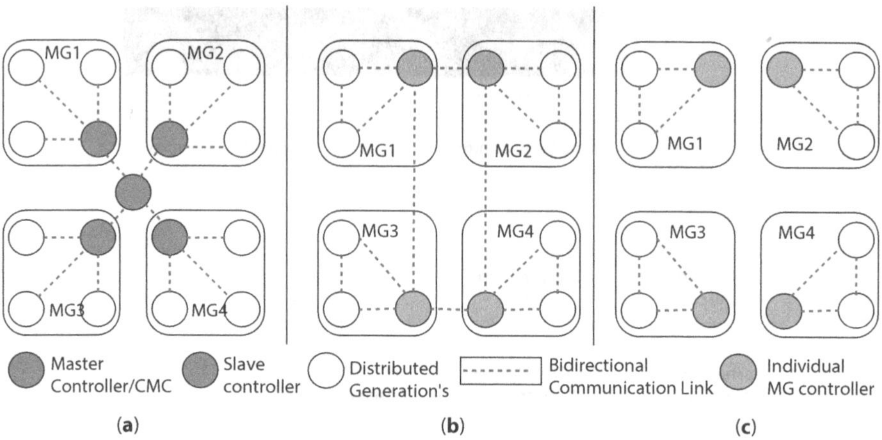

Figure 14.2 (a) Centralized controller. (b) Distributed controller. (c) Decentralized controller.

that connect CMC, SC, and DER should have high bandwidth and bidirectional characteristic, as bulk information has to be passed from the master unit to the individual DER via the slave unit and vice-versa using the single path available. Low-bandwidth links delay the speed of operation of the microgrid, which is not advisable during the time of transient and fault occurrence in the grid.

The CMC unit acts as a voltage source converter, and the SC acts as current source inverter. The VSC in the master unit is used to achieve DC bus voltage regulation and active sharing of current in all the other converters connected to it. This type of strategy has a clear observability. The major advantage of this scheme is that all the details are monitored and required protective measures are initiated by the central controller itself in case of abnormal conditions. Thereby, a strong controllability is achieved in this method. On the contrary, the main drawback of this system is single point failure. If there occurs any fault in the DCL, there is a lack of communication between the central controller and the end DERs. The system becomes completely unreliable and less insufficient due to these single point failures. In addition, when this system is applied to a large area, the requirement of the communication link also adds to the initial cost. Therefore, to achieve good efficiency, this type of centralized control should be implemented for small size microgrid where the information and datum to be gathered are confined.

14.3.2 Decentralized Control

This type of control is depicted in Figure 14.2c, which has an individual independent controller for each microgrid. There is no communication link between each microgrid as in the case of centralized control. The individual microgrid operates independently without considering the conditions of neighboring microgrid. The commonly adopted control technique in decentralized control is droop control technique [52]. This control is shown in Figure 14.3, in which all the converters are connected to the DC bus in parallel.

The principle advantage of droop control [53–58] is that it removes the circulating current flowing in the centralized control due to the presence of communication link. This control technique works in a relatively low bandwidth compared to the previous method. The droop control provides accurate and efficient current sharing among all the converters connected to the common DC bus. Therefore, selecting the droop coefficient plays a very important role in this technique. If the selected droop coefficient is very high, then it results in quick damping and accurate current sharing [59]. At the same time, it causes significant increase in the voltage deviation. Similarly if the droop coefficient selected is very low damping rate is minimized and improper current sharing is achieved among the converters.

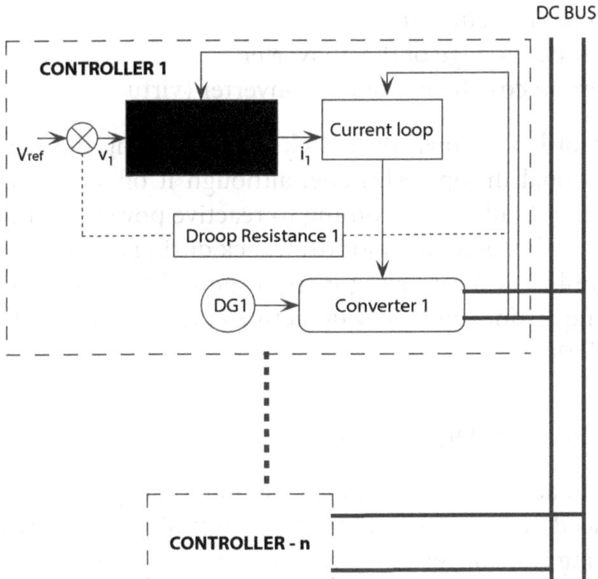

Figure 14.3 Decentralized droop control technique.

Thus, a linear trade-off should be maintained between current sharing and voltage deviations.

14.3.2.1 Droop Control With Virtual Resistance

In Figure 14.3, each converter is provided with their own droop controller that has voltage loop, current loop and a virtual resistance (R_{droop}). The virtual droop resistance is a function of the terminal voltage. This virtual resistance offers more flexibility to the system by providing wide range of control in the microgrid. Thereby, better current sharing with low bandwidth is achieved in this control. In addition, the voltage regulation is enhanced by introducing virtual droop resistance. The characteristics equations governing the controller are depicted as follows:

$$V_{ref} = V_o + (I_{ref} \times R_{droop}) \tag{14.1}$$

$$I_{ref} = (V_{ref} - V_o) / R_{droop} \tag{14.2}$$

$$R_{droop} = func(V_o) \tag{14.3}$$

where,

V_{ref} = Reference DC grid voltage
I_{ref} = Reference current
V_o = Output voltage of the converter
R_{droop} = Droop co-efficient of the converter (virtual droop resistance)

This control has its own drawback on controlling the reactive power. This conventional droop technique, although it offers active power and current sharing it fails when coming to reactive power sharing among the converters. This becomes a major drawback during connecting the microgrid to the utility and removing it from the utility grid. In addition, this droop technique does not provide active current sharing at low voltage regulations [60].

14.3.2.2 Adaptive Droop Control

In [61–64], an adaptive droop control strategy is explained to remove the drawback faced by the conventional droop technique. Here, reactive power sharing among the converters is also achieved in additional to the active power sharing. These adaptive droop control are used in both grid connected and islanded mode and has the capability of injecting the required active and reactive power to the grid. It also ensures good voltage regulation even at full load condition [65, 66]. Thus, this type of control technique can be employed for microgrids which cover large geographical area.

14.3.3 Distributed Control

Figure 14.2b depicts the structure of distributed control. This control has the advantages of both centralized and de-centralized control techniques [67, 68]. The operation of de-centralized control is based only upon the local DER measurements alone but the distributed control works on the basis of data obtained from both the local and the neighboring DER measurements. Information from all the neighboring converters is exchanged between all the converters as shown in Figure 14.4. By using distributed protocols such as consensus-based algorithm, agent-based control average of voltage is obtained and a consensus is achieved. The average consensus voltage is compared to the reference voltage and the output signal is given to the

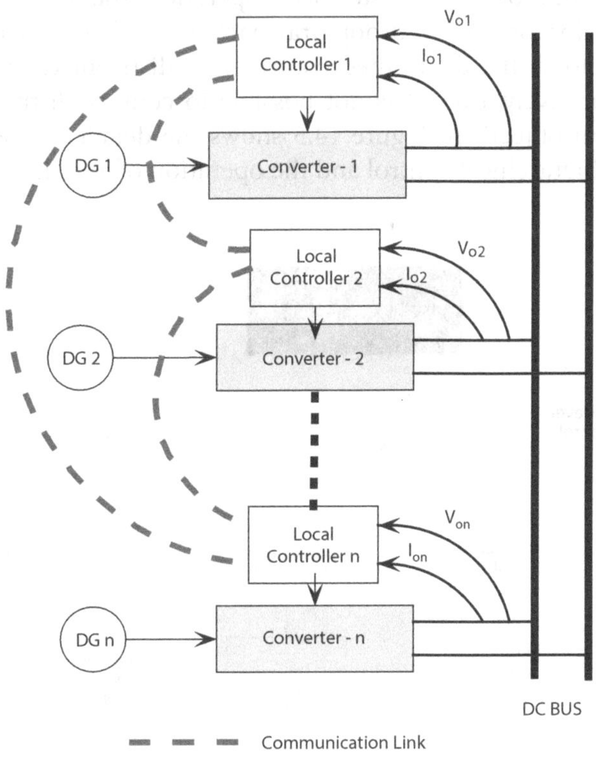

Figure 14.4 Distributed control technique.

controller. The same method is applicable for current sharing by performing average consensus current. The output of the controller accordingly removes the steady state error. The distributed system is applicable where the centralized system has more number of DER in it. This method avoids single point failure and requires low bandwidth for communication links. Even though this strategy suffers from bus voltage deviation, error in power tracking and also the complexity in achieving the consensus are very high.

14.4 Hierarchical Control

The control techniques described above are a single-layer basic control. The basic objectives of power system such as voltage control, current sharing, power sharing, power quality control, transient/dynamic stability control, synchronization issues, economic dispatch, unit commitment, losses minimization, and supervisory control cannot be governed by a single-layer control. The drawback of centralized control is that it requires high bandwidth DLC lines for basic Master-Slave operation, the decentralized control has a major load based droop drawback, i.e., voltage deviation–based variation in load, the distributed control. The distributed control is efficient comparatively, even it is not possible to control all the mentioned power system objectives. Figure 14.5 shows the detailed classification of multi-layer hierarchical control and the operation of each layer [69–75].

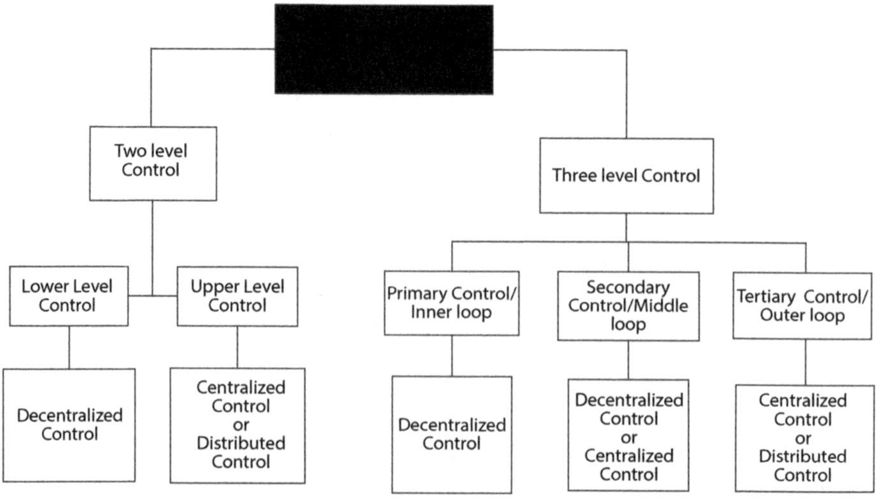

Figure 14.5 Multi-layer hierarchical control technique.

Table 14.1 Comparison chart of DC Microgrid control strategy.

Aspect	Centralized control	Decentralized control	Distributed control	Hierarchical control
Information exchange	The CMC continuously monitors all the data in the participating DERs with an individual salve controller in each microgrid. The necessary control actions required by the DER are processed by the CMC and the command is sent back through a bi-directional communication link.	Each microgrid has its own individual MG controller. All the requirements of the DERs are processed by this single controller. This type does not consider the conditions prevailing in the neighboring microgrids and acts independently based on local DER requirement alone.	Here each microgrid has its own individual MG controller. All of the individual MG controller is connected through a bidirectional communication link so that the conditions in the neighboring microgrid are also taken into account.	This method has a minimal of two layers of control in it. Two or more than two way of controlling is executed in this control. The first layer works in a decentralized way and the second layer works in distributed manner.

(Continued)

Table 14.1 Comparison chart of DC Microgrid control strategy. (*Continued*)

Aspect	Centralized control	Decentralized control	Distributed control	Hierarchical control
Scalability	The size of the communication system is highly restricted by the bandwidth of communication. Centralized control is highly efficient in case of small microgrids with less number of DERs.	The decentralized control is also similar to centralized control. Smaller scalability ensures good control within the microgrid.	The DERs interact with the neighboring microgrid with sparse communication, enhances the scalability. Future expansion is also possible in this control with good control efficiency.	This type of control can be employed for a very vast area. Increasing the geographical size of the grid has no impact on efficiency. Future expansion is also possible in this control.
Bandwidth requirement	The system has low-bandwidth digital communication links. The data from the CMC and the DER pass to and fro in the same link.	The communication link is set up within the microgrid alone and low bandwidth is sufficient to pass the information without delay.	Low bandwidth is sufficient quick operation of control.	The bandwidth requirement increases with increase in the number of layers.

(*Continued*)

Table 14.1 Comparison chart of DC Microgrid control strategy. (*Continued*)

Aspect	Centralized control	Decentralized control	Distributed control	Hierarchical control
Speed of operation	Speed of operation is low, owing to low bandwidth, communication delay, and the packet losses.	High speed of operation is possible since the control operations are performed within the microgrid. In addition, no other neighboring grid conditions are taken into account, so high speed of operation is achieved.	Speed of operation is high compared to centralized control.	High speed of operation is possible in primary layer.
Reliability	Least reliable owing to single point failure. Communication link failure in central controller makes the system more vulnerable to single point failure. Without information from the CMC, no further control operations will be performed	Good reliability on comparison with centralized control.	Highly reliable	Highly reliable

(*Continued*)

Table 14.1 Comparison chart of DC Microgrid control strategy. (*Continued*)

Aspect	Centralized control	Decentralized control	Distributed control	Hierarchical control
Continuity of service	Continuity of operation is highly affected by single point failure.	If there is a failure of controller in one microgrid, continuity of power output is lost as no other neighboring grid support the faulty grid.	System will always provide continuous power output even in case of failure of one controller.	System will always provide continuous power output even in case of failure of one controller.
Widely implemented control strategy	Master slave control	Droop control	Consensus-based algorithm	Primary layer: Droop Control Secondary layer: Consensus-based algorithm

The hierarchical control is capable of monitoring all the parameters with more layers of controls embedded in it. Even if the upper controllers fail to operate, the lower layer controller maintains uninterrupted operation of the system, and this type of control strategy is highly reliable in case of large DC microgrids. Different ideas have been proposed in many articles [76–80] explaining this hierarchical control. This chapter consolidates the proposed ideas and explains clearly about the operation of each layer in a precise manner. Basically, there are three layers in the hierarchical controller and further more layers can be extended based the requirement of the microgrid. The inner loop or the primary layer controls the fundamental requirements such as voltage control and current control and is responsible for power sharing. It also monitors the occurrence of fault and gives appropriate signal to the relays and circuit breakers. The middle loop or the secondary layer is responsible for grid synchronization and restoration. It takes full control of the point of common coupling parameters in order to maintain steady state stability after synchronization. It also monitors power quality and rate of power flow in a grid. The outer loop or the tertiary layer is an additional loop. It is responsible for planning the generator output, unit commitment, and economic dispatch. Furthermore, based on the control requirements this layer can also be utilized to check the amount of exchange of incoming and outgoing power, supervision of microgrid, etc. In some cases, the inner layer acts as lower level control and the other two layers (middle layer and outer layer) combined together acts as upper level control. The fashion at which each control layer works is shown neatly in Figure 14.5. Thus, in order to maintain all the power quality issues and to have an overall control of all the technical parameters, hierarchical control can be utilized and the efficiency obtained is quite promising on par with all other types of single-layer control. A detailed comparison chart of all the control techniques is shown in Table 14.1.

14.5 Conclusion

This chapter depicts an explicit review and provides clear guideline on the different types of technical challenges faced by the DC microgrid. The major technical problems encountered in the microgrid are as follows: difficulty experienced in suppressing the arc due to absence of zero crossing DC current, identification of proper grounding, effect of short-circuit fault current and its sensitivity toward the inverter, electromagnetic interference, and inrush currents. The solutions to minimize the issues are recommended in this chapter. Secondly, it gives a brief and clear picture on the control strategies implemented in the DC microgrid that is broadly classified into two different layers: single-layer and multi-layer control strategy. The basics of all the control strategies are elaborated and the reader can have a clear idea on the working principal of all the control layer. In addition, the place of implementation of each control is explained clearly along with the merits and de-merits of the system.

References

1. Kumar, A., Zafar Ullah Khan, M., Hussain, D.M.A., Microgrids Technology: A Review Paper. *Gyancity J. Electron. Comput. Sci.*, 3, 1, 11–20, March 2018.
2. Arfeen, Z.A., Khairuddin, A.B., Larik, R.M., Saeed, M.S., Control of distributed generation systems for microgrid applications: A technological review. *Int. Trans. Electr. Energy Syst.*, 29, e12072, 2019.
3. Arani, A.A.K., Gharehpetian, G.B., Abedi, M., Review on Energy Storage Systems Control Methods in Microgrids. *Int. J. Electr. Power Energy Syst.*, 107, 745–757, 2019.
4. Badal, F.R., Das, P., Sarker, S.K. *et al.*, A survey on control issues in renewable energy integration and microgrid. *Prot. Control Mod. Power Syst.*, 4, 1–27, 2019.
5. Choudhury, S., Mohapatra, L., Rout, P.K., A Comprehensive Review on Modeling, Control, Protection and Future Prospects of Microgrid. *International Conference on Electrical, Electronics, Signals, Communication and Optimization (EESCO)*, IEEE, pp. 1–6, 2015.
6. Mariam, L., Basu, M., Conlon, M.F., Microgrid: Architecture, policy and future trends. *Renew. Sust. Energ. Rev.*, 64, 477–48, 2016.
7. Hirscha, A., Parag, Y., Guerrero, J., Microgrids: A review of technologies, key drivers, and outstanding issues. *Renew. Sust. Energ. Rev.*, 90, 402–411, 2018.
8. Bayati, N., Hajizadeh, A., Soltani, M., Impact of Faults and Protection Methods on DC Microgrids Operation. *IEEE International Conference on Environment and Electrical Engineering*, Aug 2018.
9. Amamra, S.-A., Ahmed, H., El-Sehiemy, R.A., Firefly algorithm optimized robust protection scheme for DC microgrid. *Electr. Pow. Compo. Sys.*, 45, 10, 1141–1151, 2017.
10. Kakigano, H., Nomura, M., Ise, T., Loss evaluation of dc distribution for residential houses compared with ac systems. *Proceedings of the 2010 International Power Electronics Conference (IPEC)*, pp. 480–486, June 2010.
11. Lu, S., Wang, L., Lo, T.–M., Prokhorov, A.V., Integration of wind power and wave power generation systems using a DC microgrid. *IEEE Trans. Ind. Appl.*, 51, 4, 2753–2761, July- August 2015.
12. Mackay, L., Hailu, T., Elizondo, L.R., Bauer, P., Towards a DC distribution system-opportunities and challenges. *1st Int. Conf. on DC Microgrids (ICDCM)*, 2015.

13. Elsayed, A.T., Mohamed, A.A., Mohammed, O.A., DC microgrids and distribution systems: an overview. *Electr. Power Syst. Res.*, 119, 407–417, February 2015.
14. Jin, Z. *et al.*, Maritime DC microgrids-a combination of microgrid technologies and maritime onboard power system for future ships. *2016 IEEE 8th International Power Electronics and Motion Control Conference (IPEMCECCE Asia)*, IEEE, 2016.
15. Dragičević, T., Lu, X., Vasquez, J.C., Guerrero, J.M., DC microgrids—part I: a review of control strategies and stabilization techniques. *IEEE Trans. Power Electron.*, 31, 7, 4876–4891, July 2016.
16. Dragičević, T., Lu, X., Vasquez, J.C., Guerrero, J.M., DC microgrids—part II: a review of power architectures, applications, and standardization issues. *IEEE Trans. Power Electron.*, 31, 5, 3528–3549, May 2016.
17. Ahmed, M., Amin, U., Afta, S., Ahmed, Z., Integration of renewable energy resources in microgrid. *Front. Energy Power Eng. China*, 7, 12–29, 2015.
18. Sen, S. and Kumar, V., Microgrid control: a comprehensive survey. *Annu. Rev. Control*, 45, 118–151, 2018.
19. Sahoo, S.K., Sinha, A.K., Kishore, N.K., Control techniques in AC, DC, and hybrid AC–DC microgrid: a review. *IEEE J. Emerg. Sel. Top. Power Electron.*, 6, 2, 738–759, June 2018.
20. de Oliveira, T.R., Bolzon, A.S., Donoso-Garcia, P.F., Grounding and safety considerations for residential DC microgrids. *Paper presented at: IECON 2014-40th Annual Conference of the IEEE Industrial Electronics Society*, IEEE, pp. 5526–5532, October 2014.
21. Whaite, S., Grainger, B., Kwasinski, A., Power quality in DC power distribution systems and microgrids. *Energies*, 8, 5, 4378–4399, 2015.
22. Yuan, C., Haj-ahmed, M.A., Illindala, M.S., Protection strategies for medium-voltage direct-current microgrid at a remote area mine site. *IEEE Trans. Ind. Appl.*, 51, 4, 2846–2853, 2015.
23. Liu, Y., Chen, D., Yuan, H., Ji, L., Ma, Z., Research of dynamic optimization for the cam design structure of MCCB. *IEEE Trans. Compon. Packag. Manuf. Technol.*, 6, 3, 390–399, 2016.
24. Krishnamurthy, V. and Kwasinski, A., Effects of power electronics, energy storage, power distribution architecture, and lifeline dependencies on microgrid resiliency during extreme events. *IEEE J. Emerg. Sel. Top. Power Electron.*, 4, 4, 1310–1323, December 2016.
25. Mohammadi, J., Ajaei, F.B., Stevens, G., Microgrid Grounding Strategies, D.C., *IEEE/IAS 54th Industrial and Commercial Power Systems Technical Conference (I&CPS)*, pp. 1–6, 2018.
26. Kang, X., Nuworklo, C.E., Tekpeti, B.S., Kheshti, M., Protection of microgrid systems: a comprehensive survey. *J. Eng.*, 13, 1515–8, 2017.
27. Jayamaha, D.K.J.S., Lidula, N.W.A., Rajapakse, A.D., Protection and grounding methods in DC microgrids: Comprehensive review and analysis. *Renew. Sust. Energ. Rev.*, 120, 109631, 2020.

28. Sarangi, S., Sahu, B.K., Rout, P.K., Distributed generation hybrid AC/DC microgrid protection: A critical review on issues, strategies, and future directions. *Int. J. Energy Res.*, 44, 3347–3364, Jan 2020.
29. Sarangi, S., Sahu, B.K., Rout, P.K., A comprehensive review of distribution generation integrated DC microgrid protection: issues, strategies, and future direction. *Int. J. Energy Res.*, Dec 2020. doi = {10.1016/j.rser.2021.111401}
30. Beheshtaein, S., Cuzner, R.M., Forouzesh, M., Savaghebi, M., Guerrero, J.M., DC Microgrid Protection: A Comprehensive Review, *in: IEEE Journal of Emerging and Selected Topics in Power Electronics*, 2020.
31. Brearley, B.J. and Prabu, R.R., A review on issues and approaches for microgrid protection. *Renew. Sust. Energ. Rev.*, 67, 988–997, 2017.
32. Swathika, O.G. and Hemamalini, S., Prims aided Dijkstra Algorithm for adaptive protection in microgrids. *IEEE J. Emerg. Sel. Top. Power Electron.*, 4, 4, 1279–1286, 2016.
33. Beheshtaein, S., Savaghebi, M., Vasquez, J.C., Guerrero, J.M., Protection of AC and DC microgrids: challenges, solutions and future trends. *Paper presented at: IECON 2015—41st Annual Conference of the IEEE Industrial Electronics Society*, IEEE, pp. 005253–005260, November 2015.
34. Kou, W. and Wei, D., Fault ride through strategy of inverter interfaced microgrids embedded in distributed network considering fault current management. *Sustain. Energy Grids Netw.*, 15, 43–52, 2018.
35. Whaite, S., Grainger, B., Kwasinski, A., Power Quality in DC Power Distribution Systems and Microgrids. *Energies*, 8, 4378–4399, 2015, doi: 10.3390.
36. Kumar, D., Zare, F., Ghosh, A., DC Microgrid Technology: System Architectures, AC Grid Interfaces, Grounding Schemes, Power Quality, Communication Networks, Applications and Standardizations Aspects. *IEEE Access*, 1–1, 2017, doi: 10.1109/ACCESS.2017.2705914.
37. Alfergani, A., Alfaitori, K.A., Khalil, A., Buaossa, N., Control strategies in AC microgrid: A brief review. *2018 9th International Renewable Energy Congress (IREC)*, Hammamet, Tunisia, pp. 1–6, 2018, doi: 10.1109/IREC.2018.8362575.
38. Rawat, G.S. and Sathans, Survey on DC Microgrid Architecture, Power Quality Issues and Control Strategies. *Proceedings of the Second International Conference on Inventive Systems and Control (ICISC 2018).*
39. Roslan, M.F., Hannan, M.A., Ker, P.J., Uddin, M.N., Microgrid control methods toward achieving sustainable energy management. *Appl. Energy*, 240, 583–607, Feb 2019.
40. Zhou, Y. and Ho, C.N.-M., A Review on Microgrid Architectures and Control Methods. *International Power Electronics and Motion Control Conference (IPEMC-ECCE Asia)*, 2016.
41. Sadabadi, M.S., Shafiee, Q., Karimi, A., Plug-and-Play Robust Voltage Control of DC Microgrids. *IEEE Trans. Smart Grid*, 9, 6, 6886–6896, 2018, doi: 10.1109/TSG.2017.2728319.

42. Gupta, A., Doolla, S., Chatterjee, K., Hybrid AC–DC microgrid: systematic evaluation of control strategies. *IEEE Trans. Smart Grid*, 9, 4, 3830–3843, July 2018.
43. Shuai, Z., Fang, J., Ning, F., Shen, Z.J., Hierarchical structure and bus voltage control of DC microgrid. *Renew. Sust. Energ. Rev.*, 82, 3, 3670–3682, February 2018.
44. Eghtedarpour, N. and Farjah, E., Power control and management in a hybrid AC/DC microgrid. *IEEE Trans. Smart Grid*, 5, 3, 1494–1505, May 2014.
45. Sahoo, S.K., Sinha, A.K., Kishore, N.K., Control techniques in AC, DC, and hybrid AC–DC microgrid: a review. *IEEE J. Emerg. Sel. Top. Power. Electron.*, 6, 2, 738–759, June 2018.
46. Kaur, A., Kaushal, J., Basak, P., A review on microgrid central controller. *Renew. Sust. Energ. Rev.*, 55, 338–345, March 2016.
47. Herrrera, L., Zhang, W., Wamg, J., Stability analysis and controller design of DC microgrids with constant power loads. *IEEE Trans. Smart Grid*, 8, 2, 881–888, March 2017.
48. Srinivasan, M. and Kwasinski, A., Control analysis of parallel DC-DC converters in a DC microgrid with constant power loads. *Int. J. Electr. Power Energy Syst.*, 122, 106207, 2020.
49. Abdelaziz, M.M.A., Farag, H.E.Z., El-Saadany, E.F., Fuel-saving benefit analysis of islanded microgrid central controllers. *Electr Power Syst. Res.*, 125, 45–54, 2015.
50. Minchala-Avila, L., II, Garza-Castañón, L.E., Vargas-Martínez, A., Zhang Y., A., review of optimal control techniques applied to the energy management and control of microgrids. *Proc. Comput. Sci.*, 52, 780–7, 2015.
51. Oliveira, T.R., Silva, W.W.A.G., Donoso-Garcia, P.F., Distributed Secondary Level Control for Energy Storage Management in DC Microgrids. *IEEE Trans. Smart Grid*, 8, 2597–2607, November 2016.
52. Federico, I., Jose, E., Luis, F., Master–slave DC droop control for paralleling auxiliary DC/DC converters in electric bus applications. *IET Power Electron.*, 10, 10, 1156–1164, August 2017.
53. Huang, P.–H., Liu, P.-C., Xiao, W., Moursi, M.S.E., A novel droop-based average voltage sharing control strategy for DCMGs. *IEEE Trans. Smart Grid*, 6, 3, 1096–1106, May 2015.
54. Bidram, A., Davoudi, A., Lewis, F.L., Guerrero, J.M., Distributed Cooperative Secondary Control of Microgrids Using Feedback Linearization. *IEEE T. Power Syst.*, 28, 3, 3462–3470, Aug. 2013.
55. Shafiee, Q., Guerrero, J.M., Vasquez, J.C., Distributed Secondary Control for Islanded Microgrids—A Novel Approach. *IEEE Trans. Power Electron.*, 29, 2, 1018–1031, Feb. 2014.
56. Bidram, A., Davoudi, A., Lewis, F.L., A Multiobjective Distributed Control Framework for Islanded AC Microgrids. *IEEE Trans. Industr. Inform.*, 10, 3, 1785–1798, 2014.

57. Simpson-Porco, J., Shafiee, Q., Dorfler, F., Vasquez, J.C., Guerrero, J., Bullo, F., Secondary Frequency and Voltage Control of Islanded Microgrids via Distributed Averaging. *IEEE Trans. Ind. Electron.*, 62, 11, 7025–7038, 2015.
58. Dragicevic, T., Guerrero, J., Vasquez, J., Skrlec, D., Supervisory control of an adaptive-droop regulated dc microgrid with battery management capability. *IEEE Trans. Power. Electron.*, 29, 695–706, Feb. 2014.
59. Schiffer, J., Seel, T., Raisch, J., Sezi, T., Voltage Stability and Reactive Power Sharing in Inverter-Based Microgrids With Consensus-Based Distributed Voltage Control. *IEEE Trans. Control Syst. Technol.*, 24, 1, 96–109, 2016.
60. Su, C.-L., Lin, K.-L., Chen, C.-J., Power Flow and Generator-Converter Schemes Studies in Ship MVDC Distribution Systems. *IEEE Trans. Ind. Appl.*, 52, 1, 50–59, Jan. 2016.
61. Mahmood, H., Michaelson, D., Jiang, J., Reactive power sharing in islanded microgrids using adaptive voltage droop control. *IEEE Trans. Smart Grid*, 6, 6, 3052–3060, 2015.
62. Han, H., Liu, Y., Sun, Y., Su, M., Guerrero, J.M., An improved droop control strategy for reactive power sharing in islanded microgrid. *IEEE Trans. Power Electron.*, 30, 6, 3133–3141, 2015.
63. Peyghami, S., Mokhtari, H., Blaabjerg, F., Decentralized Load Sharing in a Low-Voltage Direct Current Microgrid With an Adaptive Droop Approach Based on a Superimposed Frequency. *IEEE J. Emerg. Sel. Top. Power Electron.*, 5, 3, 1205–1215, Sept. 2017.
64. Nasirian, V., Davoudi, A., Lewis, F.L., Distributed adaptive droop control for DC microgrids. *IEEE T. Energy Conver.*, 29, 4, 1147–1152, Dec. 2014.
65. Meng, L., Dragicevic, T., Vasquez, J.C., Guerrero, J.M., Tertiary and Secondary Control Levels for Efficiency Optimization and System Damping in Droop Controlled DC–DC Converters. *IEEE Trans. Smart Grid*, 6, 6, 2615–2626, Nov. 2015.
66. Farasat, M., Mehraeen, S., Arabali, A., Trzynadlowski, A., GA-based optimal power flow for microgrids with DC distribution network, in: *2015 IEEE Energy Conversion Congress and Exposition (ECCE)*, pp. 3372–3379, 2015.
67. Ahmed, M., Meegahapola, L., Vahidnia, A., Datta, M., Stability and Control Aspects of Microgrid Architectures – A Comprehensive Review. *IEEE Access*, 8, 144730–144766, 2020.
68. Bharath, K.R., Krishnan, M.M., Kanakasabapathy, P., A Review on DC Microgrid Control Techniques, Applications and Trends. *Int. J. Renew. Energy Res.*, 9, 3, 1328–1338, 2019.
69. Meng, L., Shafiee, Q., Trecate, G.F., Review on control of DC microgrids and multiple microgrid clusters. *IEEE J. Emerg. Sel. Top. Power Electron.*, 5, 3, 928–948, September 2017.
70. Wang, P., Xiao, J., Setyawan, L., Hierarchical Control of Hybrid Energy Storage System in DC Microgrids. *IEEE Trans. Ind. Electron.*, 62, 8, 4915–4924, Aug. 2015.

71. Meng, L., Dragicevic, T., Vasquez, J., Guerrero, J., Sanseverino, E.R., Hierarchical control with virtual resistance optimization for efficiency enhancement and State-of-Charge balancing in DC microgrids, in: *2015 IEEE First International Conference on DC Microgrids (ICDCM)*, pp. 1–6, 2015.
72. Wang, P., Lu, X., Yang, X., Wang, W., Xu, D.G., An Improved Distributed Secondary Control Method for DC Microgrids with Enhanced Dynamic Current Sharing Performance. *IEEE Trans. Power Electron.*, 31, 9, 6658–6673, 2016.
73. Kumar, J., Agarwal, A., Agarwal, V., A review on overall control of DC microgrids. *J. Energy Storage*, 21, 113–138, Feb. 2019.
74. Yadav, M., Pal, N., Saini, D.K., Microgrid control, storage, and communication strategies to enhance resiliency for survival of critical load. *IEEE Access*, 8, 169047–169069, 2020.
75. Ekanayake, U.N. and Navaratne, U.S., A Survey on Microgrid Control Techniques in Islanded Mode. *J. Electr. Comput. Eng.*, 2020, Article ID 6275460, 8, 2020.
76. Abhishek, A., Ranjan, A., Devassy, S., Verma, B., Ram, S., Dhakar, A., Review of Hierarchical Control Strategies for DC Microgrid. *IET Renew. Power Gen.*, 14, 10, 1631–1640, 2020.
77. Jin, C., Wang, P., Xiao, J., Tang, Y., Choo, F., Implementation of Hierarchical Control in DC Microgrids. *IEEE Trans. Ind. Electron.*, 61, 4032–4042, 2014, doi: 10.1109/TIE.2013.2286563.
78. Mohamed, A. and Mihet-Popa, L., Hierarchical Control for DC Microgrids, in: *Energy Management of Distributed Generation Systems*, July 13th 2016, DOI: 10.5772/63986.
79. Nahata, P., La Bella, A., Scattolini, R., Ferrari Trecate, G., Hierarchical Control in Islanded DC Microgrids With Flexible Structures. *IEEE Trans. Control Syst. Technol.*, 29, 6, 2379–2392, 2021, doi: 10.1109/TCST.2020.3038495.
80. Al-Ismail, F.S., DC Microgrid Planning, Operation, and Control: A Comprehensive Review. *IEEE Access*, 9, 36154–36172, 2021.

15

Maximizing Power Generation of a Partially Shaded PV Array Using Genetic Algorithm

Alice Hepzibah A.*, Premkumar K., Shyam D. and Aarthi B.

Department of Electrical and Electronics Engineering, Rajalakshmi Engineering College, Chennai, India

Abstract
Renewable energy is replaced in most of the places to preserve the environment and the atmosphere. The release of CO_2, SO_2 in utilization of energy leads to various hazards in the environment and creates health issues in human being. The demand of maximum power generation without affecting the atmospheric conditions is required. Pollution caused by use of conventional fuels and their depletion needs new technologies for saving the earth. Solar energy poses challenges in attaining maximum power, which, otherwise, is beneficial when the global warming is concerned. From the photovoltaic (PV) system, it is able to take the maximum power from the each and every array of the PV system. An innovative two-stage MPPT technique is used to overwhelm the demerit for maximum power extraction. The method proposed is robust in addressing shifting ecological settings and array features and has least negative effect on the associated electrical network. The implementation of the proposed system is economical and affordable, according to the researchers. The research work was solved in MATLAB/Simulink.

***Keywords*:** ANFIS, single-ended primary-inductor converter (SEPIC), PV, artificial neural networks (ANN), MPPT

**Corresponding author*: alicehepzibah.a@rajalakshmi.edu.in

A. Chitra, V. Indragandhi, and W. Razia Sultana (eds.) Smart Grids and Green Energy Systems, (231–246)

15.1 Introduction

Fundamental disadvantage of utilizing sustainable power sources is low not up to the imprint execution boundaries. The appeal of utilizing sun-powered photovoltaic (PV) frameworks is expanding quickly as a result of its very much planned circumstance for existing circulation frameworks. Energy created by SPV frameworks is overwhelmed by working voltage of module or an exhibit. Execution improvement of SPV Systems should be possible by following the greatest power point for it. Furthermore, in the current estimation of PV exhibit, the voltage esteem is expanded or diminished to move toward the most extreme in power point.

15.2 Literature Review

Single-stage full-connect inverter has restricted current wave at the DC interface and power at the alternating current results [1]. The twofold line-recurrence (2 ω) power wave of single-stage semi-Z-source (qZS) PV inverter exceptionally influences the entire framework's plan and execution [2]. The proposed inverter accomplishes low–DC-input voltage swell and sinusoidal current. It can accomplish solidarity power factor without huge uninvolved segments in DC transport. For example, smoothing inductors and electrolytic capacitors. The principal activities of the proposed inverter are exhibited experimentally [3]. The base vitality stockpiling needed to detach the power swell from the DC port is introduced [4]. Least capacitance required for converters vitality use proficiency of business PV siphoning frameworks can be altogether improved [5]. The PV frameworks are frequently liable to concealing in viable activity, which regularly results in multi-peak power-voltage (P-V) attributes [6]. Most existing MPPT control techniques are neighbor-hood ideal arranged, in view of the slope search [7]. The over voltages in low-voltage (LV) feeders with high infiltration of PVs are normally forestalled by restricting the feeders [8].

There is an expanding center around lessening costs and improving proficiency for PV cells and modules just as finding a more proficient way to deal with the item producing [9]. The cell plan and innovation on huge territory business grade Czochralski Si [10]. X-beam diffraction shows two In GaN players with indium creations of 14.8% and 16.8%. Photoluminescence tops were seen at 2.72 and 2.67 eV. The structure of GA-ANFIS comprises of various summed up outspread premise work (GRBF) units [11]. The spiral premise capacities are unpredictably dispersed as hyper-patches in the information yield space [12–14]. Brief background about the proposed planning PV frameworks is getting better known for age of intensity [15]. However, there are numerous difficulties in establishment and usage of these frameworks. One of the most significant issues is non-consistency of irradiance on the components of PV set that is called confuse issue [16, 17]. Reconfiguration of modules requires countless switches; so for applications like PV plants, there is another arrangement that isolates modules into two isolated sets [18, 19]. In this chapter, we center on reconfiguration. Modules in an exhibit are introduced on the top of a home or a vehicle [20]. Hence, consider reconfiguration ability for every module and pick the best association design for each PV module for ideal productivity.

15.3 Proposed System Design

This chapter shows an improved power quality in single stage littler scale structure. The proposed system uses sun situated PV show. This reasonable power sources are consolidated using only a solitary stage voltage source converter VSC. Test results of proposed littler scale cross section exhibits that the grid voltage is kept up relentless, while the structure is following a sudden change in burdens and under irregular passageway of sun arranged imperativeness source.

Current produced from the PV exhibit is legitimately relative to the sun-powered insolation. In Figure 15.1, PV created current is represented by the accompanying conditions:

$$I_{bv} = I_m - I_{sh} \tag{15.1}$$

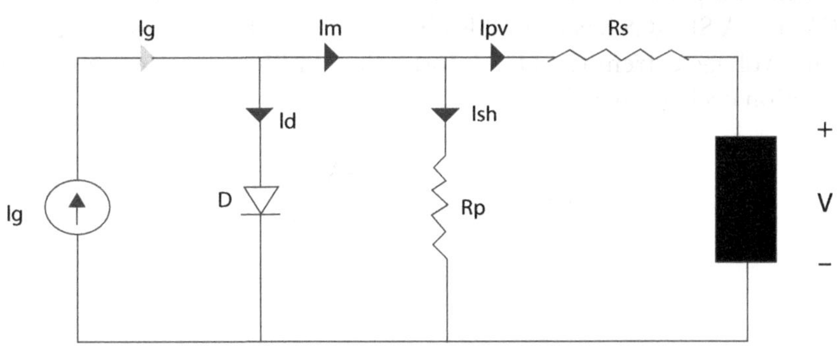

Figure 15.1 Equivalent model of PV cell.

where I_{sh} is the current through shunt resistor given by

$$I_{sh} = \left(\frac{V + I_{pv} R_s}{R_p} \right) \quad (15.2)$$

$$I_m = I_g - I_d \quad (15.3)$$

where Ig is the generated PV current proportional to solar insulation given by

$$I_g = (I_{g,n} + K_1 \Delta T) \frac{G}{G_n} \quad (15.4)$$

where Id is the diode current given by

$$I_d = I_0 \left(\exp\left(\frac{(V + I_{pv} R_s)}{v_t n} \right) - 1 \right) \quad (15.5)$$

where I_0 is the reverse saturation current of the diode.

15.4 Design of SEPIC Converter

A single-ended primary-inductor converter (SEPIC) is a sort of DC-DC converter permitting voltage at its yield to be more noteworthy than, not exactly, or equivalent to that of its info. The yield voltage is constrained by controlling the obligation pattern of a semiconductors or IGBT or MOSFET. A SEPICS has favorable circumstances of having the yield with a similar voltage extremity as the information. SEPIC converter working the obligation cycle is given by

$$D_{max} = \frac{V_{OUT} + V_D}{V_{IN(min)} + V_{OUT} + V_D} \quad (15.6)$$

15.5 Comparison of Different Optimization Tools

15.5.1 Fuzzy Logic Control

Fuzzy is a fundamental control framework that depends on the degrees of condition of the information and the yield. It essentially deals with the rule of if-else-the, for example, in the event that An AND B, at that point Z. Uncertain rationale deals with same standard on the idea on choosing the yield based on presumptions. It takes a shot at a premise of sets. Each set speaks to some etymological variable characterizing the conceivable condition of a set. Every conceivable state and degrees of progress of the state are an aspect of the set, contingent on which the yield is anticipated.

The factors of the set are the condition of the information sources and the degrees of changes of the information. The participation of the yield relies upon the rationale of AND activity. For a multi-input framework, the factors can likewise be the various information sources. The yield can be the conceivable aftereffect of the AND activity between the factors which is shown in Figure 15.2.

15.5.2 ANFIS Model

Versatile organization–based fluffy deduction framework of ANFIS is a straight mix of the information factors included by a consistent term. The last yield is the weighted normal of each standard's yield. The separated films of this ANFIS construction area is delineated in the following, where

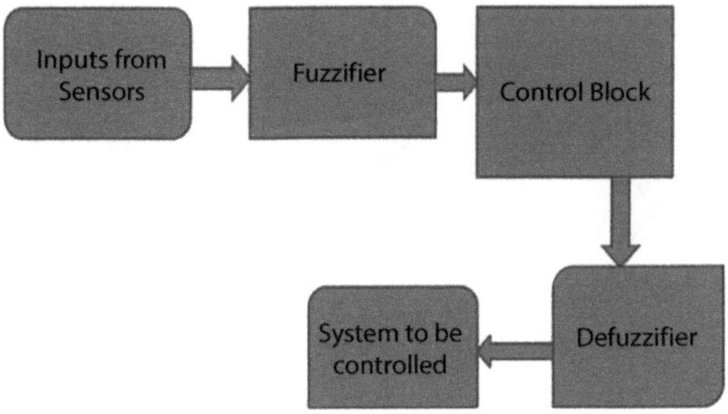

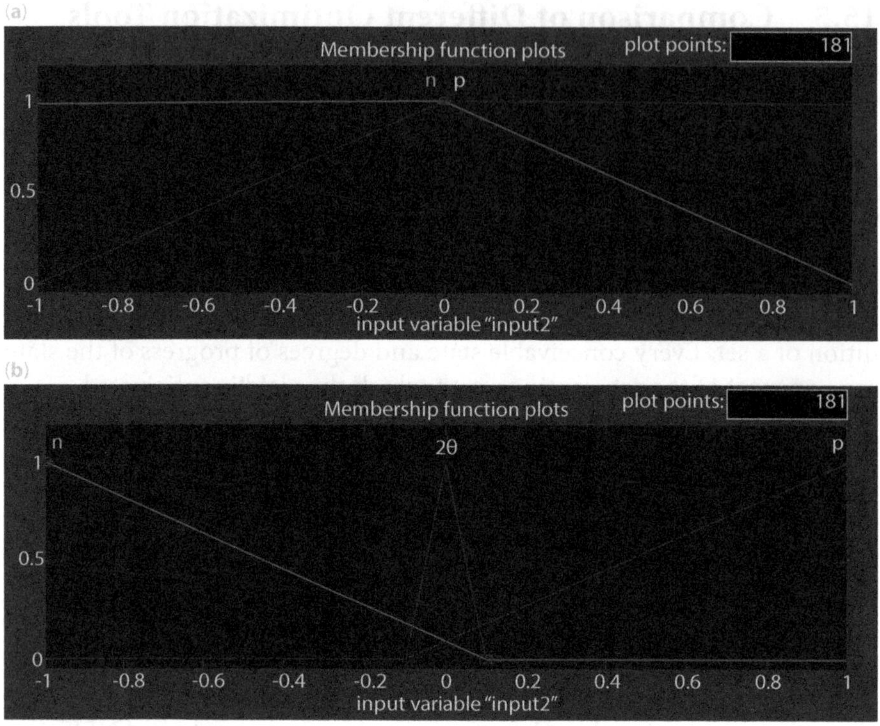

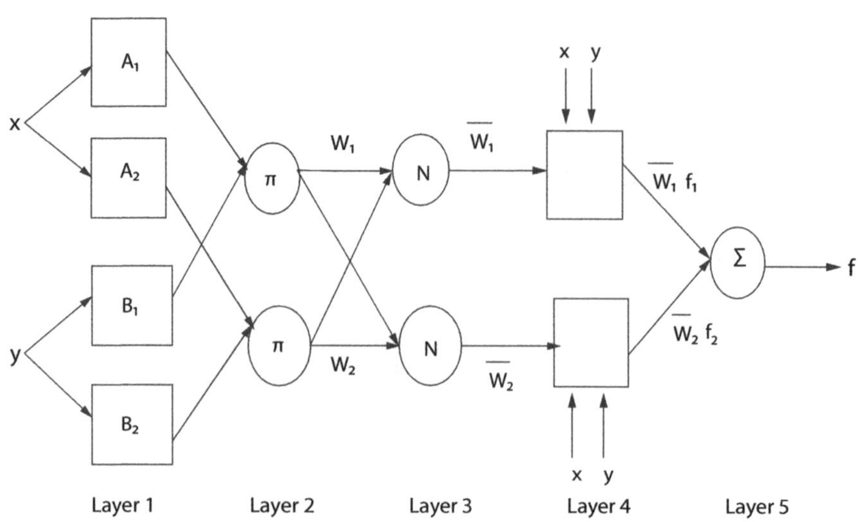

Figure 15.4 Block diagram of ANFIS model.

x is that the input to node i, and Ai is that the linguistic variable related to this node operate, Every hub in this layer is a fixed hub which ascertains the terminating quality of a standard.

The yield of every hub is the result of the apparent multitude of approaching signs to it and is given by

$$O_i^1 = \mu_{A_i}(x) \tag{15.7}$$

$$O_i^2 = w_i = \mu_{A_i}(x) \times \mu_{B_i}(y), \quad i = 1,2 \tag{15.8}$$

$$O_i^4 = \overline{w_i} f_i = \overline{w_i}(p_i x + q_i y + r_i), \quad i = 1,2 \tag{15.9}$$

$$\mu_{A_i}(x) = \frac{1}{1 + \left[\left(\frac{x - c_i}{a_i}\right)^2\right]^{b_i}} \tag{15.10}$$

$$\mu_{A_i}(x) = \frac{1}{1 + \left[\left(\frac{x - c_i}{a_i}\right)^2\right]^{b_i}} \tag{15.11}$$

Each hub in this layer is a fixed hub. The yield from the ith hub is the standardized terminating quality given in Equations (15.7) to (15.11). This layer contains just one fixed hub that computes the general yield as the summation of every single approaching sign. In the ANFIS structure, it is seen that given the estimations of reason boundaries, the last yield can be communicated as follows:

$$\begin{aligned} f &= \frac{w_1}{w_1 + w_2} f_1 + \frac{w_2}{w_1 + w_2} f_2 \\ &= \overline{w_1} f_1 + \overline{w_2} f_2 \\ &= \left(\overline{w_i} x\right) p_i + \left(\overline{w_1} y\right) q_1 + \left(\overline{w_1}\right) r_1 + \left(\overline{w_2} x\right) p_2 + \left(\overline{w_2} y\right) q_2 + \left(\overline{w_2}\right) r_2. \end{aligned} \tag{15.12}$$

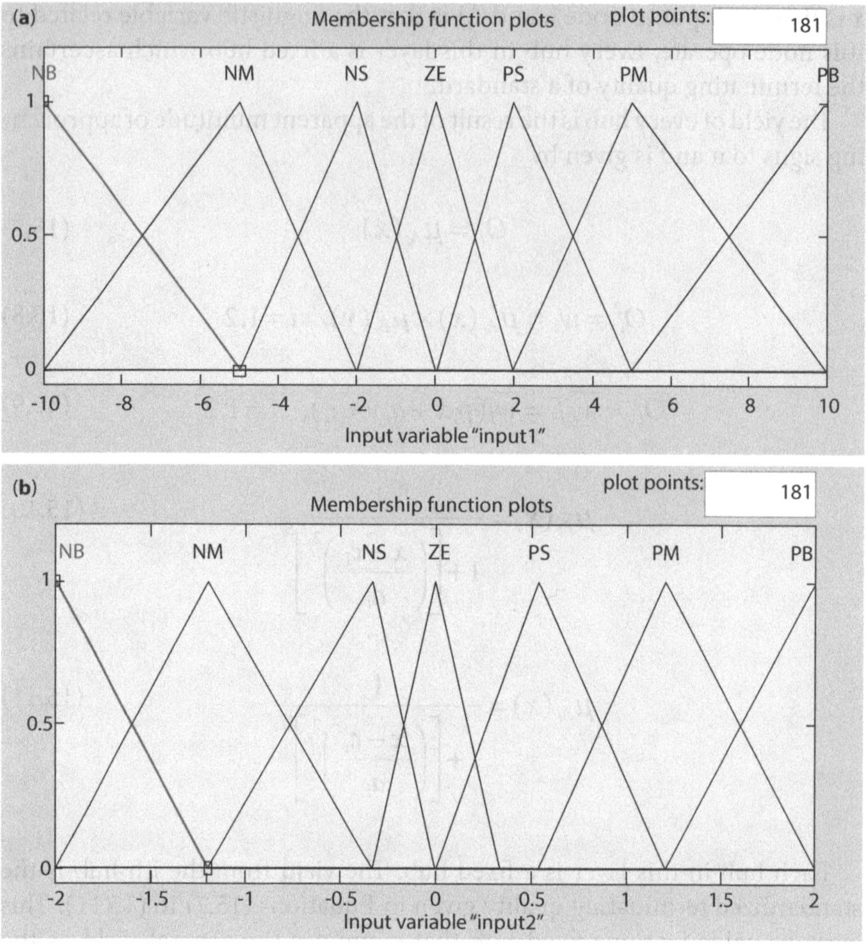

Figure 15.5 ANFIS structure (a) input 1 (b) input 2.

In the forward go of the learning calculation, ensuing boundaries are recognized by the least squares gauge. In the retrogressive pass, the blunder signals are the subsidiaries of the squared mistake as for every hub yield. In this retrogressive pass, the reason boundaries are refreshed by the slope plunge calculation.

15.5.3 Genetic Algorithm

In 1975, J. Holland initially presented genetic algorithm (GA). It is sustained the Darwinian advancement philosophy survival of the fittest. It is

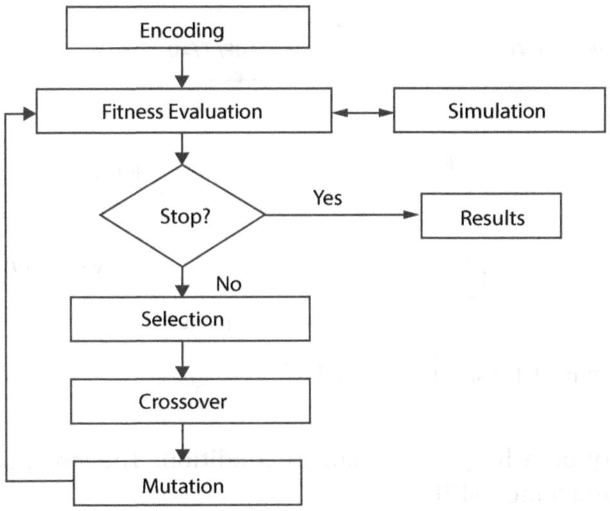

Figure 15.6 Flow diagram of genetic algorithm.

often empowered as associate improvement search procedure that uses the principles of genetic science and selection by modeling potential solutions to a research downside as strings of zeroes and ones. Every purpose in an exceedingly parameter or resolution house is often encoded by GA rule in an exceedingly binary bit string that is named the body. When the creation of a generation every of those points is additionally allotted, a fitness worth that is given by the evaluated worth of the target perform at that time. This fitness worth is then accustomed produce a replacement population which might be repeatedly evolved toward a higher fitness worth within the next sequential generations by upgrading the complete population.

15.5.4 Incremental Conductance Method (INC)

The technique abuses the supposition of the proportion of progress in yield conductance is equivalent to the negative yield Conductance Quick conductance. The MPPT manages the PWM control sign of the DC-to-DC help converter. The pinnacle intensity of the module lies at above 98% of its steady conductance at MPP. The above condition could be written as far as cluster voltage V and exhibit current. In this strategy, steady conductance was planned upheld a perception of P-V characteristic work. This calculation was created in 1993 and was planned to beat some disadvantage of P&O calculation. IC attempts to improve the following time and to deliver

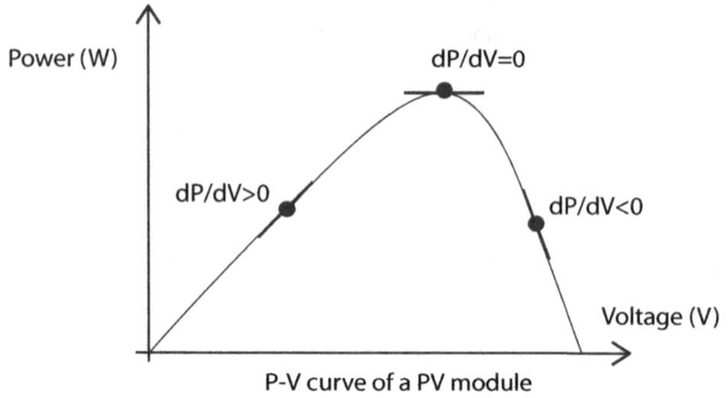

Figure 15.7 PV module using INC method.

more vitality on a huge light changes condition. The stream diagram of gradual conductance MPPT.

PV frameworks are regularly liable to concealing in down to earth activity. This frequently results in multi-peak P-V qualities. In the event that MPP lies on right aspect, dI/dV < - I/V, thus, the PV voltage ought to be limited to accomplish the MPP. Usage IC on a microcontroller made a ton of stable presentation once it contrasted with P&O.

$$\begin{aligned}\frac{dP}{dV} &= \frac{d(\mathrm{V.I})}{dV} = I\frac{dV}{dV} + V\frac{dI}{dV} \\ &= I + V\frac{dI}{dV}\end{aligned} \tag{15.13}$$

$$\begin{aligned}&\frac{dI}{dV} = -\frac{I}{V} \\ &\frac{dP}{dV} > 0 \text{ then } \mathrm{V_p} < \mathrm{V_{mpp}} \\ &\frac{dP}{dV} = 0 \text{ then } \mathrm{V_p} = \mathrm{V_{mpp}} \\ &\frac{dP}{dV} < 0 \text{ then } \mathrm{V_p} > \mathrm{V_{mpp}}\end{aligned} \tag{15.14}$$

15.6 Single-Phase Inverter

Inverter is characterized as an electrical gadget that changes over the immediate flow source into the exchanging flow source. DC yield of the battery is kicked or supported predictable with the need at that point changed over into air conditioning utilizing a DC air conditioning inverter. The primary wellspring of electrical power is the battery which is a DC source. Inverter work is to change a DC input voltage to a symmetric AC yield voltage of wanted greatness and recurrence. The yield voltage waveforms of ideal inverters ought to be sinusoidal, and handy inverters are non-sinusoidal and contain certain sounds. Switches S1, S2, S3, and S4 are unidirectional, for example, they lead current one way. At the point when switches S1 and S2 are turned on at the same time for a span 0, the info voltage Vin shows up over the heap and the current streams from direct a toward b, Q1 - Q2 àON, Q3 → Q4 → OFF.

15.7 Simulation Results

Sun-powered cell is basically a semiconductor unit presented to lightweight. PV cell boundaries along with impede, circuit voltage, power, and strength are changed in view of power and temperature varieties. The diode fundamentally based PV cell demonstrating has increased expanded

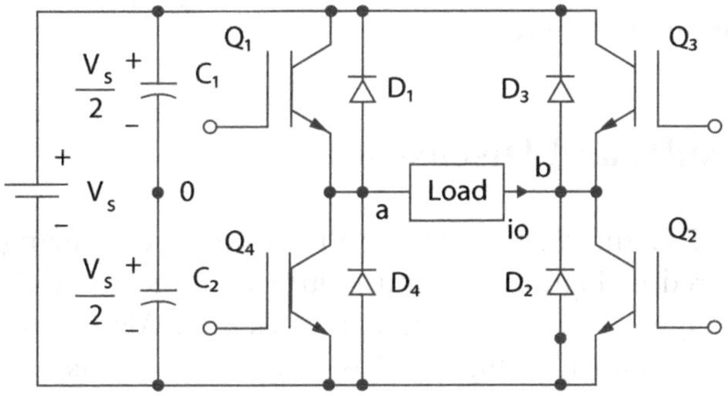

Figure 15.8 Single phase inverter.

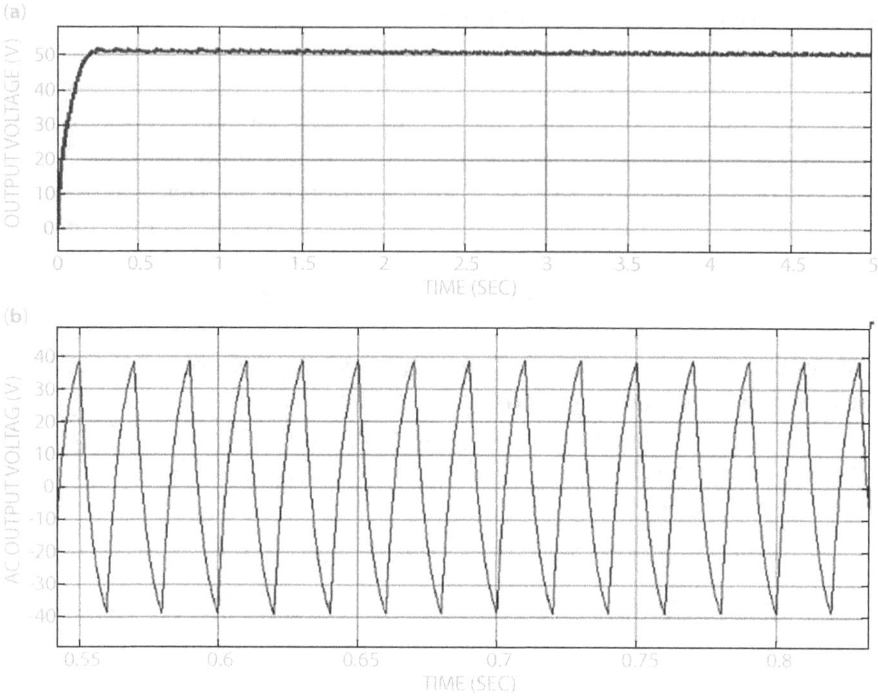

fascination in light of the graphical interface of programming resembles MATLAB/Simulink family.

15.8 Results and Discussion

The Output DC interface voltage and PV voltage utilizing fuzzy system is appeared in Figure 15.3 and Figure 15.4 and its ANFIS structure appeared in Figure 15.5. The flowgraph of Genetic Algorithm for the proposed system is shown in Figure 15.6. The PV model using Incremental conductance method is shown in Figure 15.7. Figure 15.8 provides the structural diagram of Single-phase bridge inverter. The output DC voltage utilizing GA-based PID appeared in Figure 15.9. Table 15.1 provides the Comparative analysis of different optimization techniques such asFuzzy, ANFIS, and GA-PID. Among this, GA-PID produces better results when compared to Fuzzy and ANFIS networks. This chapter focuses only on the peak overshoot and settling time.

Table 15.1 Comparison of Fuzzy, ANFIS, and GA-PID.

Parameters	**Peak overshoot**	**Settling time**
Fuzzy	0.35V	3 s
ANFIS	0.35V	3 s
GA-based PID	0.1V	0.3 s

15.9 Conclusion

The system proposed successfully presents maximum power extraction from solar panel with use of INC-based MPPT along with GA-based PID controller to keep the output voltage from the SEPIC converter constant. Analysis is done with different controllers, and the results have been compared. The SEPIC converter has the advantage of same polarity output. The GA-based PID controller shows better results in case of ripples and settling time. The parameters used are Fuzzy, ANFIS, and GA-based PID, and study of their performances is shown. GA-based PID gives a peak overshoot of 0.1 V and settling time of 0.3 s compared to a peak overshoot of 0.35 V and settling time of 3 s in fuzzy systems and ANFIS. Thus, GA-based PID systems proved to be much better than fuzzy systems and ANFIS for attaining maximum power in solar systems.

References

1. Almaktoof, A., Shebani, N., Elfallah, A., GUI-PV Application Tool for Teaching Performance of PV System using MATLAB-Graphical User Interface Environment. *2021 IEEE 1st International Maghreb Meeting of the Conference on Sciences and Techniques of Automatic Control and Computer Engineering MI-STA*, pp. 447–451, 2021, doi: 10.1109/MI-STA52233.2021.9464367.
2. da Rocha, N.M., Coelho, R.F., Passos, J.C., Martins, D.C., Suggestion of associating a PV MPPT algorithm based on temperature control with a PV cooling system. *3rd Renewable Power Generation Conference (RPG 2014)*, pp. 1–6, 2014, doi: 10.1049/cp.2014.0890.
3. Adam, K.B. and Miyauchi, H., Design of PV-DC mini grid using PI-based fuzzy MPPT control considering load demand. *TENCON 2017 - 2017 IEEE Region 10 Conference*, pp. 2990–2995, 2017, doi: 10.1109/TENCON.2017.8228374.

4. Patil, R., Fuzzy based MPPT technique for PV system. *2017 International Conference on Energy, Communication, Data Analytics and Soft Computing (ICECDS)*, pp. 1455–1459, 2017, doi: 10.1109/ICECDS.2017.8389686.
5. Shyam, D., Jai Ganesh, R., Premkumar, K., Nazar Ali, A., Sivamani, D., Narendiran, S., Symmetrical and Asymmetrical Multilevel Inverter with configurational parameters for power quality applications. *2020 IEEE 7th Uttar Pradesh Section International Conference on Electrical, Electronics and Computer Engineering (UPCON)*, pp. 1–5, 2020, doi: 10.1109/UPCON50219.2020.9376505.
6. Siddula, S., Analysis of Fuzzy Logic Based MPPT Using Incremental Conductance Technique for PV Cell. *2020 International Conference on Smart Technologies in Computing, Electrical and Electronics (ICSTCEE)*, pp. 180–185, 2020, doi: 10.1109/ICSTCEE49637.2020.9277172.
7. Nazar Ali, A., Jayabharath, R., Udayakumar, M.D., An ANFIS Based Advanced MPPT Control of a Wind-Solar Hybrid Power Generation System. *Int. Rev. Model. Simul. (I.RE.MO.S.)*, 7, 4, 638–643, 2014.
8. Anand, R. and Nazar Ali, A., A Single phase Five level Inverter for Grid Connected Photovoltaic System by employing PID Controller. *Afr. J. Sci. Res.*, 6, 1, 306–315, 2011.
9. Gupta, A., Kumar, P., Pachauri, R.K., Chauhan, Y.K., Performance analysis of neural network and fuzzy logic based MPPT techniques for solar PV systems. *2014 6th IEEE Power India International Conference (PIICON)*, pp. 1–6, 2014, doi: 10.1109/POWERI.2014.7117722.
10. Amara, K. *et al.*, Improved Performance of a PV Solar Panel with Adaptive Neuro Fuzzy Inference System ANFIS based MPPT. *2018 7th International Conference on Renewable Energy Research and Applications (ICRERA)*, pp. 1098–1101, 2018, doi: 10.1109/ICRERA.2018.8566818.
11. Deopare, H. and Deshpande, A., Fuzzy based incremental conductance algorithm for PV system. *2016 International Conference on Automatic Control and Dynamic Optimization Techniques (ICACDOT)*, pp. 683–687, 2016, doi: 10.1109/ICACDOT.2016.7877673.
12. Bouakkaz, M.S., Boukadoum, A., Boudebbouz, O., Attoui, I., Boutasseta, N., Bouraiou, A., Fuzzy Logic based Adaptive Step Hill Climbing MPPT Algorithm for PV Energy Generation Systems. *2020 International Conference on Computing and Information Technology (ICCIT-1441)*, pp. 1–5, 2020, doi: 10.1109/ICCIT-144147971.2020.9213737.
13. Ramkumar, R., Sivamani, D., Preetha, K., Shyam, D., On line monitoring and control of grid-interfaced PV system using zigbee. *IOP Conference Series: Materials Science and Engineering*, vol. 937, 2020.
14. Shyam, D., Sivamani, D., Jaiganesh, R., Narendiran, S., Solar PV Fed Parallel SEPIC Converter for Highly Efficient Multilevel Inverter Integration. *Int. Transaction Power Energy Syst.*, 1, 1–17, 2021.

15. Hepzibah, A.A. and Premkumar, K., ANFIS current–voltage controlled MPPT algorithm for solar powered brushless DC motor based water pump. *Electr. Eng.*, 102, 421–435, 2020.
16. Sivamani, D., Shyam, D., Nazar Ali, A., Premkumar, K., Narendiran, S., Alexander, S.A., Solar Powered Battery Charging System Using Optimized PI Controller for Buck Boost converter. *IOP Conference Series: Materials Science and Engineering*, vol. 1055, 2021.
17. Ganthia, B.P., Pradhan, R., Das, S., Ganthia, S., Analytical study of MPPT based PV system using fuzzy logic controller. *2017 International Conference on Energy, Communication, Data Analytics and Soft Computing (ICECDS)*, pp. 3266–3269, 2017, doi: 10.1109/ICECDS.2017.8390063.
18. Kumar, J., Rathor, B., Bahrani, P., Fuzzy and P&O MPPT Techniques for Stabilized the Efficiency of Solar PV System. *2018 International Conference on Computing, Power and Communication Technologies (GUCON)*, pp. 259–264, 2018, doi: 10.1109/GUCON.2018.8674909.
19. Fitriyanah, D.N. and Abadi, I., Fuzzy Logic Control Design of Mobile PV Using Bacterial Foraging Optimization. *2018 International Seminar on Intelligent Technology and Its Applications (ISITIA)*, pp. 215–220, 2018, doi: 10.1109/ISITIA.2018.8711161.
20. Pachauri, R.K. and Chauhan, Y.K., *Fuzzy logic controlled MPPT assisted PV-FC power generation for motor driven water pumping system*, 2014.

16

Investigation of Super-Lift Multilevel Inverter Using Water Pump Irrigation System

Johny Renoald Albert[1], Premkumar K.[2], Vanchinathan K.[3], Nazar Ali A.[2*], Sagayaraj R.[4] and Saravanan T.S.[2]

[1]*Department of Electrical and Electronics Engineering, Vivekanandha College of Engineering for Women, Namakkal, Tamil Nadu, India*
[2]*Department of Electrical and Electronics Engineering, Rajalakshmi Engineering College, Chennai, Tamil Nadu, India*
[3]*Department of Electrical and Electronics Engineering, Velalar College of Engineering and Technology, Erode, Tamil Nadu, India*
[4]*Department of Electrical and Electronics Engineering, Muthayammal Engineering College, Tamil Nadu, India*

Abstract

The most significant development in DC-to-DC power conversion systems is the positive output voltage-lift method. The topology exists in two sorts such as super-lift and double-lift technique. Double-lift converter functions successfully DC-to-DC converter. The developed approach employs the invention of a super-lift multilevel method, which may minimize the number of recurring components when compared to the traditional sort of weird inverter. In this point, a new approach of single-source super-lift multilevel inverter (SLMLI) is proposed with concentrator solar module, which can perform the necessary voltage conversion applied in water pump irrigation using agricultural field. Designing a new SLMLI applied with PSO algorithm is also capable of determining the desired value of harmonics value performed with necessary power conversion. Solar water pumping systems are integrated in motor load; it has series connected to generate the required system load voltage, which is supplied by the availability of solar irradiation. Results present modified SLMLI with submersible water pump that helps to save the fuel and electricity cost,

**Corresponding author*: nazarali.a@rajalakshmi.edu.in

A. Chitra, V. Indragandhi, and W. Razia Sultana (eds.) Smart Grids and Green Energy Systems, (247–262)

when compared to equivalent use of a diesel-operated water pump. The performances at various operating conditions are analyzed using MATLAB/Simulink and hardware-based result.

Keywords: PV (photovoltaic), PSO (particle swarm optimization), SLMLI (super-lift multilevel inverter), submersible water pump, THD (total harmonic distortion)

16.1 Introduction

The concentrator solar photovoltaic (SPV) energy is one of the best alternate remedies or solutions for the rapid reduction of nonrenewable resources. The SPV energy is free and enormous, highly beneficial, and economically viable to many industrial applications [1]. Nonrenewable resources such as coal, petroleum, and natural gas are the primary sources of energy, but they will be depleted in a few decades. It also leads to environmental problems such as CO_2 emissions and the environmental impact [2]. Nowadays, water pumping is drawing huge attention among all the applications of PV module. To enhance the efficiency of concentrator SPV system, it is operated based on standard test condition [3]. Direct current converters are used to convert one level to another DC required level. Double-lift converters are mainly focused to enhance the DC-to-DC output voltage [4]. The positive output voltage at any required level attainments depends upon positive input DC source voltage. Voltage-lift performance is a standard scheme that is widely applied in electronic circuit design [5]. The multilevel converter employs power semi-conversion devices from lower DC voltage sources to generate an almost sinusoidal waveform of the staircase [6]. The main objective of the proposed system is to reduce the complexity of the multilevel inverter with reduced switches and diode sizes called as super-lift Luo inverter [7]. Inversion technology operates on vital mostly used in residential, domestic, electrical vehicles, and renewable energy systems. The proposed super-lift converter is put forth for PV concentrator module to apply water pump irrigation system. PV concentrate water pump system can operate all kinds of electric water pumps with applications ranging from irrigation to residential requirements [8]. A stand-alone PV system is provided for water pumping irrigation technology that has been constantly developed since the early 1980s.

In India, there are three primary solar water pumping configurations. Brushless DC pump, DC positive displacement pump, and AC centrifugal pump are all examples of brushless DC pumps. Submersible irrigation

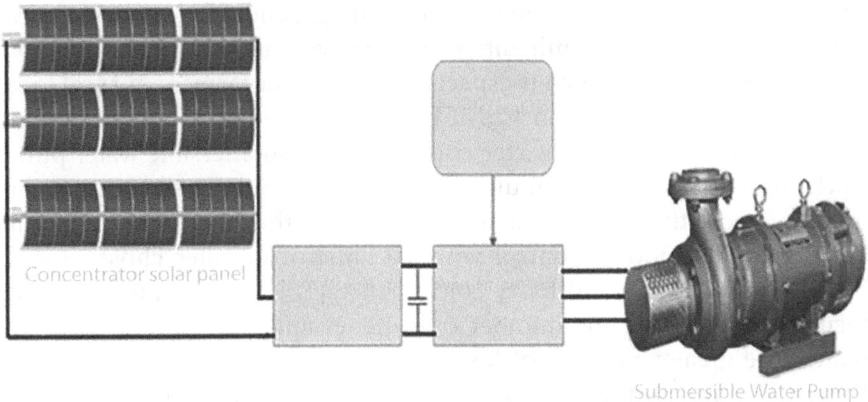

water pumps can also be used in conjunction with drip irrigation systems to maximize the benefits of this setup [9]. Submersible pumps have the benefit of being totally submerged, including the engine. Water pumps are also more efficient since it is simply had to push water and do not have to suck it. The sum amount of solar array size necessary to operate the connected pump is known as a standard concentrator solar water pumping system. A number of solar modules are linked in series in each solar array. The integrated AC submersible pumps are connected to pull the water and feed it into delivery pipes, and the electrical energy from the complete array is controlled and directly adjusted by them [10]. These low-power submersible pumps make it possible to create a low-cost solar pumping system for a deep well. Figure 16.1 shows the suggested approach, which requires a 0.5-kWp solar water pump for the water pumping system. The unit is capable of extracting and pumping 20,000 to 25,000 liters of water per day from a source up to 3–5 meter deep. MATLAB/Simulink was used to assess the performance and measurement overall system output, and the prototype module for freestanding solar PV systems uses hardware-based results.

16.2 Proposed System Configuration

Small and marginal farmers are at danger economically due to a sparse electricity associated with the increased regularity of mountain rain and the prevalence of expensive diesel pumping. A complex mix of issues, such as global warming, competitive land usage, and a lack of basic infrastructure, is posing new difficulties to India's vast agricultural population. Then, there is power, which is especially difficult for farmers in rural locations. The configuration consists of PV module feeding the power to Luo converter, and super-lift inverter employed with submersible water pump work out with MATLAB/Simulink platform [11]. Submersible pumps are used to choose the outfall area and also connect the hose up to the pump. This prototype module control with PIC microcontroller chosen for its PWM control ability and hasty convergence [12]. This chapter generally focused on reducing the number of switches and DC sources used. We reduce the greater number of losses; voltage dropped across the switch. Design, control, and operation of each stage are elaborate individually as following section.

16.3 Design of Concentrator SPV Array

To determine the amount of water used, multiply the width of the yard by the length of the yard in feet to obtain the number of square feet. The water is driven by a 0.5-kW, 50-Hz submersible water pump that is powered by the SPV array. To compensate for motor and inverter losses, the SPV array rating is somewhat greater than the pump motor rating [13]. A SPV is created by linking 72 cell SPV modules in series with an open circuit voltage of 24 V and 72 cell SPV modules in sequence with a short circuit current of 1.5 A. As shown in Table 16.1, the voltage and current achieve their MPP at approximately 81% of open circuit voltage and 90% of ISC, correspondingly. There are various sub-series circuits in positive output super-lift converters [14, 15]. Stage by stage, the circuits' voltage gains grow in geometrical series. Figure 16.2 depicts a basic positive output double-lift or re-lift converter. When switch S is turned on, the comparable circuits are shown in Figure 16.2a, and when switch S is turned off, the corresponding circuits are shown in Figure 16.2b. When the primary switch S is switched off, the inductor L1 and capacitor C1 raise the output voltage V 0 significantly from the source voltage, as shown in the equation below. When compared to a traditional inverter, the super-lift multilevel inverter (SLMLI) has benefits such as modularity, fewer components, the absence of additional clamping diodes, a lower number of switches, and the ability to quickly alter the number of output voltage levels.

Figure 16.3 depicts a 15-level super-lift inverter [16–20]. The period of switches turning ON and OFF in SLMLI can only be done once per

Table 16.1 Design of PV array.

S. no.	Parameter	Range
1	Voltage at maximum power point	24 V
2	Total Power at MPP	0.5 kW
3	Current at MPP, $I_{mp} = P_{mp}/V_{mp}$ $I_{mp} = P_{mp}/V_{mp}$	0.08 A
4	No. of series connected module, $N_{Ser} = V_{mp}/V_{mpp}$ $N_{Ser} = V_{mp}/V_{mpp}$	cells

cycle. As a result, it simply addresses the switching loss problem. Each SLMLI produces seven different voltage outputs: (+E, 0, and–E), (+1E, 0, and –1E), (+2E, 0, and –2E) up to (+7E, 0, and –7E), etc. However, SLMLI produces stairway output voltage waveform that is measured. The output voltage waveform of a 15-level SLMLI value is seven times the input DC voltage [14]. There is a new way of achieving multilevel outputs using the SLMLI band inverter, which switches 10 capacitors in seven levels to achieve: E, 2E, 3E, 4E, 5E, 6E, and 7E. When duty cycle (k) = 6/7 (≈ 0.857) in 15-level SLMLI. Therefore, the operation of ON status is shown in Figure 16.3.

Each super-lift converter is nourished with the estimation of DC output voltage. The single-phase voltage of the inverter depends upon the band switching capacitor. The 15-level yield waveforms acquired by various switching combinations and switching pattern appeared in Table 16.2. Multistage centrifugal pumps that operate in a vertical position are known as electric submersible pumps. The load torque and motor speed of a water pump have a non-linear parabolic relationship. Liquids that are accelerated by the impeller lose their kinetic energy in the diffuser in which kinetic energy is converted to pressure. A mechanical coupling at the pump's base connects the pump shaft to the gas separator or protector.

An optional thrust bearing is used to support part of the axial forces in the pump but most of the thrust bearing is absorbed. The water pump has a parabolic nonlinear connection between load torque and speed, i.e., brake horsepower is the power to drive the pump [21–41].

$$bhp = \omega T \tag{16.1}$$

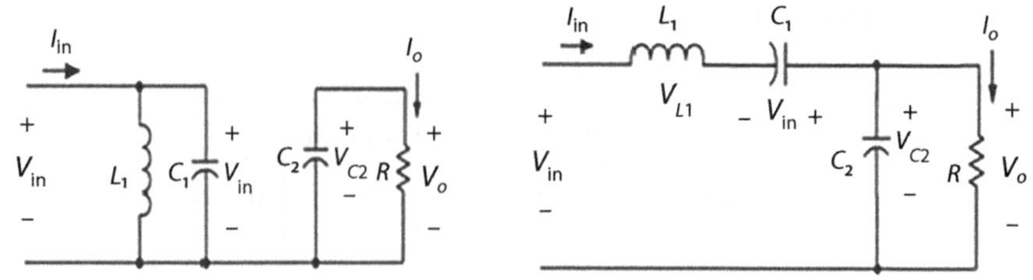

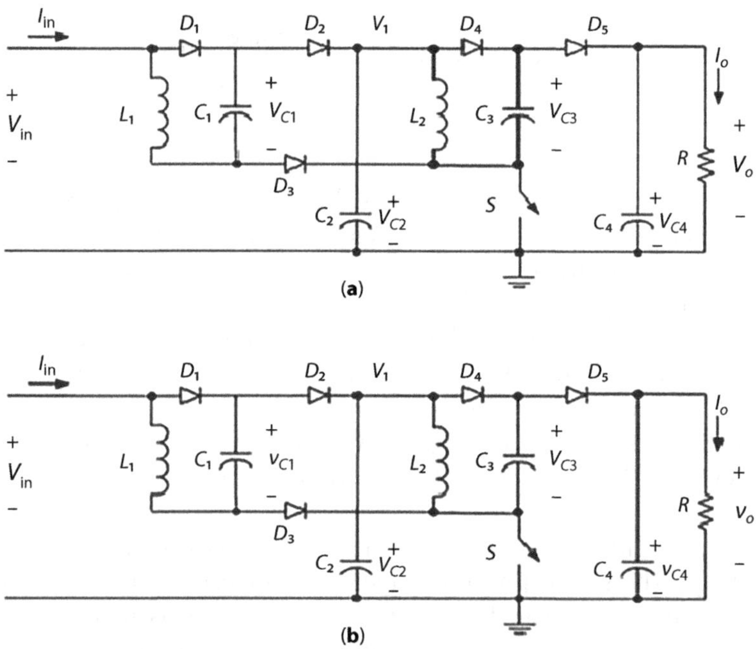

Figure 16.2 Double-lift circuit Luo converter. (a) Equivalent circuit for switch S is ON. (b) Switch S is OFF.

where is the shaft angular velocity and T is the shaft torque. If there were no losses, then P_w P_w and brake horse power would be equal, but, of course, P_w P_w is actually less and the efficiency of the power is defined as

$$\eta = \frac{P_w}{bhp} = \frac{\rho gQH}{\omega T} \tag{16.2}$$

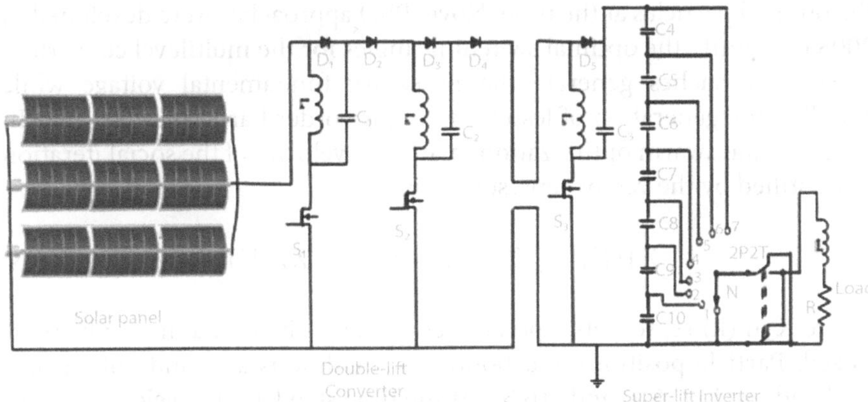

16.4 Principle of Particle Swarm Optimization

PSO was learning from the problem-solving situation. In PSO, each solution is referred to as a "bird" in the search space. It is referred to as a "particle" by us. All particles have fitness values that are assessed by the fitness function in order to be optimized, as well as velocities that direct the particles' flight. The particles move through the problem space by following the optimal particles at the time. Novel PSO approaches were developed in 2008 to identify the optimal switching angles for the multilevel converter.

The approaches generate the necessary fundamental voltage while avoiding the generation of lesser and higher order harmonics [16–20]. In fundamental swarm optimization, matching velocity of the social iteration is identified by the vector representation.

$$X1(K) = [X2(K), X2(K).....Xn(K)] \tag{16.3}$$

where X_n (k) represents a parameter of the problem that has to be optimized. Particle positions and boundary conditions are randomly generated and use in (16.4) and (16.5). At the time step k+1, the velocity of i_th particle is given by

$$\begin{aligned} V\_i\ (k+1) = wV\_i\ (k)+C1rand(x)\ ((P\_i\ (best)(k)-X\_i\ (k))/\\ \Delta t+C2rand(x)\ ((Gbest)(k)-X\_i\ (k))/\Delta t \end{aligned} \tag{16.4}$$

where w is the component of inertia; C1 is the social and C2 the cognitive; Pi (best) is the best location ever for a particle; G best is the best position ever obtained during movement of the whole social population of a particular particle; Rand (x) is a uniform positive random number from 0 to 1. The new location of each particle is determined by following one iteration:

$$X\_i\ (k + 1) = X\_i(k) + V\_i\ (k)\ \Delta t \tag{16.5}$$

The present speed is set to zero, and the new speed is assessed according to Equation (16.4), but a new location is evaluated using equation with the difference of the term (16.5). The particle swarm algorithm starts from the creation and assignment of the first particles [21]. The goal function is evaluated at every particle location and the best (lowest) function value and the optimal position is determined. It selects new speeds, based on the present speed, the optimal locations for the particles, and best places in its neighborhood. It then updates the particle positions iteratively to identify the optimal angle of switch in SLMLI and minimize THD percent in sequence. PSO changes the particle location on an iterative basis and its iterations continue until optimum values of the method are achieved, and then, a stop condition may be used.

16.5 Result and Discussion

The flowgraph of Genetic Algorithm for the proposed system is shown in Figure 16.4. The simulation results of Output voltage and current for the proposed 15-level super-lift Luo converter is appeared in Figure 16.5 and Figure 16.6. The THD results of Output voltage and current for the proposed 15-level super-lift Luo converter from simulation is shown in Figure 16.7 and Figure 16.8. The hardware model of concentrated SPV array with proposed converter is shown in Figure 16.9. The comparative analysis was carried out using simulation and hardware which is shown in Table 16.3.

The configuration of the proposed concentrated SPV module fed water pumping system is designed and simulated using MATLAB. The performance of the water pumping system using double-lift converter and SLMLI is evaluated using the simulation package, which gives the 15-level waveform for staircased Luo converter output voltage and current, as shown in Table 16.4. The output of SLMLIs includes harmonics, which may be

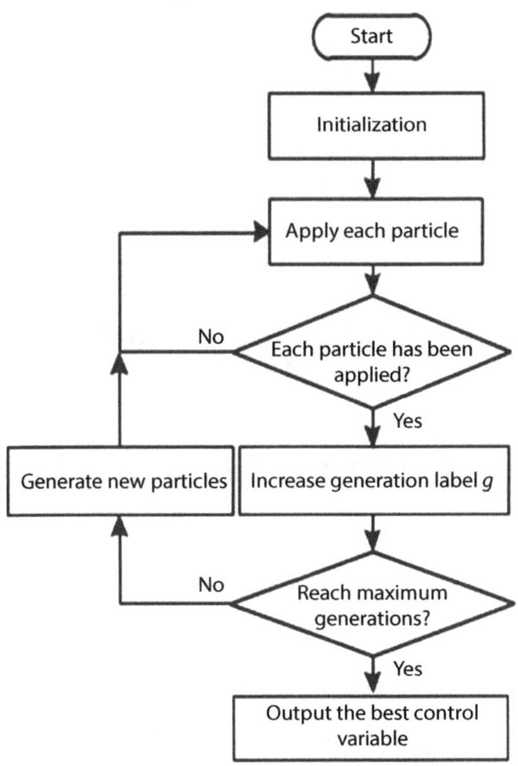

Figure 16.4 Flow chart for PSO algorithm.

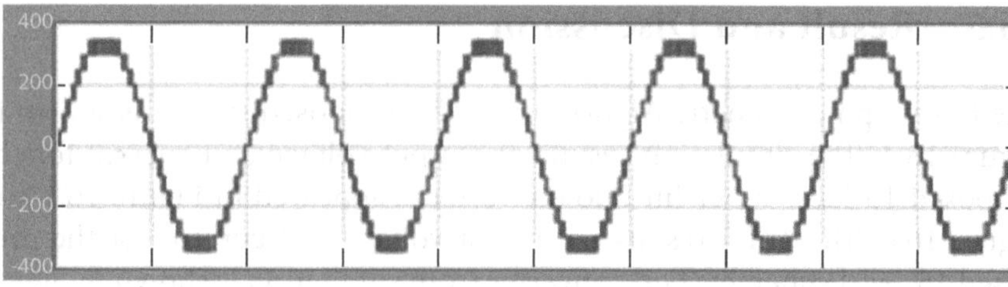
400
200
0
-200
-400

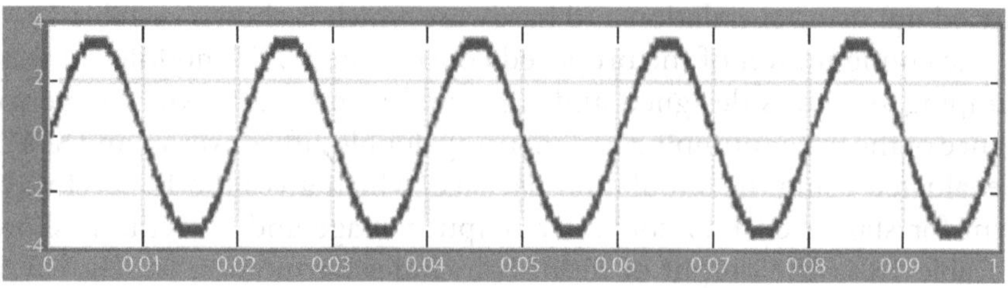
4
2
0
-2
-4
0
0.01
0.02
0.03
0.04
0.05
0.06
0.07
0.08
0.09
1

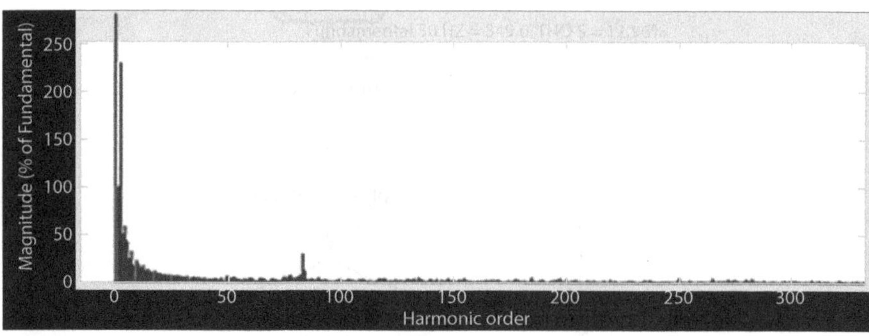
Magnitude (% of Fundamental)
250
200
150
100
50
0
0
50
100
150
200
250
300
Harmonic order

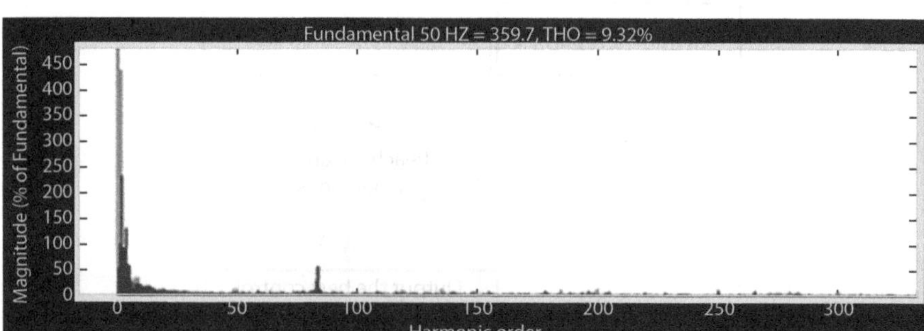
Fundamental 50 HZ = 359.7, THO = 9.32%
Magnitude (% of Fundamental)
450
400
350
300
250
200
150
100
50
0
0
50
100
150
200
250
300
Harmonic order

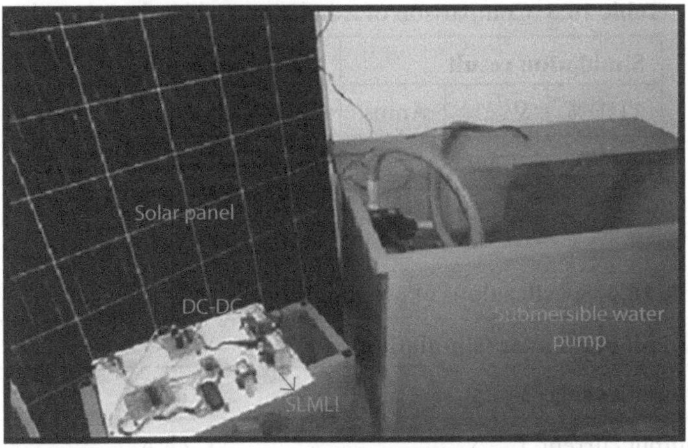
Solar panel
DC-DC
Submersible water pump
SLMLI

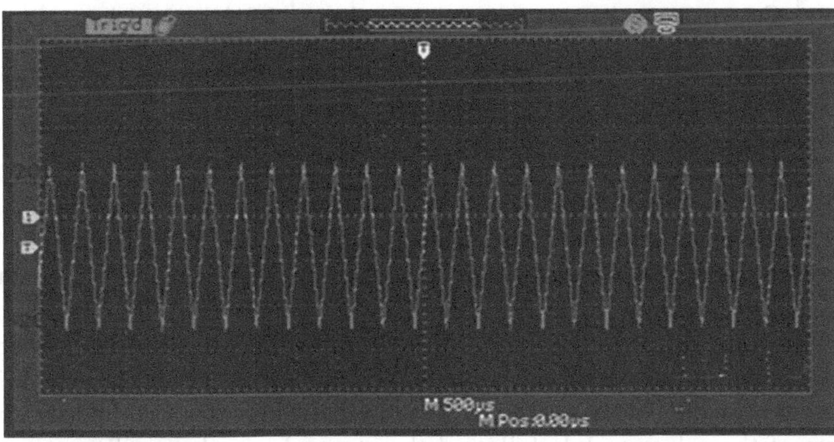
M 500μs
M Pos:0.00μs

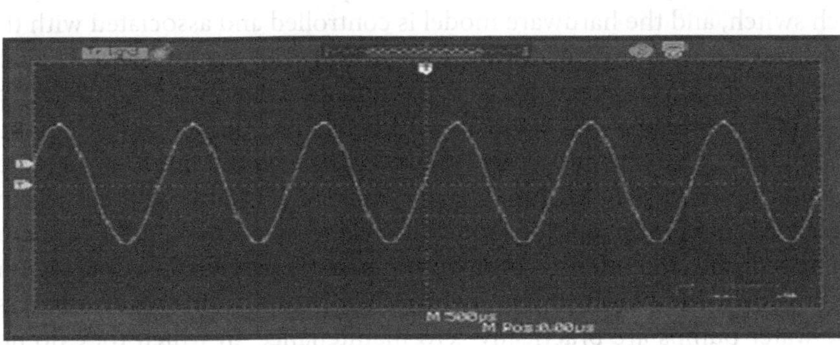
M 500μs
M Pos:0.00μs

Table 16.3 Comparison of simulation and hardware result.

Simulation result			Hardware result		
THD%	V_{rms}	Amps	THD%	V_{rms}	Amps
9.32%	387 V	2.4 A	24.12%	240 V	1.8 A

Table 16.4 Specifications of proposed converter.

Circuit parameter (simulation)	
Output voltage, V_o	387 V
Output current, I_0	2.4 A
Voltage at maximum power point, V_{mpp} V_{mpp}	0.81*387 = 313.47 V
Current at maximum power point, I_{mpp} I_{mpp}	0.9*2.4 = 2.16 A
Capacitor (C_1 – C_{11}) C_1 – C_{11})	1,000e–6
Load resistance (RL)	100 Ω

measured using total harmonic distortion (THD). Using FFT analysis, it is a measure of the shape similarity between the waveform and the fundamental component.

Harmonic distortions are minimized in this case in relation to the magnetic phase output voltage. Except for the highest peak voltage, which is regulated by the PSO algorithm with a low THD percentage, other parameters of the solar module followed the atmospheric change. The output buffer is provided to an isolated circuit (optical isolator IC MCT2E) for each switch, and the hardware model is controlled and associated with the assistance of a PIC microcontroller child [22]. The output buffer is operated with an output frequency of 50 Hz and a carrier switching frequency of 10,000 Hz. Figures 16.10 and 16.11 demonstrate the submersible pump load output of 240 V and 1.8 A, respectively. The output line voltage waveform of the 15-level SLMLI is virtually sinusoidal, indicating that the submersible water pump method is effective. Concentrated PV (CPV) systems are specifically intended to provide water for irrigation in rural regions without access to power supply energy. Their main advantages of submersible water pumps are practically zero maintenance, in which they do not require fuel, contaminate, and finally are straight forward to install.

16.6 Conclusion

Agriculture today suffers from multiple crises, and one of the worst crises is pertaining irrigation and water resources. The proposed system gives the basic principle of super-lift converters that is good for a small yard or for watering individual plants compare with diesel water pump. SPV panels are a low maintenance and environmentally friendly option for pumping irrigation water. The SLMLI has the simple structure and straight forward operation procedure, and easy control is used to construct higher level output voltage with reduced lower THD value. The proposed super-lift inverter with less number of semiconductor switches with minimum THD with the optimization control techniques such as PSO by using MATLAB/Simulink attains the value of THD 9.32% (simulation), and the appropriate prototype model THD is 24.12%. The drip irrigation must reduce water losses, without adding a substantial amount of head to the irrigated pumping system. It is also utilized to save installation space and costs.

References

1. Albert, J.R. and Stonier, A.A., Design and development of symmetrical super-lift DC–AC converter using firefly algorithm for solar-photovoltaic applications. *IET Circ. Device. Syst.*, 14, 3, 261–269, 2020. https://doi.org/10.1049/iet-cds.2018.5292.
2. Vanaja, D.S., Albert, J.R., Stonier, A.A., An experimental investigation on solar PV fed modular STATCOM in WECS using intelligent controller. *Int. Trans. Electr. Energy Syst.*, 13, e12845, 2021. https://doi.org/10.1002/2050-7038.12845.
3. Albert, J.R. and Vanaja, D.S., *Solar Energy Assessment in Various Regions of Indian Sub-continent, Solar Cells*, 1–14, 2021, DOI: 10.5772/intechopen.87735.
4. Kaliannan, T. *et al.*, Power quality improvement in modular multilevel inverter using for different multicarrier PWM. *EJECE*, 05, 02, 19–27, 2021, ISSN: 2736-5751. DOI: http://dx.doi.org/10.24018/ejece.2021.5.2.315.
5. Dhivya, M. and Renoald, A.J., Fuzzy grammar based hybrid split-capacitors and split inductors applied in positive output Luo-converters. *Int. J. Sci. Res. Sci. Eng. Technol.*, 3, 327–332, 2017. http://ijsrset.com/ IJSR SET173174.
6. Albert, J.R. *et al.*, Solar roadways-the future rebuilding infrastructure and economy. *Int. J. Electr. Electron. Res.*, 4, 2, 14–19, 2016. https:// research publish.com/journal-details/IJEER.
7. Santhiya, K. *et al.*, Real time speed control of three phase induction motor by using lab view with fuzzy logic. *JSET*, 5, 2, 21–27, 2018. http://jset.sasapublications.com/wp-content/uploads/2018/04/1841019.pdf.

8. Albert, J.R. *et al.*, Micro grid connected solar PV employment using for battery energy storage system. *J. Xidian Univ.*, 15, 3, 85–97, 2021. https://doi.org/10.37896/jxu15.3/010.
9. Kumarasamy, V., Ramasamy, V.K., Chinnaraj, G., Systematic design of multi-objective enhanced genetic algorithm optimized fractional order PID controller for sensorless brushless DC motor drive. *Circuit World*, 2021. https://doi.org/10.1108/CW-07-2020-0137
10. Vanchinathan, K. and Valluvan, K.R., A metaheuristic optimization approach for tuning of fractional-order PID controller for speed control of sensorless BLDC motor. *J. Circuit. Syst. Comp.*, 27, 08, 1850123, 2018. https://doi.org/10.1142/S0218126618501232.
11. Vanchinathan, K., Valluvan, K.R., Gnanavel, C., Gokul, C., Design methodology and experimental verification of intelligent speed controllers for sensorless permanent magnet Brushless DC motor. *Int. Trans. Electr. Energy Syst.*, 9, e12991, 2021. https://doi.org/10.1002/2050-7038.12991.
12. Vanchinathan, K. and Valluvan, K.R., Tuning of fractional order proportional integral derivative controller for speed control of sensorless BLDC motor using artificial Bee colony Optimization technique, in: *Intelligent and Efficient Electrical Systems*, pp. 117–127, Springer, Singapore, 2018, https://doi.org/10.1007/978-981-10-4852-4_11.
13. Rajasekaran, P. and Vanchinathan, K., Improved performance of four switch three phase brushless DC motor using speed-current control algorithm. *Int. J. Comput. Appl.*, 68, 11, 1–7, 2013. DOI: 10.5120/11620-5237.
14. Vanchinathan, K. and Valluvan, K.R., A study of sensorless BLDC motor drives and future trends. *Asian J. Res. Soc. Sci. Humanities*, 6, 9, 1863–1887, 2016. DOI: 10.5958/2249-7315.2016.00912.6.
15. Chandrasekaran, G., Kumarasamy, V., Chinraj, G., Test scheduling of core based system-on-chip using modified ant colony optimization. *J. Eur. Systèm. Autom.*, 52, 6, 599–605, 2019.
16. Chandrasekaran, G., Karthikeyan, P.R., Kumar, N.S., Kumarasamy, V., Test scheduling of system-on-chip using dragonfly and ant lion optimization algorithms. *J. Intell. Fuzzy Syst.*, (Preprint), 1–13, 2021.
17. Vanchinathan, K. and Valluvan, K.R., Improvement of time response for sensorless control of BLDC motor drive using ant colony optimization technique. *Int. J. Appl. Eng. Res.*, 10, 55, 6, 2015.
18. Gnanavel, C. and A. Alexander, S., Experimental validation of an eleven level symmetrical inverter using genetic algorithm and queen bee assisted genetic algorithm for solar photovoltaic applications. *J. Circuit. Syst. Comp.*, 23, 13, 1850212–23, 2008.
19. Vanchinathan, K., Sathiskumar, P., Selvaganesan, N., A Metaheuristic optimization algorithm-based speed controller for brushless DC motor: Industrial case study, in: *Metaheuristic Algorithms in Industry 4.0*, pp. 189–216, CRC Press, 2021.

20. Gnanavel, C., Kamalamoorthy, N., Prabhu, V., Assessment among Single and three Phase 14–Echelon cascaded multilevel inverter. *Int. J. Sci. Res. Publ.*, 3, 5, 1–8, 2013.
21. Vanchinathan, K. and Selvaganesan, N., Adaptive fractional order PID controller tuning for brushless DC motor using artificial bee colony algorithm. *Results Control Optimization*, 4, 100032, 2021.
22. Gnanavel, C., Khanna, P.S., Rajavelan, M., Vanchinathan, K., Gokul, C., Real time monitoring and controlling of boiler drum parameters using Internet of Things (IoT). *Test Eng. Manage.*, 82, 3, 2195–2198, 2020.
23. Kalavalli, C., Sachinamreiss, G.N., Ali, A.N., Performance enhancement of boost converter fed permanent magnet synchronous machine. *IOP Conference Series: Materials Science and Engineering*, vol. 937, p. 012015, 2020.
24. Udayakumar, M.D., Gowthaman, K.S., Prabhu, A., Ali, A.N., HERIC Inverter-A SEPIC based transformer-less converter design and simulation for isolated standalone PV system. *IOP Conference Series: Materials Science and Engineering*, vol. 937, p. 012031, 2020.
25. Vijayalakshmi, S., Sivaraman, P.R., Karthick, R., Ali, A.N., Implementation of a new Bi-directional switch multilevel inverter for the reduction of harmonics. *IOP Conference Series: Materials Science and Engineering*, vol. 937, p. 012026, 2020.
26. Subramanian, A.T.S., Priyadharsini, S., Palaniyappan, S., Ali, A.N., A novel IOT based domestic automation system for load monitoring and efficient control. *IOP Conference Series: Materials Science and Engineering*, vol. 937, p. 012028, 2020.
27. Praveen, A.A.A., Kumaran, M.M., Ali, A.N., Minimization of power factor penalty charges for non-linear domestic loads with IOT technology. *IOP Conference Series: Materials Science and Engineering*, vol. 937, p. 012011, 2020.
28. Sivamani, D., Ramkumar, R., Nazarali, A., Shyam, D., Design and implementation of highly efficient UPS charging system with single stage converter. *Mater. Today: Proc.*, 45, 2, 1809–1819, 2021, 10 October 2020.
29. Paveethra, S.R., Kalavalli, C., Vijayalakshmi, S., Ali, Dr. A. N., Shyam, D., Evaluation of voltage stability of transmission line with contingency analysis. *Int. J. Sci. Technol. Res.*, 9, 02, 4018–22, 2020.
30. Ali, A.N., Sivamani, D., Jaiganesh, R., Pradeep, M., Solar powered air conditioner using BLDC motor. *IOP Conference Series: Material and Science Engineering*, vol. 623, Oct 2019.
31. Venkatesh, V., Ali, A.N., Jaiganesh, R., Indragandhi, V., Extraction and conversion of exhaust heat from automobile engine in to electrical energy. *IOP Conference Series: Material and Science Engineering*, vol. 623, Oct 2019.
32. Shyam, D., Premkumar, K., Thamizhselvan, T., Ali, A.N., Priya, M.V., Symmetrically modified laddered H-bridge multilevel inverter with reduced configurational parameters. *Int. J. Eng. Adv. Technol.*, 9, 1, 5525–5532, Oct 2019.

33. Ali, A.N., Jayabharath, R., Udayakumar, M.D., An ANFIS based advanced MPPT control of a wind-solar hybrid power generation system. *Int. Rev. Model. Simul. (IREMOS)*, 7, 4, 638–643, 2014.
34. Ali, A.N., Jayabharath, Dr. R., Priyadharshini, R.S., A Single phase high efficient transformer less inverter for PV grid connected power system using ISPWM technique. *Int. J. Appl. Eng. Res. (IJAER)*, 10, 9, 7489–7496, 2015.
35. Ali, A.N. and Jeyabharath, R., Ride through strategy for a three-level dual Z-source inverter using TRIAC. *Circuits Syst.*, 7, 11, 3911, 2016.
36. Ali, A.N. and Jayabharath, R., Performance enhancement of hybrid wind/photovoltaic system using Z source inverter with Cuk-sepic fused converter. *Res. J. Appl. Sci. Eng. Technol.*, 7, 19, 3964–3970, 2014.
37. Ali, A.N., Gowri, N., Scholor, P.G., Power factor correction based bridgeless single switch SEPIC converter fed BLDC motor. *Adv. Nat. Appl. Sci.*, 10, 3, 190–198, 2016.
38. Ali, A.N., Jeyabharath, R., Udayakumar, M.D., Cascaded multilevel inverters for reduce harmonic distortions in solar PV applications. *Asian J. Res. Soc. Sci. Humanities*, 6, 11, 703–705, 2016.
39. Subramanian, A.T.S., Sabarish, P., Ali, A.N., *A Power factor correction based canonical switching cell converter for VSI fed BLDC motor by using voltage follower technique*, IEEE xplore Digital Library, pp. 1–8, 2017.
40. Ganesh, R.J., Ali, A.N., Kodeeswaran, S., Karthikeyan, B., Nagarajan, L., IoT based water management system for highly populated residential buildings. *Int. J. of Disaster Recovery And Business Continuity*, 11, 01, 4018–22, Apr 2020.
41. Ali, A.N., Ganesh, R.J., Sivamani, D., Shyam, D., Solar powered highly efficient seven-level inverter with switched capacitors. *IOP Conference Series: Material and Science Engineering*, vol. 906, p. 17578981, Aug. 2020, doi: 1757899X.

17

Analysis of Load Torque Characteristics for an Electrical Tractor

Gade Chandra Sekhar Reddy, Sujay Deole, Mandar More, Razia Sultana W.* and Chitra A.

School of Electrical Engineering Vellore Institute of Technology, Vellore, India

Abstract

Due to decrease in manpower, farm machineries are placing vital role in agriculture. Tractor is kind of farm machinery that reduces the manpower and increases the work efficiency. This chapter presents the design and analysis of the load torque characteristics of electric tractor. Tractor is an off-road vehicle unlike conventional automobiles. Tractor is used in various agricultural applications like plowing, tillage, and transport. In such kind of field applications, torque requirement is more. Electric motor is used as drive train in E-tractor so that the selected motor should be able to meet the load requirement. Finding the load torque, power, and its variations according to field application must choose suitable electric motor. This works aims to design and analyze the load toque characteristics and its dynamics for different tillage applications. In addition, this work also aims to determine load toque for goods transportation. Load torque is modeled in MATLAB Simulink.

***Keywords*:** Electrical tractor, motor, grade ability, torque

17.1 Introduction

Nowadays, the world's population increases explosively day by day; with this, it will require more amount of food to feed them. Hence, the agricultural production should have to increase to fulfill this requirement [1, 2]. Thus, the farmer needs to improve productivity in their farm, while increasing the production energy input needs to be taken into account.

**Corresponding author*: wraziasultana@vit.ac.in

A. Chitra, V. Indragandhi, and W. Razia Sultana (eds.) Smart Grids and Green Energy Systems, (263–284)

Agricultural equipment in farm should be ecofriendly to nature. The continuous hike in the prices of fuel and the hazardous effects on environment caused by the combustion of fossil fuels has made it very important for researchers to come up with an alternative energy source to run the vehicles [3–5]. One such alternative is the development and usage of electric vehicles, which do provide high sustainability as compared to the conventional fuel vehicles [6]. Recently, electrification in agricultural field increases in various parts such as implemented electrification, engine auxiliaries, and energy storage for electric drives, traction drives; in this, we are going to design traction drives, which can be useful in agricultural field [7].

Tractors are the essential part in farming aspect; in modern time, due to electrification in agricultural field, E-tractors were developed; so for this, E-tractor should be design in such way that it will work on electrical energy to improveme in their performances and also to increase efficiency. Farmers use tractors in agricultural field for various such as heavy and light duty works; it is used to carry out various agricultural activities such as plowing, tilling, and transportation, etc., [8–10] so draft force requirement in agricultural is more, so it has to be fulfilled by vehicle that we are going to design; normal E-vehicle could not handle such large draft force so E-tractor needs to be designed in such way that it will handle this draft force.

The traction power provided by the motor needs to meet driving requirement of the tractors. In this work, we are going to calculate the total tractive force required to pull the implement in field. Accordingly, by calculating various parameters, we will select a suitable motor drive for application.

17.2 Methodology

- Estimation of resistive forces acting on the vehicle.
- Estimation of agricultural implement draft force for E-tractor.
- Estimation of load torque at wheels.
- Simulation of load torque characteristics for an E-tractor using MATLAB simulation.

17.2.1 Traction Resistive Forces

When vehicle is running in a particular direction, there are some forces that will act in opposite direction to the vehicle movement, which is known as resistive forces [11–14]. Main function of these forces is to reduce the performance of the vehicle. These all forces are negative forces. These forces should be considered to obtain maximum torque required; these forces are as follows:

- Rolling resistance force
- Grade resistance
- Aerodynamic force
- Acceleration force

17.2.2 Calculation of Rolling Resistance Force

Rolling resistance is corresponding to rolling motion between the wheels and the surface motion of the vehicle. Figure 17.1 illustrate the concept of the rolling resistance.

$$Frr = GVW * Crr \tag{17.1}$$

where GVW is the gross vehicle weight and Crr is the coefficient of rolling resistance.

Rolling resistance is depends on various parameter like GVW, Crr, roughness of surface of motion, contact area, and tyre parameters. A rated tyre is better than G rated tires by this fuel consumption reduces up to 7.5%. If tyre having flat surface, then contact with ground is more and F_{rr} is increases, whereas tyre having point contact with ground there, F_{rr} decreases.

17.2.3 Calculation of Grade Resistance

Component of vehicle weight acts parallel to an inclined surface then grade resistance acts on vehicle shown in Figure 17.2. When vehicle is traveling up a grade, grade resistance is positive. When traveling upgrade,

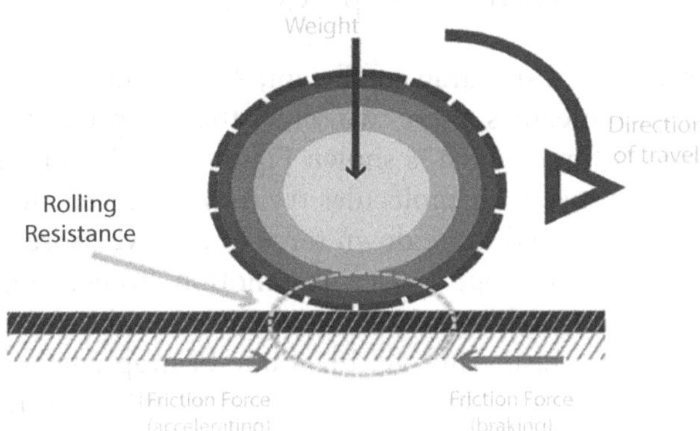

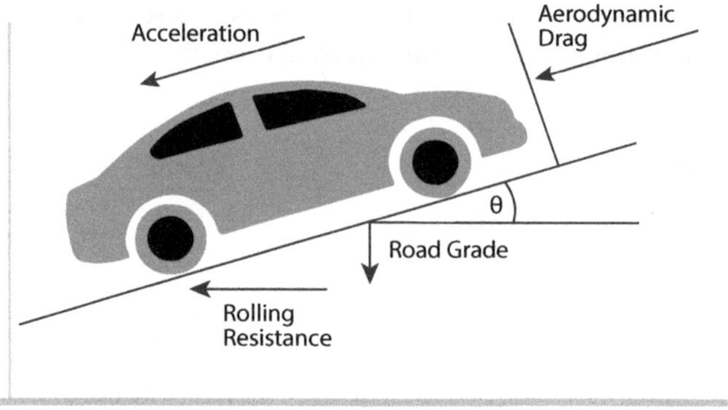

Figure 17.2 Representation of grading force acting on a vehicle.

grade resistance is negative. More inclination means more grade so grade resistance is positive. The grade resistance acting on the vehicle can be calculated as follows:

$$GR = GVW * SIN\theta \tag{17.2}$$

where GR = grade resistance and Sinθ = grade or inclination angle.

17.2.4 Calculation of Aerodynamic Force

When vehicle is moving in particular direction, air molecules oppose the motion of vehicle. Due to this, force resisting its motion known as aerodynamic force. This force referred to as aerodynamic drag.

Aerodynamic force is calculated as

$$Faero = 0.5 * \rho * \mathrm{Cd} * \mathrm{A} * v^2 \tag{17.3}$$

where ρ = air density; Cd = drag coefficient; A = frontal area; υ = velocity.

Because of air molecules, there two forces that are produced, which are drag force and lift force which is shown Figure 17.3. When vehicle moving in a particular speed, air molecules try to oppose vehicle motion and lift the vehicle so we need more down force to achieve more acceleration, whereas drag component opposes vehicle motion. To get drag coefficient of vehicle, we have to take some test like CFD simulation, wind tunnel test, and coast down test. After getting value of drag coefficient, we can calculate drag coefficient. Cd value is mainly depending on frontal area of the vehicle. In case of air density, during rainy season, air density is more because the density of air and density of water are also considered.

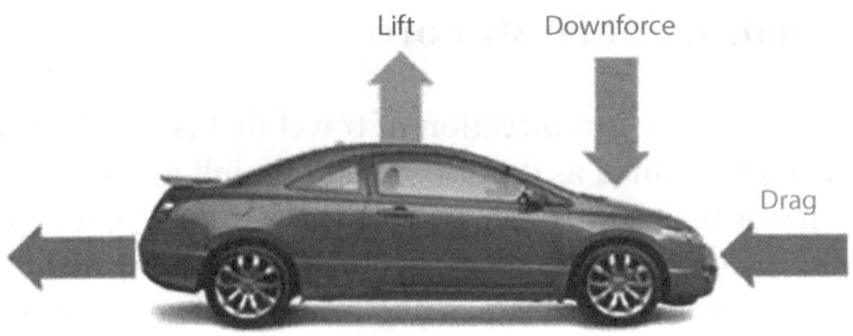

17.2.5 Calculation of Acceleration Force

The acceleration force is mainly depending on Newton second law of motion. It is depend upon total force acting on an object and mass of object.

Acceleration force is calculated as follows:

$$Fa = m * a \tag{17.4}$$

where m is the mass of the vehicle and a is the required acceleration.

17.2.6 Contribution of Total Running Resistances

The following Table 17.1 shows the resistance force contribution of each vehicle component in percentage of total running resistance.

Table 17.1 Percentage contribution of vehicle components to total running resistance.

Running resistances	
Components	Percentage contribution to total running resistance
Drive train components	5–10
Brake pads	4–5
Wheel hub	2–4
Rolling resistance	35–45
Aerodynamic resistance	45–75

17.3 Dynamics of Draft Force

Total force parallel acted in direction of travel that is required to pull the implement, which defined as draft force [7]. To pull many seeding implements and minor tillage tools operated at shallow depths, it is primarily a function of the width of the implement and the speed at which it is pulled. For tillage tools operated at deeper depths, draft also depends upon soil texture, depth, and geometry of the tool. Few farm implements which are majorly used are shown in Figures 17.4–17.7 [15–17]. Typical draft requirements can be calculated as follows:

$$D = Fi\,[A + B * S + C * S^2] * W * T \tag{17.5}$$

where

D is implemented draft;
F is a dimensionless soil texture adjustment parameter;
i = 1 for fine, 2 for medium and 3 for coarse textured soils;
A, B, and C are machine-specific parameters;
S is speed, km/h;
W is implemented width, (m) or number of rows or tools;
T is tillage depth, (inch) for major tools.

Table 17.2 shows the parameters from ASABE standards [7].

To calculate the draft force, we have taken a different implements like sweep plow, field cultivator, seed planter, and disc harrow (tandem).

E-Tractors can be pulled to a variety of farm implements for plowing, planting, and harvesting crops and also used for dragging materials and individual transportation. Tractors are put to use accordingly size of land and tillage depth. For that, we calculate the draft force, by considering the speed of the vehicle constant with different tillage depth. By taking constant parameter A, B, and C from Table 17.3 and take the speed as 7 kmph width of implement 3-m tillage depth as 4, 8, 12, 16, and 20 inches [18, 19]. From the above result, the graph between tillage depth vs. draft force can be drawn.

In Figure 17.8, it presented the calculated tillage depth vs. draft force. We observe that variations of draft force by changing tillage depths for agricultural implements.

Now, we calculate the draft force for by considering the tillage depth of the vehicle constant with different speed. By taking constant parameters A, B, and C from Table 17.3 and taking the tillage depth as 5 inches, speed of the vehicle is 0, 2, 4, 6, 8, and 10 kmph.

From the above result, the graph between speeds vs. draft force is shown in Figure 17.9.

Table 17.2 Parameters of ASABE.

	SI units					English units			Soil parameters		
		Machine parameters			Width units	Machine parameters					
Implement	**Width units**	**A**	**B**	**C**		**A**	**B**	**C**	**F1**	**F2**	**F3**
Major Tillage Tools											
Narrow Point	tools	226	0	1.8	tools	129	0	2.7	1	0.7	0.45
30-cm winged point	tools	294	0	2.4	tools	167	0	3.5	1	0.7	0.45
Chisel Plow	m	652	0	5.1	ft	113	0	2.3	1	0.7	0.45
5-cm straight point	tools	91	5.4	0.0	tools	52	4.9	0.0	1	0.85	0.65
7.5-cm sweep	tools	107	6.3	0.0	tools	61	5.8	0.0	1	0.85	0.65
Sweep plow	m	390	19.0	0.0	ft	53	4.6	0.0	1	0.88	0.78
Tandem primary tillage	m	309	16	0.0	ft	68	5.2	0.0	1	0.88	0.78
Tandem secondary	m	216	11.2	0.0	ft	37	3.2	00	1	0.88	0.78

(Continued)

Table 17.2 Parameters of ASABE. (*Continued*)

Implement	SI units				English units				Soil parameters		
	Width units	Machine parameters			Width units	Machine parameters					
		A	B	C		A	B	C	F1	F2	F3
Disc harrow offset	m	364	18.8	0.0	ft	62	5.4	0.0	1	0.88	0.78
Disc gang	m	124	6.4	0.0	ft	21	1.8	0.0	1	0.88	0.78
Coulters smooth	tools	55	2.4	0.0	tools	31	2.5	0.0	1	0.88	0.78
Coulter's bubble	tools	66	3.3	0.0	tools	37	3.0	0.0	1	0.88	0.78
Rod weeder	m	210	10.7	0	ft	37	3.0	0.0	1	0.85	0.65
Rotary hoe	m	600	0.0	0.0	ft	41	0.0	0.0	1.0	1.0	1.0
Land plane	m	8,000	0.00	0.0	ft	550	0.0	0.0	1.0	1.0	1.0
Pneumatic drill	m	3,700	0.0	0.0	ft	250	0.0	0.0	1.0	1.0	1.0
Field cultivator	tools	46	2.8	0.0	tools	26	2.5	0.0	1.0	0.88	0.78
Coil tin harrow	m	250	0.0	0.0	ft	17	0.0	0.0	1.0	1.0	1.0
Disc bedder	rows	185	9.5	0.0	rows	106	8.7	0.0	1.0	0.88	0.78

(*Continued*)

Table 17.2 Parameters of ASABE. (*Continued*)

	SI units				English units				Soil parameters		
		Machine parameters			Width units	Machine parameters					
Implement	**Width units**	**A**	**B**	**C**		**A**	**B**	**C**	**F1**	**F2**	**F3**
Hoe drill primary tillage	m	6,100	0.0	0.0	ft	420	0.0	0.0	1.0	1.0	1.0
Hoe drill secondary tillage	m	2,900	0.0	0.0	ft	200	0.0	0.0	1.0	1.0	1.0

Figure 17.4 Sweep plow.

Figure 17.5 Field cultivator.

Figure 17.6 Seed planter.

Figure 17.7 Disk horrow (tandem).

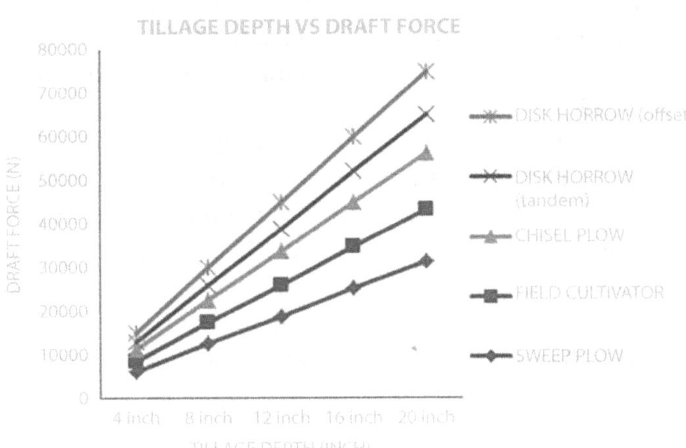

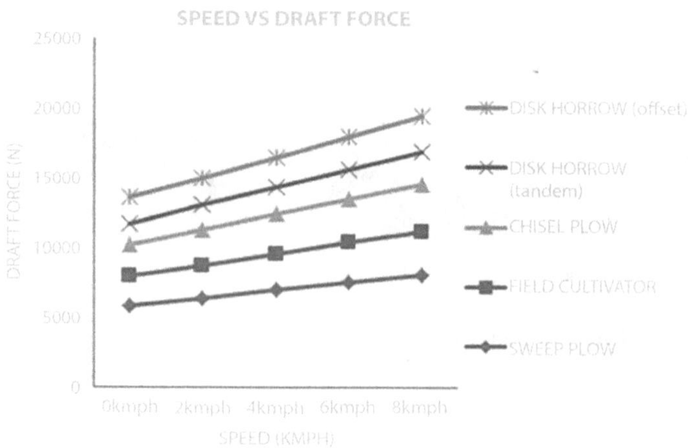

17.4 Power Train Calculation

The purpose of drivetrain is decisive to determine the power and torque requirements of the tractor. During operation of tractor in field requires large amount of power to drag various agricultural implements. It is a necessity to calculate power required before selecting motor specification and battery. The different parameters like torque, efficiency, power, and rpm are required for desired performance of E-tractor. In this case, we are taking two different applications. For field application, we do not consider aerodynamic force because the speed of E-tractor during field application is very low. The grading force is also not considered in case of field application and calculations performed only leveled field.

For calculating power, we are taking parameters into considerations, as shown in Table 17.3.

The force required to pull a tillage tool through the soil is called its draft. The draft force is located at the point where the tool is attached to the power unit called hitch. Standard formulae from the American Society of Agricultural and Biological Engineers (ASABE) are used for the calculation of draft force. The major farm implements and their parameters considered for the powertrain sizing is listed in Table 17.4.

We consider values for machine parameters are as follows:

A = 91, B = 5.4, C = 0;
Fi = 0.65, W = 3 m, T = 8 inches;
S = 5 kmph.

Table 17.4 Major farm implements used considered for powertrain sizing.

	SI units				English units				Soil parameters		
	Width	Machine parameters			Width	Machine parameters					
Implement	**units**	**A**	**B**	**C**	**units**	**A**	**B**	**C**	**F1**	**F2**	**F3**
Major Tillage Tools											
Narrow point	tools	226	0	1.8	tools	129	0	2.7	1	0.7	0.45
30-cm winged point	tools	294	0	2.4	tools	167	0	3.5	1	0.7	0.45
Chisel Plow	m	652	0	5.1	ft	113	0	2.3	1	0.7	0.45
5-cm straight point	tools	91	5.4	0.0	tools	52	4.9	0.0	1	0.85	0.65
7.5-cm sweep	tools	107	6.3	0.0	tools	61	5.8	0.0	1	0.85	0.65

17.4.1 Calculations for Field Applications

Acceleration force:

Fa = 550 * (5 * 5/18)/5 (assuming 0 – 5 km/h in 5 s)
= 152.77 N

Draft force:

D_f = 0.65[91 + 5.4950 + 0] * 3 * 8
= 1503.84 N

Rolling resistance:

Fr = 550 * 9.8 * 0.5
= 2695 N

Total tractive force required:

Ft = *Fa* + *Fr* + *Df*
= 152.77 + 2695 + 1503.84 = 4351.61 N

Assuming a 20% higher than required for extra power:

Fta = 1.2 * *Ft* = 1.2 * 4351.64 = 5221.932 N

Power required:

$$(Fta * \upsilon)/(transmission\ efficiency * motor\ efficiency)$$
$$= (5221.939 * (5 * 5/18)/(0.94 * 0.85) = 5.9 \text{ kW}$$

Wheel torque required:

Fta * *r* = 5221.932 * 0.35 = 1969 Nm
(r = 0.35 radius of wheel)

Wheel RPM:

N = υ * (5/18) * (60/2 * 0.35 * π)
=37.89 RPM

Gear Ratio:

motor speed/wheel rpm = 45

Motor Torque:

Wheel torque/Gear torque = 49.74 Nm

For transport application, aerodynamic force is considered because for transport application speed of E-tractor is more as compare to the field application. Grade resistance is also taking part in case of transport application.

17.4.2 Calculation for Transport Applications

For transportation application, NEDC (New European Drive Cycle) [17] gives the velocity input

Frr = 550 * 9.8 * 0.5 = 2695 N
GR = 550 * 9.8 * *sin*12 = 1120.64 N
Aerodynamic force = 0.5 * 1.225 * 0.2 * 1.23 * 12.25 = 1.8457 N
Fa = 550 * (5 * 5/18)/5 = 152.77 N
Ft = 2695 + 1120.64 + 1.845 + 152.77 = 3970.255 N
Torque required at wheel = 3970.255 * 0.35 = 11343.58 N

17.5 MATLAB Simulation and Result

For transportation application, NEDC gives the velocity input.

Figures 17.10 and 17.11 are the simulation diagrams for the field application; from the above simulation models, we obtained output power that is 5.9 kW.

Figure 17.12 shows the simulation diagram for the transportation application. From simulation, we obtained various drive cycle graphs for speed, motor torque, and power with respect to the time. Figure 17.13 shows the modified NEDC drive cycle with reduction speed ranges applied to tractor for transport application in field. Figure 17.14 shows the power requirement of the applied NEDC drive cycle and Figure 17.15 give the motor torque required for the applied velocity cycle.

Drive cycle shows variation of speed from 0 to 7 kmph with respect to time.

As simulations results, we ensure that maximum amount of power required for E-tractor is 3.65 kW. For a transportation application, whereas, the maximum amount of power required for field application is 5.9 kW. From above power requirement, we select motor in a such way that it can able to produce the requirements of load torque and load power.

17.6 Motor Specifications

Motor details are selected from the VEM group company that is a leading motor manufacturer, which is listed in Table 17.5.

17.7 Conclusion and Discussion

From the above result, we calculated load torque and load power from which we can select a suitable motor that is required; so, we have to determine the load torque characteristics and analyze the draft force requirement for different parameters.

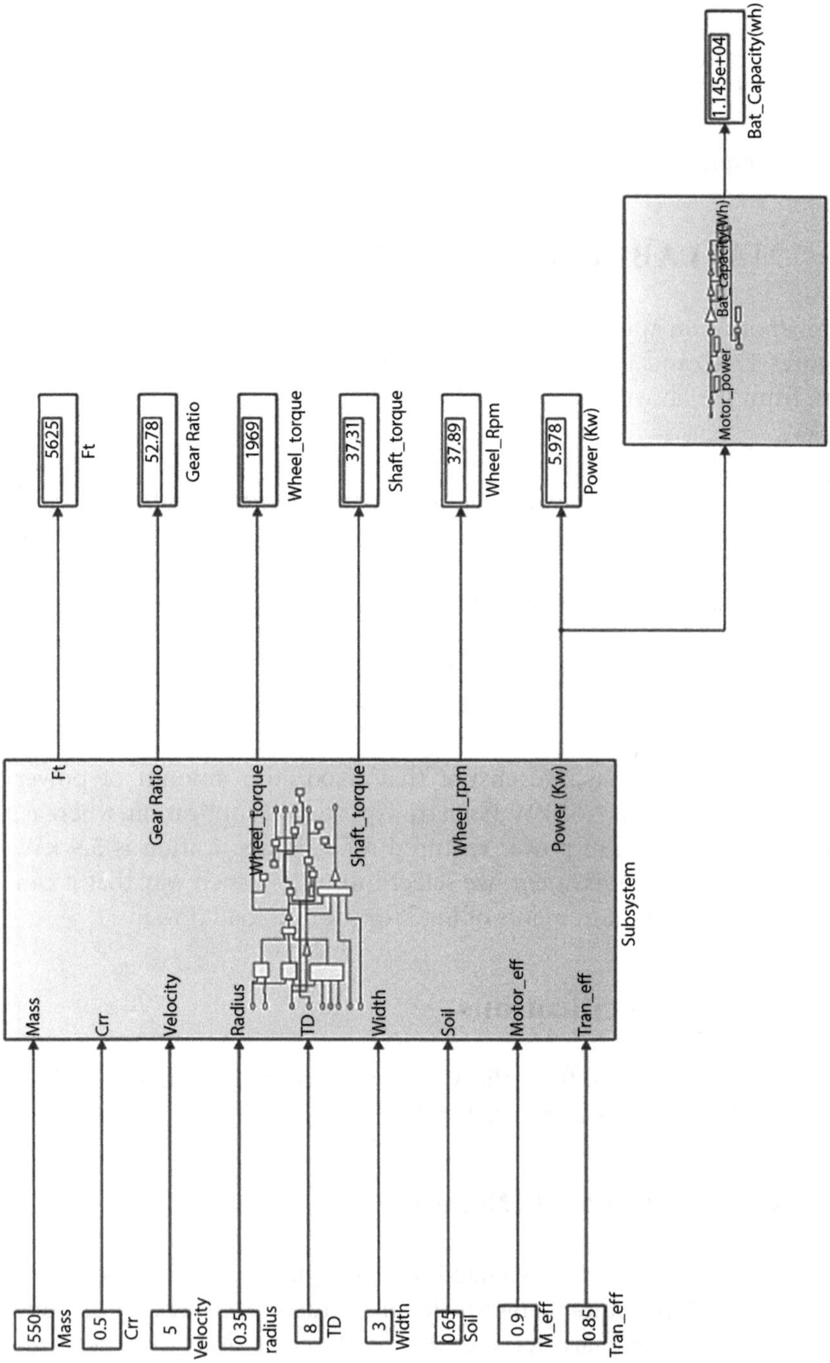

Figure 17.10 MATLAB simulation diagram for powertrain sizing.

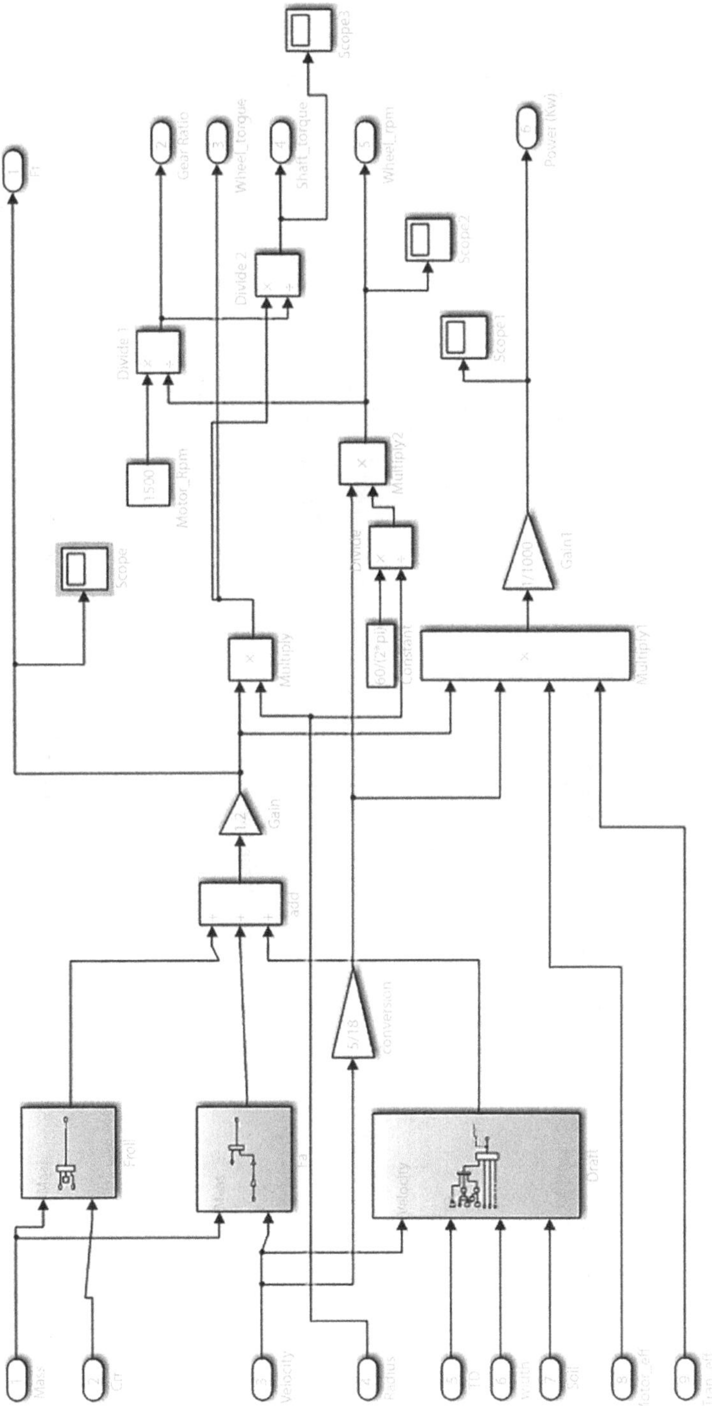
Mass
Crr
Velocity
Radius
T.D
Width
Soil
Motor_eff
Tran_eff
Froll
Fa
Draft
add
1.2
Gain
5/18
conversion
Multiply
Scope
Ft
60/(2*pi)
Constant
Divide
Multiply2
Multiply1
1/1000
Gain1
1500
Motor_Rpm
Divide 1
Divide 2
Scope1
Scope2
Scope3
Gear Ratio
Wheel_torque
Shaft_torque
Wheel_rpm
Power (kw)

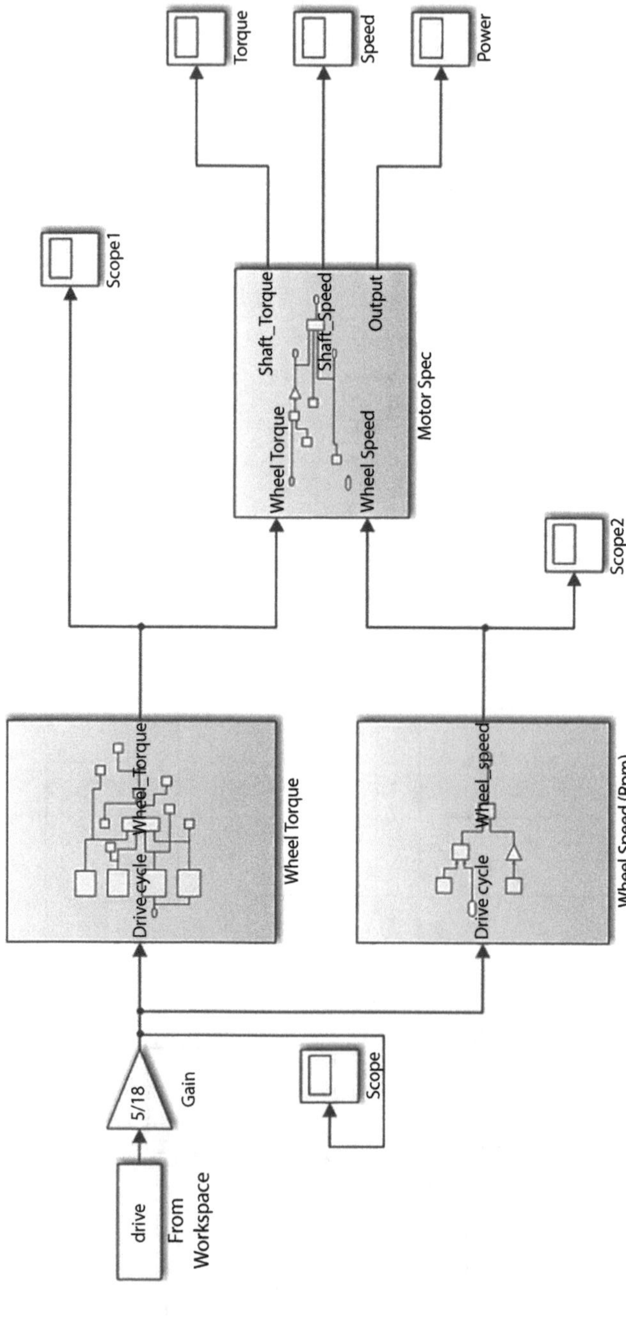

Figure 17.12 MATLAB simulation diagram for calculating torque and power for transport application.

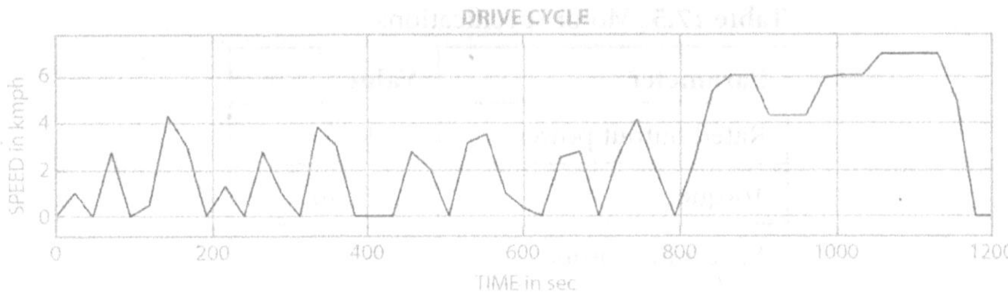
DRIVE CYCLE
SPEED in kmph
0
2
4
6
0
200
400
600
800
1000
1200
TIME in sec

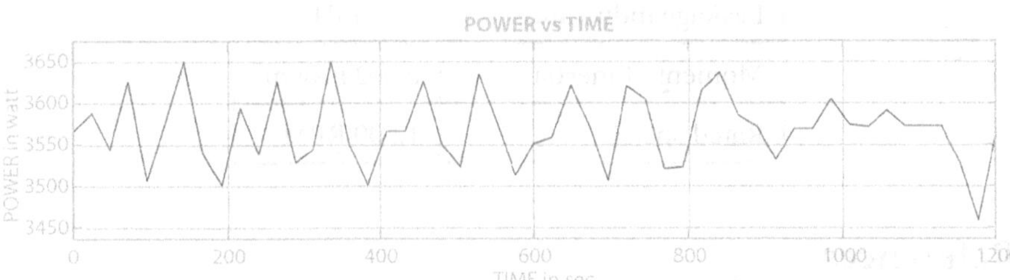
POWER vs TIME
POWER in watt
3450
3500
3550
3600
3650
0
200
400
600
800
1000
1200
TIME in sec

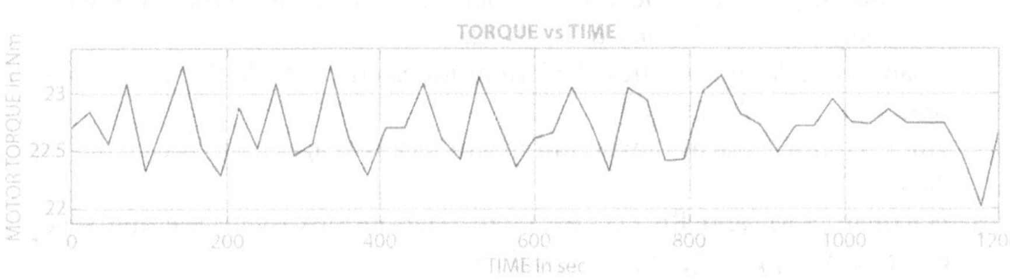
TORQUE vs TIME
MOTOR TORQUE in Nm
22
22.5
23
0
200
400
600
800
1000
1200
TIME in sec

Table 17.5 Motor specifications.

Parameter	Value
Rated output power	7.5 kW
Torque	42.8 Nm
Maximum torque	60 Nm
Voltage	347 V
Current	14 A
Efficiency	94%
Phase resistance	0.48 Ω
Magnetizing inductance	6.09 mH
Leakage inductance	2.37 mH
Moment of inertia	0.0024 Kg m^2
Rated speed	1,500 RPM

References

1. Askari, M. and Khalifahamzehghasem, S., Draft force inputs for primary and secondary tillage implements in a clay loam soil. *World Appl. Sci. J.*, 21, 12, 1789–1794, 2013.
2. Das, A., Jain, Y., Agrewale, M.R.B., Bhateshvar, Y.K., Vora, K., Design of a Concept Electric Mini Tractor, in: *2019 IEEE Transportation Electrification Conference (ITEC-India)*, IEEE, pp. 1–5, 2019.
3. Tabbache, B., Djebarri, S., Kheloui, A., Benbouzid, M., A power presizing methodology for electric vehicle traction motors, *International Review on Modelling and Simulations*, 6, 1, 29–32, 2013.
4. Chauhan, S., Motor torque calculations for electric vehicle. *Int. J. Sci. Technol. Res.*, 4, 8, 126–127, 2015.
5. Hunt, Donnell. Farm power and machinery management. Waveland Press, 2008.
6. Harrigan, T.M. and Rotz, C.A., Draft relationships for tillage and seeding equipment. *Appl. Eng. Agric.*, ll, 6, 773–783, 1995.
7. American Society of Agricultural and Biological Engineers (ASABE) Standards. *Agricultural Machinery Management Data. Test*, 9, 2011.
8. Troncon, D., Alberti, L., Bolognani, S., Bettella, F., Gatto, A., Electrification of agricultural machinery: A feasibility evaluation, in: *2019 Fourteenth*

International Conference on Ecological Vehicles and Renewable Energies (EVER), IEEE, pp. 1–7, 2019.

9. Baek, S.Y., Jung, Y.J., Siddique, M.A.A., Nam, K.C., Kim, T.B., Choi, C.H., Kim, Y.J., Development of simulation model for electric all-wheel-drive tractor, in: *2019 ASABE Annual International Meeting*, American Society of Agricultural and Biological Engineers, p. 1, 2019.
10. Buning, E.A., Electric drives in agricultural machinery-approach from the tractor side. *J. Agric. Eng.*, 47, 3, 30–35, 2010.
11. Amongo, R.M.C., Quilloy, E.P., Ranches, M.A.F., Larona, M.V.L., Madlangbayan, M.S., Development of an electric hand tractor (e-Tractor) for agricultural operations, in: *IOP Conference Series: Earth and Environmental Science*, vol. 542, IOP Publishing, p. 012027, 2020.
12. Chen, Y., Xie, B., Du, Y., Mao, E., Powertrain parameter matching and optimal design of dual-motor driven electric tractor. *Int. J. Agric. Biol. Eng., 12, no. 1*, 12, 1, 33–41, 2019.
13. Petrecca, G., *Energy Conversion and Management*, Springer, Berlin, Germany, Springer, 2014.
14. Sandeep, V. and Shastri, S., Analysis and design of PMBLDC motor for three wheeler electric vehicle application, in: *E3S Web of Conferences*, vol. 87, EDP Sciences, p. 01022, 2019.
15. Matache, M.G., Cristea, M., Găgeanu, I., Zapciu, A., Tudor, E., Carpus, E., Popa, L.D., Small power electric tractor performance during ploughing works. *INMATEH-Agric. Eng.*, 60, 1, 2020.
16. Brenna, M., Foiadelli, F., Leone, C., Longo, M., Zaninelli, D., Feasibility proposal for heavy duty farm tractor, in: *2018 International Conference of Electrical and Electronic Technologies for Automotive*, IEEE, pp. 1–6, 2018.
17. Zhang, X., Design theory and performance analysis of electric tractor drive system. *Int. J. Eng. Res. Technol. (IJERT)*, 2278–0181, 2017.
18. Bin, X., Hao, L., Zheng-He, S., En-Rong, M., Powertrain system design of medium-sized hybrid electric tractor. *Inf. Technol. J.*, 12, 23, 7228, 2013.
19. Shishkov, A.N., Sychev, D.A., Bychkov, A.E., Yu Sidorenko, N., The DET-400 tractor traction electric drive. *Russ. Electr. Eng.*, 85, 10, 610–612, 2014.

18

Comparison of Wireless Charging Compensation Topologies of Electric Vehicle

M. Rajalakshmi[1] and W. Razia Sultana[2]*

[1]Research Scholar, Vellore Institute of Technology, Vellore, Tamil Nadu, India
[2]Faculty of School of Electrical Engineering, Vellore Institute of Technology, Vellore, Tamil Nadu, India

Abstract

In automobile sector, due to increase in the demand and price of fuel of IC engines, the investments turned toward the green technologies the electric vehicle. To reduce the pollution, the government also promotes the electric vehicle usage by giving lots of subsidies. The major criteria that affect the usage of electric vehicle are the charging time and the charging station availability. Lots of research is undergoing to reduce the charging time and to increase the range of electric vehicle. The charging technology is one of the other blooming technologies to promote the electric vehicle utilization. Among different charging methods, the wireless charging plays an important role in development of electric vehicle to popularize the EV. This chapter discusses about the different wireless charging topologies; from that, the recent technology of inductive charging with compensation network is discussed, SS and PS topologies are simulated, and the results are compared.

***Keywords*:** Wireless charging, compensation topology, electric vehicle

**Corresponding author*: wraziasultana@vit.ac.in

A. Chitra, V. Indragandhi, and W. Razia Sultana (eds.) Smart Grids and Green Energy Systems, (285–300)

18.1 Introduction

The electrification of transportation has been concluding for several years for conserving energy and controlling environmental pollution. Electric vehicle wireless power transfer technology is the latest one that gives a great solution to many obstacles like different charging plugins in charging technology [1]. This is done by the inductive charging method that is the oldest technology that had been developed 30 years before.

Different countries deploying different charging plugins related to their standards [2]. There is a huge difference between them. To overcome this, the wireless technology is the only solution that provides universal adaptability for the charging of the electric vehicle. The only thing is the design of the system with high efficiency to transfer the power effectively. The efficient design of the system gives better and fast charging of the battery of the electric vehicle [3]. This chapter focuses on studying and implementing various varieties of wireless charging techniques effectively into electric vehicles, to extend their easy use.

The wireless transfer technology had already been well developed in the railway system several years ago, in which the train runs on the laid track and the power is transferred using conductors from pantograph sliders. Figure 18.1 depicts the components of the simple electric vehicle with some compensation network in the transmitter and receiver sides.

From the grid, the power is converted into DC using AC-to-DC converter, then that power is again converted into high-frequency AC power, then the high-frequency transformer is the inductor, and capacitor combination of the power is transmitted from the transmitter side. From the

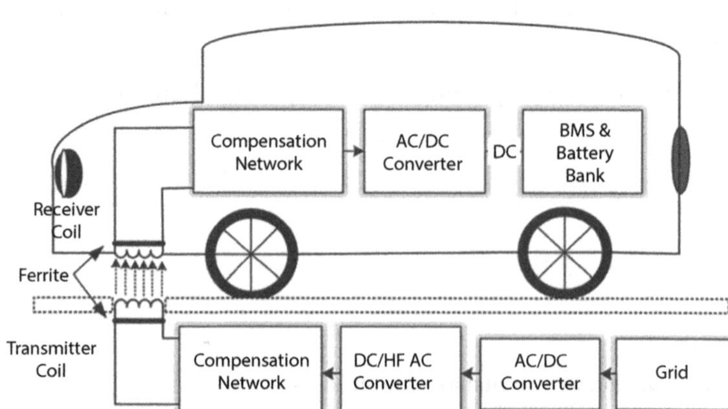

Figure 18.1 Wireless electric vehicle charging system (WEVCS).

receiver side through the compensation network again, the AC power is converted into DC power; then, it is connected to the battery through the battery management system and then that battery is connected to the DC motor that is connected to the wheel of the electric vehicle.

The main consideration in an electric vehicle is the high power and range of the vehicle compared to the IC engine vehicle. The range of the vehicle is decided by the capacity of the battery pack placed in that vehicle. Fifty percent of the cost of the vehicle is occupied by the major part of the electric vehicle is battery technology that is the heart of the electric vehicle. Even though the government is offering many subsidies for promoting the electric vehicle, the major concern that affects the utilization of the electric vehicle is the availability of the charging station.

The energy density of the gasoline engine is about 12,000 watt-hour/kg and the range would be 300 miles. To reach this stage, the electric vehicle has to move many steps forward [4]. The quasi-dynamic charging system provides charging of the electric vehicle during traffic signals and their scheduled stoppings. This dynamic charging is achieved by using the supercapacitor for fast charging and discharging instead of battery for energy storage [5–10].

18.2 Types of Electric Vehicle Wireless Charging Systems (EVWCS)

The wireless charging technology is classified based on the component used for transmitting and receiving [11] the power as mentioned in Figure 18.2.

18.2.1 Capacitive Wireless Charging System (CWCS)

In this technology, the power is transferred from the transmitter to the receiver through the bias current that causes the electrical field to vary. However, the magnets or coils as receivers and transmitters are used in this communication, and capacitors are used for the wireless transfer of power. The AC voltage is first applied, the coefficient of the correction circuit to enhance the efficiency of the energy, to keep up the voltage level, and reduce power transmission losses. It is a tool for generating high frequency, variable voltage, and with a high level of frequency of the variable is that the current through the transfer of the plate, which causes the event of an oscillating force field and current; so as to compensate, they need the plate by electrostatic induction.

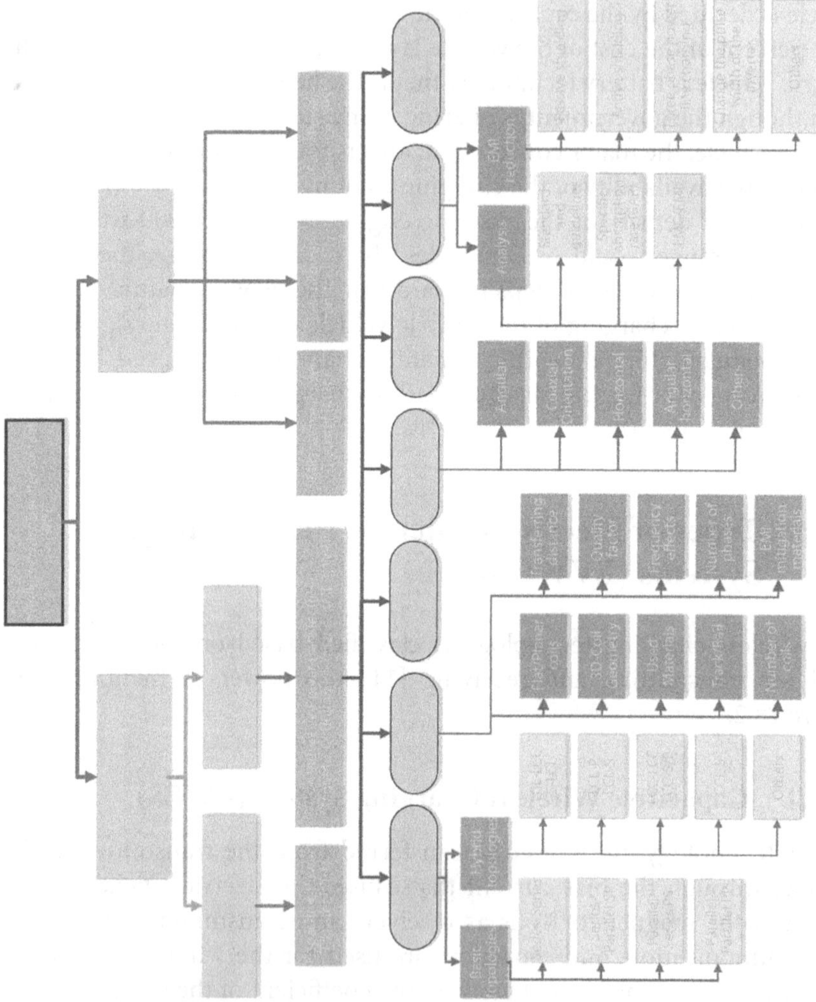
Angular
Coaxial orientation
Horizontal
Angular horizontal
Other
Transferring distance
Quality factor
Frequency effects
Number of phases
EMI mitigation materials
Flat/Planar coils
3D Coil Geometry
Used Materials
Track/Pad
Number of coils
Analysis
EMI reduction
Basic topologies
Hybrid topologies

18.2.2 Permanent Magnet Gear Wireless Charging System (PMWC)

In this method, the transmitter and receiver both are having armature winding and synchronized perpetual magnets. When the power is entering the transmitter side, it gets energized and the interaction between the magnets and the current-carrying conductor in the transmitter makes it rotate and that induces the current in the receiver coil. Presently, the change in the collector perpetual attractive field makes the alternating current flowing in the receiver side and makes it act as a generator, and this is transferred to the battery through the power electronics interface.

18.2.3 Inductive Wireless Charging System (IWC)

In this method, the power is transport based on faradays law of induction. At the purpose when the primary AC supply is applied to the transmitter coil, it makes an AC attractive field that goes through the recipient curl, and this attractive field moves electrons within the receiver coil causing AC power yield. This AC yield is corrected and sifted to charge the EV's energy stockpiling framework. The measure of power moved relies upon frequency, shared inductance, and distance among transmitter and recipient curl. The working recurrence of IWC is between 19 and 50 kHz.

18.2.4 Resonant Inductive Wireless Charging System (RIWC)

Type of wireless charging has the highest functionality efficiency; so by functioning at resonance, even with more vulnerable attractive fields, we are able to send an identical measure of force as in IWC. The force is moved to significant distances without wires. Max move of control over the air happens when the transmitter and receiver coils are tuned, i.e., the 2 curls' full frequencies should be coordinated. Hence, to urge great full frequencies, extra pay networks within the arrangement and equal blends are added. This extra remuneration network alongside progress in thunderous frequency likewise lessens the additional misfortunes.

18.3 Classification of Compensation Topologies

The compensation topology is classified into two major groups, and it is divided into several groups based on the connection of the inductor and capacitor [12]. Figure 18.3 describes the classifications of the compensation topology. They are divided into basic and hybrid topologies.

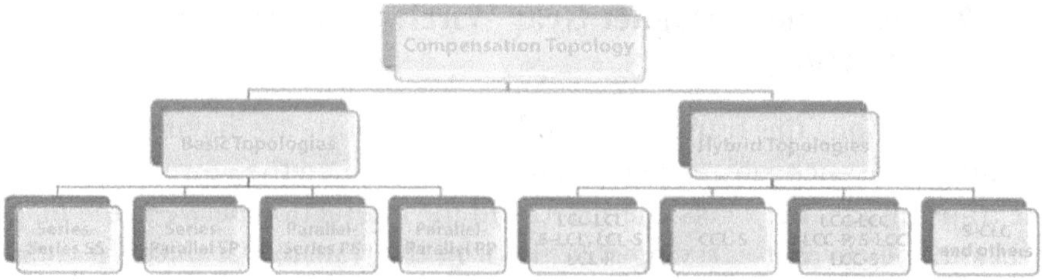

The basic topology is classified as series and parallel based on the connection of inductor and capacitor.

Based on the application, the topology has to be selected. SS topology is nothing but the inductor and capacitor connected in series at the primary and secondary side [13]. PP topology is nothing but the inductor and capacitor connected in series at the primary and secondary sides. SP and PS topologies comprise the series and parallel combination of inductor and capacitor at either the primary and secondary side. The SS combination is suitable for the distributed winding, and the PP combination is suitable for concentrated winding. SS topology is suitable for higher voltages, and PP is suitable for higher currents. In SP topology, the capacitance in the series side affects the input-to-output voltage ratio of the overall system efficiency that affects the ability factor. The SS and the SP topologies find most applications [14, 15]. However, the PS and PP topologies are difficult in design and analysis, and so, it has bulky formulas to find the capacitance of the primary side of the converter. Figures 18.4 and 18.5 depict the equivalent circuit diagram of the basic topologies and hybrid topologies.

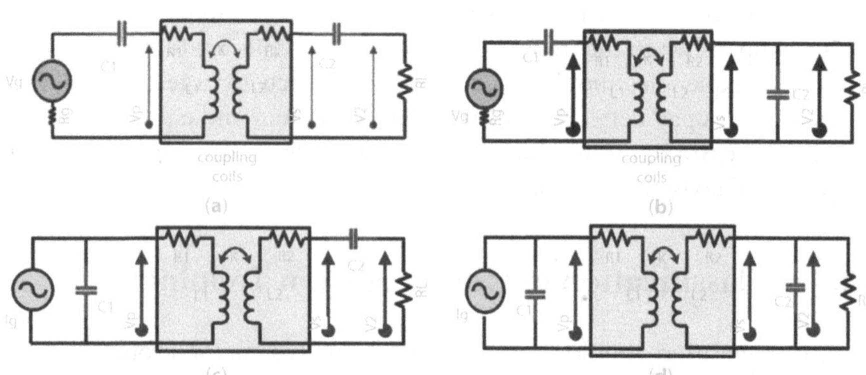

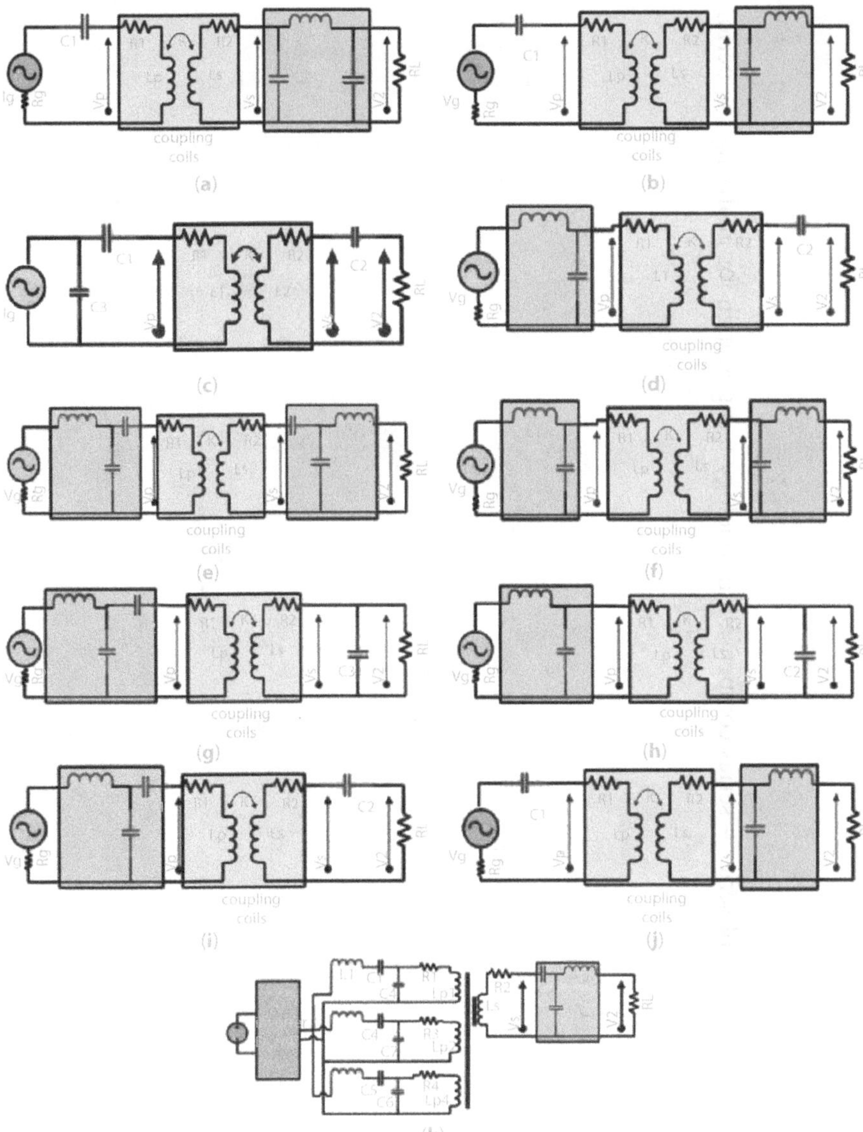
(a)
(b)
(c)
(d)
(e)
(f)
(g)
(h)
(i)
(j)
(k)

18.4 Simulation Diagram

18.4.1 Series-Series

The Figure 18.6 shows the simulation of series – series compensation topology using Matlab for analysing the performance of the circuit.

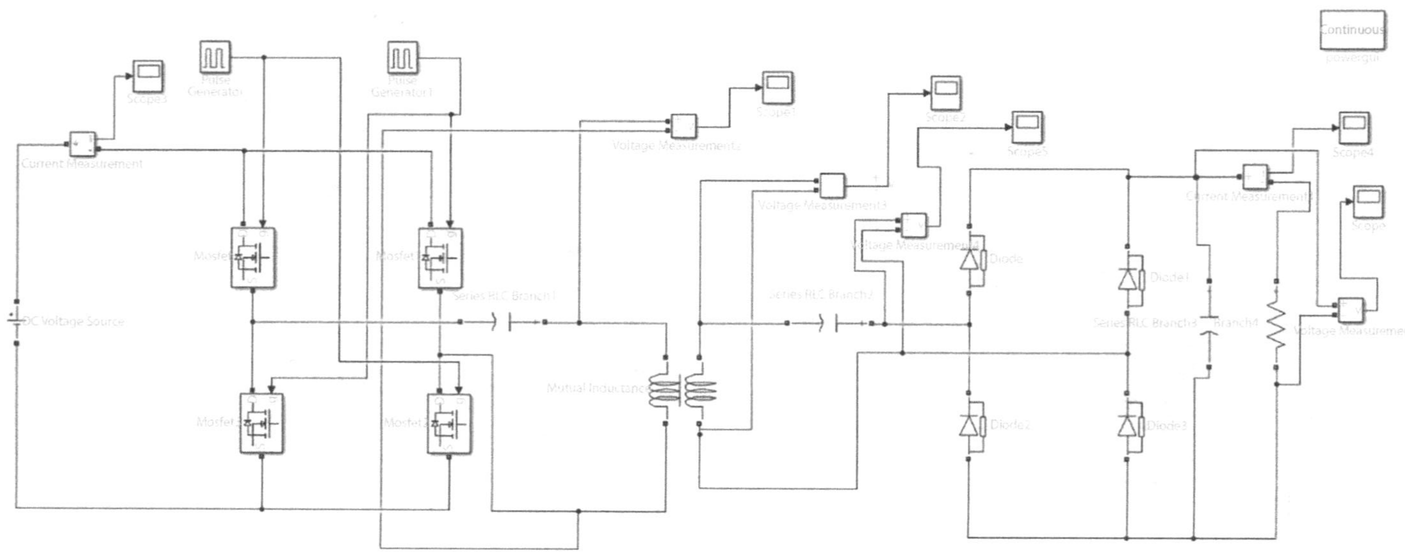

18.4.2 Parallel-Series

The following Figure 18.7 depicts the simulation of the Parallel-series compensation topology using Matlab for AC input and for analyzing the performance.

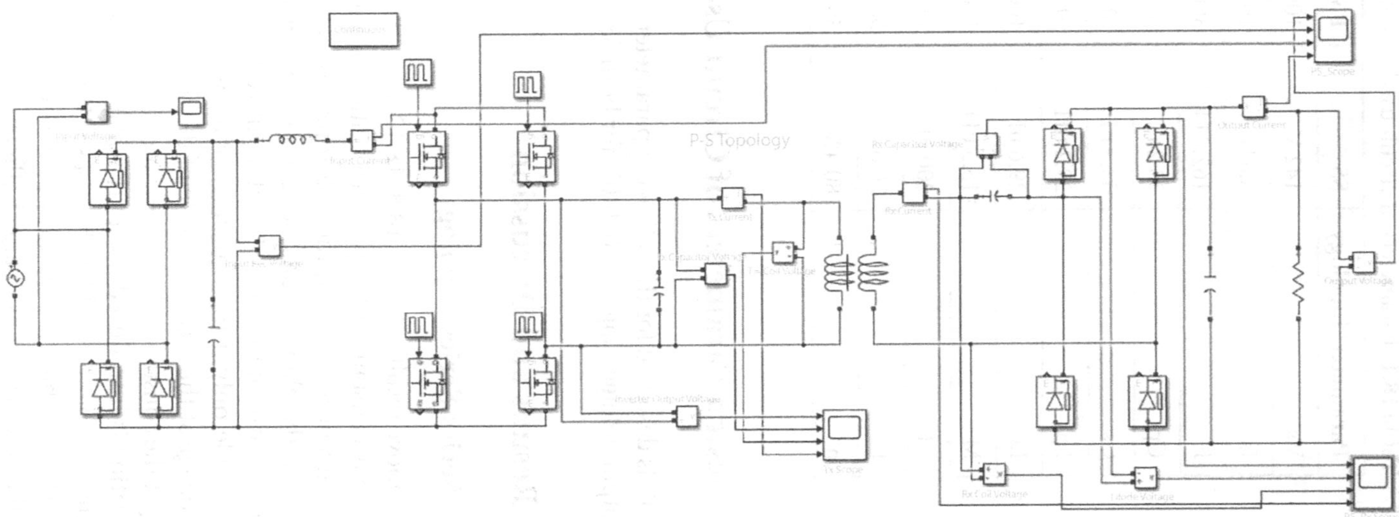

Table 18.1 Parameters of circuit used in simulation.

Parameter/Topology	SS	PS
Vs	142 V	100 V, 60 Hz
Is	8.631 A	10 Ω
F	162.84 kHz	300 mHz
Cp	33.2 nF	[0 Ω 39 μH]
Cs	33.3 nF	[0 Ω 28 μH]
Lp	29.2 uH	[0 Ω 7.6 μH]
Ls	29.6 uH	100 nF
M	15.3 uH	150 nF
Cf	200 μF	11.13 μF
R	20.6	56 μH
η	80.4%	55.6%

18.5 Design Parameters of Circuit Used in Simulation

The circuit is designed for the following parameters as shown in Table 18.1 to get output voltage range as 50-140V for both SS and PS topologies.

18.6 Results and Discussion

18.6.1 Series-Series Topology

From the above results, Figures 18.8–18.12 it can be observed that the output voltage is constant. If the primary voltage is maintained constant and the load current is constant, then SS topology acts as a constant voltage source. From the above results, we can observe that the output voltage is maintained constant for the constant input voltage and fixed load current. This is suitable only if the transformer turns ratio is kept constant. The output voltage at the secondary side of the transformer is not pure square wave to get the constant voltage at the output side of the rectifier. When the compensation is introduced, the pure square wave can be obtained.

Lithium-ion batteries, like those used in electric vehicles, have two charging stages, which are typically provided using CC-CV topology. This ensures that the battery is charged efficiently and is suitable for different

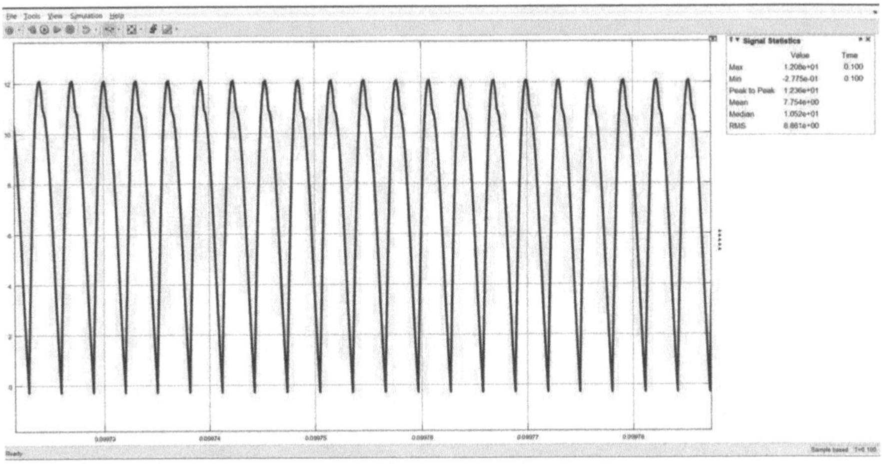

Figure 18.8 Input current waveform.

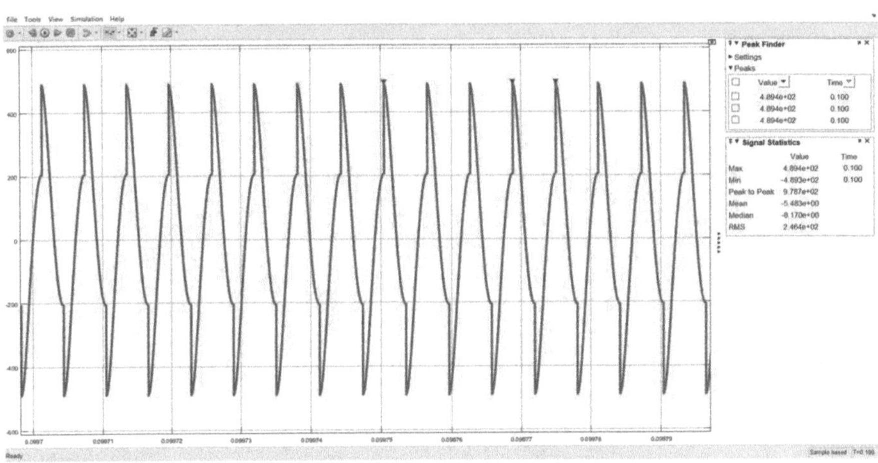

Figure 18.9 Transmission coil voltage after compensation.

conditions. The SS topology is a better alternative to the SP topology, as it provides better compensation scheme. However, for the constant voltage source, the system frequency is not equal to resonant frequency, which depends on the mutual coupling of the primary and secondary winding.

The model's simulation results are discussed in this section. The parameters of the circuit are tuned to provide the ideal output and input voltage and resistance. The results of the S-parameters analysis and transient

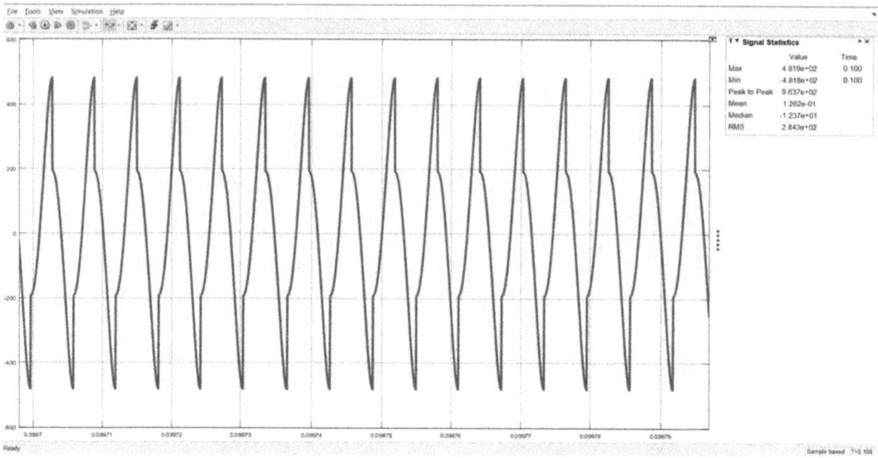

Figure 18.10 Receiver coil voltage before compensation.

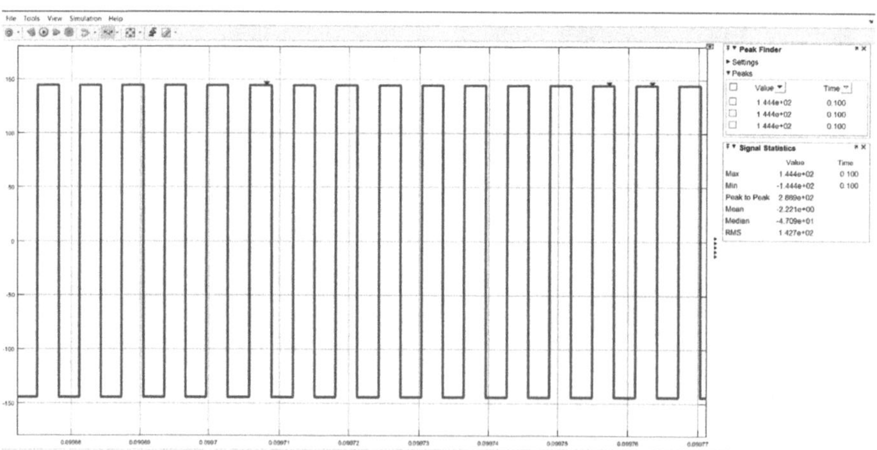

Figure 18.11 Receiver coil voltage after compensation.

analysis of the SP combined topology model show that its power transfer efficiency is higher than that of SS topology model.

18.6.2 Parallel-Series Topology Waveforms

From the above results, Figures 18.13–18.16 the DC-DC efficiency of the system with an SS compensation topology is greater than that of the PS

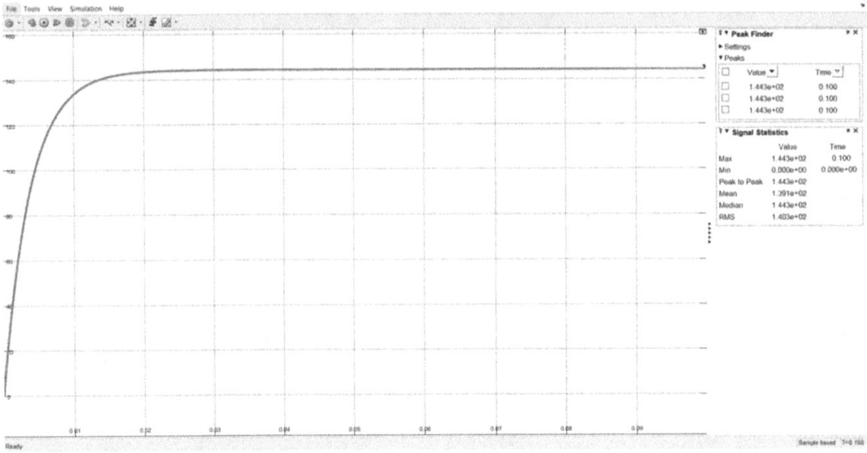

Figure 18.12 Output voltage waveform.

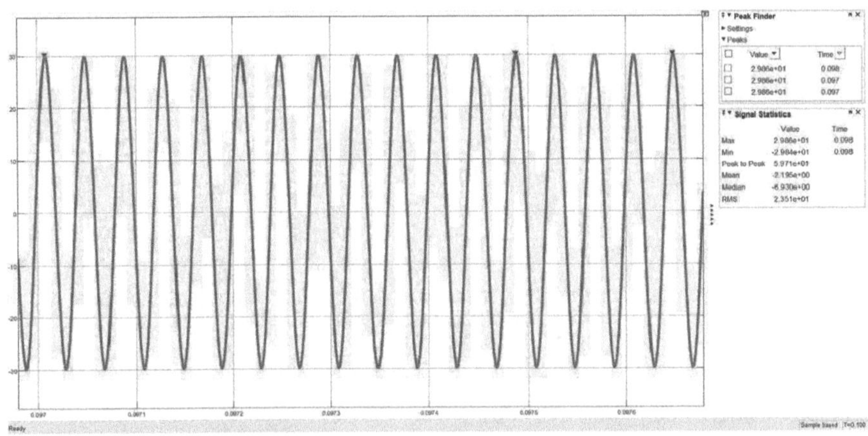

Figure 18.13 Receiver coil voltage before compensation.

configuration due to the lower loss of the converter. The system's output power is also proportional to the battery voltage and the efficiency of the converter. The voltage on C2 without the L1 is decreased by 40%, which makes it possible to charge the battery. In addition, without the L1, the highest efficiency of the receiver is still 94%. Since the output current is continuous in both SS and SP topology, the control of battery charging is so easy with that. There is not much difference in efficiency with L1 filter inductor; it can be removed to reduce the size of the converter.

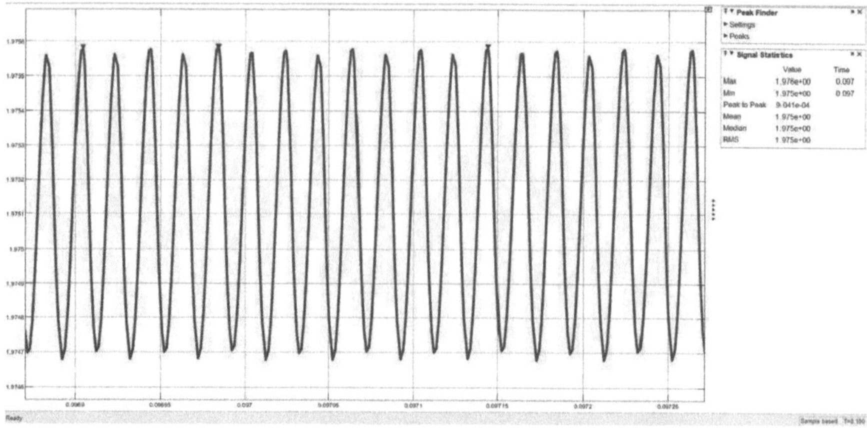

Figure 18.14 Transmission coil voltage before compensation.

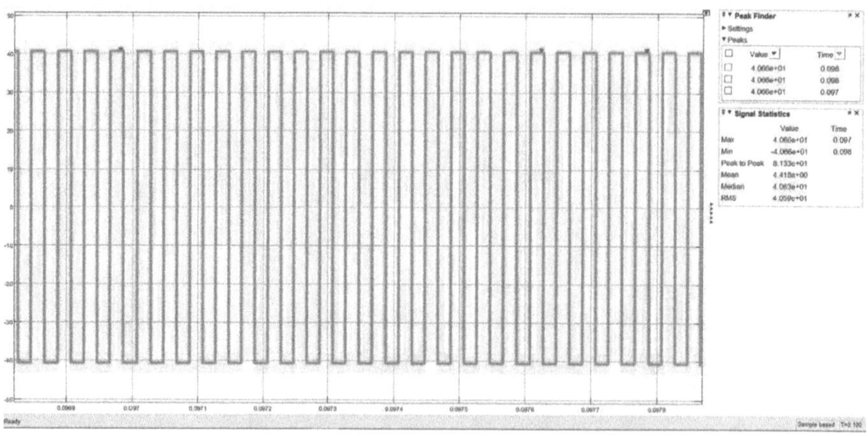

Figure 18.15 Transmission coil voltage after compensation.

18.7 Conclusion

The performance analysis of SS and PS topologies is done in MATLAB to compare their performance with each other. From that, the conclusion is that the SS is the best suitable topology for wireless charging topology compared to PS topology. For charging EV, the PS is suitable for high-input and low-output power, but SS can be applicable for all kind of applications with dynamic load. Other hybrid topologies that include more inductances and capacitances increase the stray resistance, and so it affects the efficiency

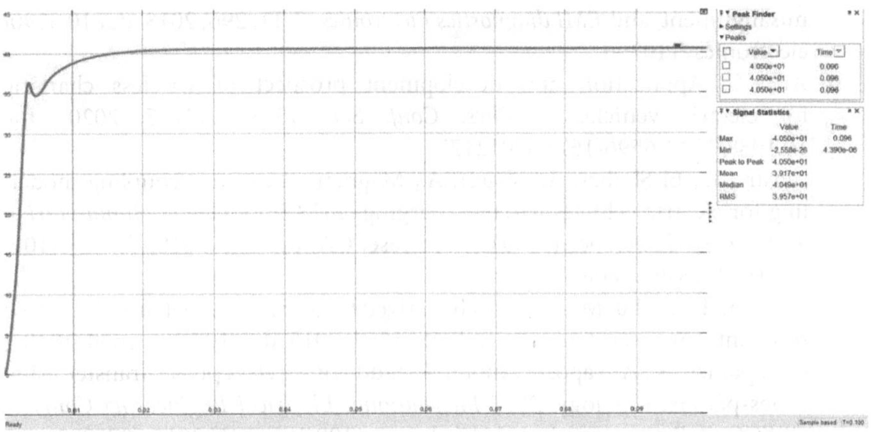

Figure 18.16 Output voltage waveform.

by increasing the copper loss than SS topology especially for high-power applications. The other factor that supports the SS topology is the complexity in control, which is simple in that compared to other hybrid topologies. Hence, the final conclusion comes as the SS topology is best among other topologies based on control and cost.

References

1. Shevchenko, V., Husev, O., Strzelecki, R., Pakhaliuk, B., Poliakov, N., Strzelecka, N., Compensation topologies in IPT systems: Standards, requirements, classification, analysis, comparison and application. *IEEE Access*, 7, 120559–120580, 2019. doi: 10.1109/ACCESS.2019.2937891.
2. Kushwaha, B.K., Rituraj, G., Kumar, P., Mathematical model of series-series compensation for contactless power transfer system. *IECON 2015 - 41st Annual Conference of the IEEE Industrial Electronics Society*, Yokohama, pp. 001321–001326, 2015. doi: 10.1109/IECON.2015.7392283.
3. Hua, J., Wang, H.-Z., Zhao, Y., Zou, A.-L., LCL resonant compensation of movable ICPT systems with a multi-load. *J. Power Electron.*, 15, 1654–1663, 2015. doi: 10.6113/JPE.2015.15.6.1654.
4. Vatsala, Ahmad, A., Alam, M.S., Chaban, R.C., Efficiency enhancement of wireless charging for electric vehicles through reduction of coil misalignment. *2017 IEEE Transportation Electrification Conference and Expo (ITEC)*, Chicago, IL, pp. 21–26, 2017. doi: 10.1109/ITEC.2017.7993241.
5. Houran, M.A., Yang, X., Chen, W., Magnetically coupled resonance WPT: Review of compensation topologies, resonator structures with

misalignment, and *EMI diagnostics Electronics*, 7, 11, 296, 2018. doi:10.3390/electronics7110296.
6. An, Z., Application and development prospect of wireless charging for electric vehicles. *J. Phys. Conf. Ser.*, 1578, 012175, 2020. doi: 10.1088/1742-6596/1578/1/012175.
7. Ayisire, E., El-Shahat, A., Sharaf, A., Magnetic resonance coupling modelling for electric vehicles wireless charging. *2018 IEEE Global Humanitarian Technology Conference (GHTC)*, San Jose, CA, pp. 1–2, 2018. doi: 10.1109/GHTC.2018.8601806.
8. Li, H.-Y., Li, X., Lu, M., Hu, S., A linearized large signal model of an LCL-type resonant converter. *Energies.*, 8, 1848–1864, 2015. doi: 10.3390/en8031848.
9. Cheapanich, S., Anyapo, C., Intani, P., Study of wireless power transfer using series-parallel topology. *2017 International Electrical Engineering Congress (iEECON)*, Pattaya, pp. 1–4, 2017. doi: 10.1109/IEECON.2017.8075802.
10. Ravikiran, V. and Keshri, R.K., Comparative evaluation of S-S and P-S topologies for wireless charging of electrical vehicles. *IECON 2017 - 43rd Annual Conference of the IEEE Industrial Electronics Society*, Beijing, pp. 5324–5329, 2017. doi: 10.1109/IECON.2017.8216922.
11. Bukya, R., Mangu, B., Jayaprakash, A., Ramesh, J., A study on current-fed topology for wireless resonant inductive power transfer battery charging system of electric vehicle. *2020 International Conference on Power Electronics & IoT Applications in Renewable Energy and its Control (PARC)*, Mathura, Uttar Pradesh, India, pp. 415–421, 2020. doi: 10.1109/PARC49193.2020.236644.
12. Cheapanich, S., Anyapo, C., Intani, P., Study of wireless power transfer using series- parallel topology. *2017 International Electrical Engineering Congress (iEECON)*, Pattaya, pp. 1–4, 2017. doi: 10.1109/IEECON.2017.8075802.
13. Ishihara, M., Umetani, K., Hiraki, E., Strategy of topology selection based on quasi-duality between series–series and series–parallel topologies of resonant inductive coupling wireless power transfer systems. *IEEE Trans. Power Electron.*, 35, 7, 6785–6798, July 2020. doi: 10.1109/TPEL.2019.2956732.
14. Fang, C., Song, J., Lin, L., Wang, Y., Practical considerations of series-series and series-parallel compensation topologies in wireless power transfer system application. *2017 IEEE PELS Workshop on Emerging Technologies: Wireless Power Transfer (WoW)*, Chongqing, pp. 255–259, 2017. doi:10.1109/WoW.2017.7959404.
15. Shevchenko, V., Husev, O., Strzelecki, R., Pakhaliuk, B., Poliakov, N., Strzelecka, N., Compensation topologies in IPT systems: Standards, requirements, classification, analysis, comparison and application. *IEEE Access*, 7, 120559–120580, 2019. doi: 10.1109/ACCESS.2019.2937891.

19

Analysis of PV System in Grid Connected and Islanded Modes of Operation

Aditya Ghatak*, Tushar Pandit, Chitra A. and Razia Sultana W.

School of Electrical Engineering, Vellore Institute of Technology (VIT), Vellore, India

Abstract

Microgrids are crucial to the integration of renewable energy systems, and one such system is the photovoltaic (PV)–based energy system. Microgrids can be operated in grid connected mode, where they share power with the grid. However, if there is a certain disturbance in the grid, then the local distributed generation system is protected by operating the microgrid in the islanded mode, i.e., by disconnecting it from the grid. This chapter proposes a simulation model of a PV-based microgrid system operated in grid connected and islanded modes. A Maximum Power Point Tracking algorithm is used for the control of the DC side converter, whereas the AC side control system maintains the DC bus voltage for the grid connected mode, and the system performance is analyzed. For the islanded mode, critical and non-critical loads are varied, and the results observed. Total Harmonic Distortion is also calculated as a means of power quality analysis.

***Keywords*:** Photovoltaic system, microgrid, MPPT, grid-connected mode, islanded mode, total harmonic distortion

19.1 Introduction

The term microgrid can be defined as a system that is self-sufficient in providing energy to a small geographical area [1]. It can have its own distributed energy sources to produce power. Microgrids are independent, i.e., they can be separated from the central grid in case of any disturbances

**Corresponding author*: adityaghatak2000@gmail.com

A. Chitra, V. Indragandhi, and W. Razia Sultana (eds.) Smart Grids and Green Energy Systems, (301–316)

and operate independently to provide energy to its particular area [2]. Microgrids provide various advantages such as improving the power quality, reducing costs and losses incurred in the transmission lines and providing a solution to reduce wasteful emissions [3].

Photovoltaic (PV) systems provide a robust and eco-friendly nature of electric power generating systems. Its integration with microgrids helps in the creation of an independent energy system [4, 5]. In grid-connected mode of operation, the microgrid administers power to the central grid providing constant voltage at a suitable frequency [6]. In the islanded mode of operation, the microgrid is separated from the central grid and provides power to its load from its local sources [7]. These operations can become complicated when the distributed energy sources and the main grid system increase in number, making the interaction between them very complex [8]. Microgrids can operate at optimal condition and provide rational distribution of energy when the variables such as active power output, bus voltage of inverter, and current source are regulated correctly [9]. Authors in [10] talked about various modelings and control strategies of microgrids, concluding Networked Control System to be the most superior control method.

Authors in [11] discussed about a smart distributed control system for microgrids in order to achieve higher efficiency in control of power system. Authors in [12] incorporated use of intelligent control in microgrids in its two modes with the use of multiagent system technique. Authors in [13] proposed a more centralized approach in the energy management of the microgrids with more focus in the islanded mode. Authors in [14] discussed use of fuzzy controller for the regulation of frequency variations in the power system and optimum utilization of wind farms incorporated in microgrids. Authors in [15] give a comparative study of the energy storage systems included in microgrids along with analysis of various control systems that are currently in use.

19.2 Grid Connected Mode

The block diagram of the grid connected PV system is shown in Figure 19.1. It consists of a PV array and a DC-DC converter in the DC side, with the Maximum Power Point Tracking (MPPT) algorithm being used for the DC-DC converter control. The DC and AC sides are linked by a DC bus, whose voltage is maintained constant. An inverter converts the DC to AC and supplies it to the grid. Table 19.1 gives the specifications for the model simulated on MATLAB. The control strategies for both the DC side and the AC side have been discussed in detail.

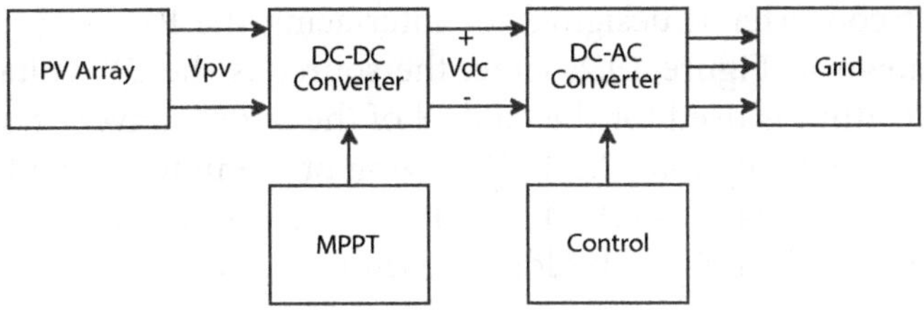

Figure 19.1 Block diagram of PV system in grid connected mode.

Table 19.1 Model specifications for PV in grid connected mode.

Parameter	Specifications
Load Rated Power	50 kW
Grid Voltage	415 V
Grid Frequency	50 Hz
Switching Frequency	5 kHz
DC Bus voltage	800 V
Maximum Power of PV Array	213.5 W

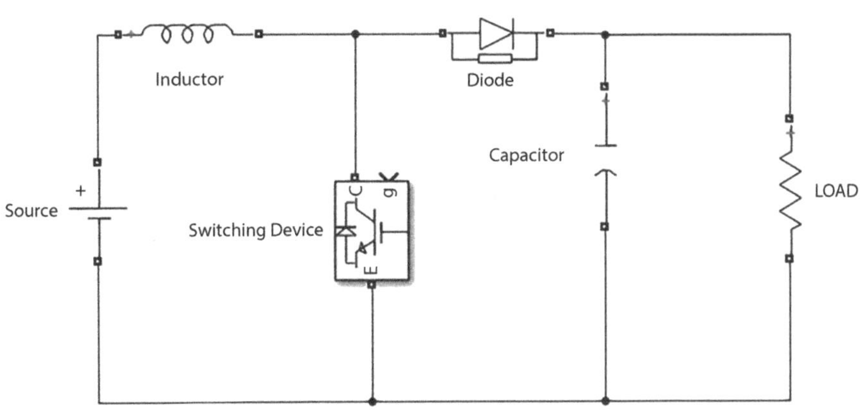

Figure 19.2 Circuit diagram of boost converter.

A boost converter is designed for interfacing the PV array at the DC side, as shown in Figure 19.2. Here, the source is the PV voltage, and a MPPT algorithm is used for the control of the boost converter, which has been discussed in Section 19.2.1. The sizing of the inductor and capacitor are done as per Equations (19.1) and (19.2), and the values are highlighted in Table 19.2 [16]. A signal builder is used to replicate the irradiation for

Table 19.2 Specifications for design of Boost converter.

Parameter	Specifications
Input voltage from PV array (maximum power point voltage)	406 V
Duty ratio	0.4925
Boundary inductance calculated	0.1623 mH
Boost inductance	1.623 mH
Minimum capacitance calculated	0.769 mF
Boost capacitance	1 mF

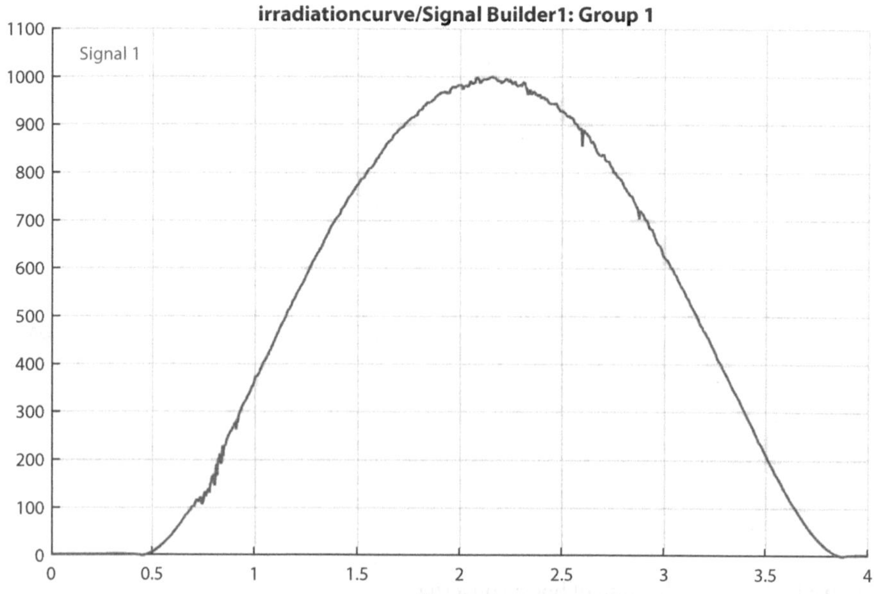

Figure 19.3 Irradiation curve for PV input.

the PV array. The curve shown in Figure 19.3 is generated in MATLAB in order to be used as an input signal to the PV array.

$$L_f = \frac{0.1V^2}{2\pi fP_a} \tag{19.1}$$

$$C_f = \frac{0.05\,P_a}{2\pi fV^2} \tag{19.2}$$

where L_f = minimum inductance, C_f = minimum capacitance, V = grid voltage, f = frequency, and Pa = power per phase.

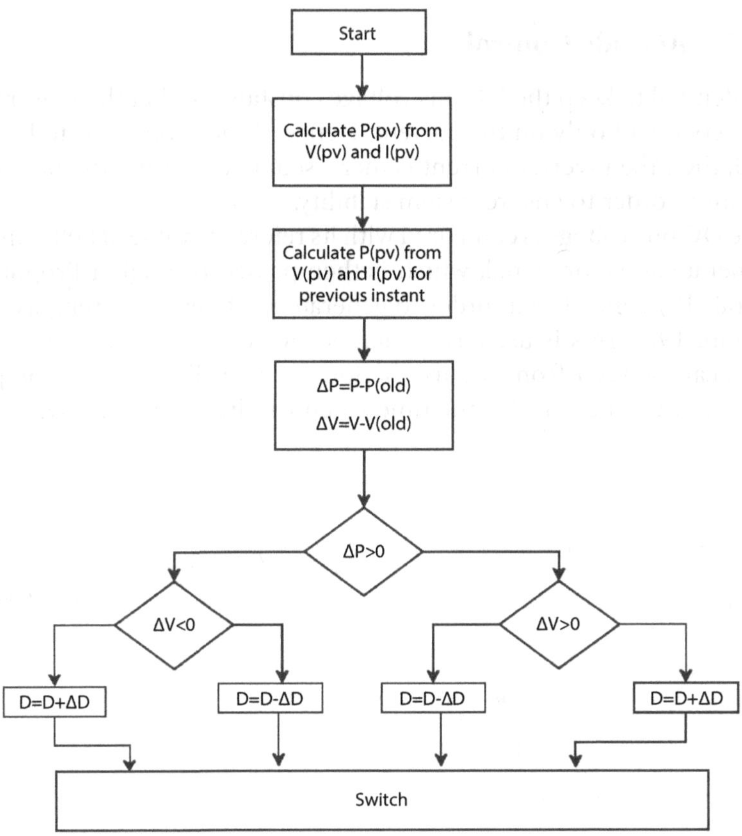

Figure 19.4 Flowchart of MPPT algorithm.

19.2.1 DC Side Control

The MPPT algorithm is used for the control of the boost converter. The MPPT is used in order to ensure that maximum power from the PV array can be supplied to the grid. The algorithm locates the maximum power point by finding the points where the derivative of power with respect to voltage (dP/dV) is zero. A small change in the duty ratio is applied to the system, because of which the power changes, and the change in power is observed. If it is seen that the power increases because of this disturbance, then the duty ratio is increased in that direction only. When the power peak is reached, it starts to decrease again, and the duty ratio is then changed in the opposite direction. A flow chart explaining the working of the MPPT algorithm is shown in Figure 19.4. The MPPT algorithm is implemented using the function generator block in MATLAB, with an initial disturbance of $125*10^6$ in order to vary the power and get the maximum power point.

19.2.2 AC Side Control

It is intended to keep the DC bus voltage constant, so that the power transfer is dependent only on the current. Hence, if more power is to be transferred, then the inverter current is increased, maintaining the bus voltage constant in order to ensure system stability.

The DC bus voltage is compared with its reference value of 800 V in order to generate an error signal, which is then passed through a Proportional Integral (PI) controller in order to generate a reference current, as shown in Figure 19.5. This is used as a reference to regulate active power transfer. As can be seen from the irradiation curve in Figure 19.3, the power transferred will be zero for the time when irradiation is zero. During this

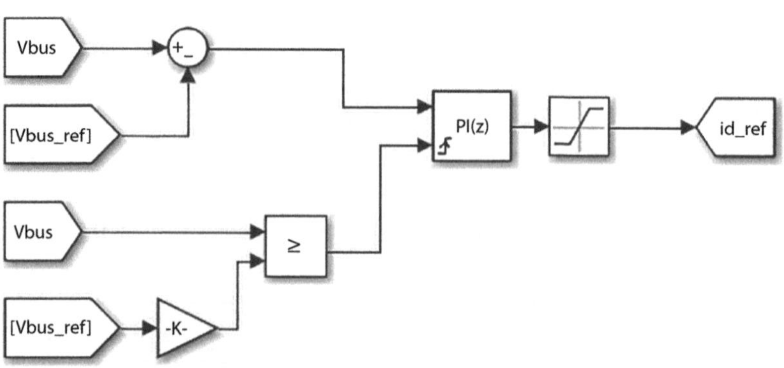

Figure 19.5 Control loop to generate d-axis reference current.

time, the PI controller will continue to function. As the power and, consequently, the DC bus voltage starts increasing gradually, it may be so that the PI controller is also saturated, and the DC bus voltage undergoes an unregulated increase. Due to this reason, an additional condition is added, so that the PI controller resets to its original operating conditions whenever the DC bus voltage exceeds 840 V.

Three phase-two phase current transformations are used in order to obtain independent control of active and reactive power flow. The wt required for this transformation is obtained by the use of a simple Phase-Locked Loop (PLL).

Two PI control loops that are used to generate modulation signals for the d- and q-axis currents are shown in Figures 19.6 and 19.7, respectively. The d-axis current, which is obtained from the control loop shown in Figure 19.5, regulates active power transfer, and the q-axis current regulates

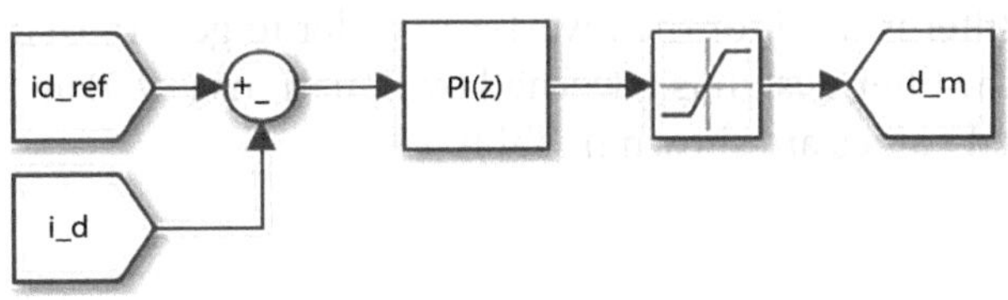

Figure 19.6 Circuit to generate d-axis modulation signal.

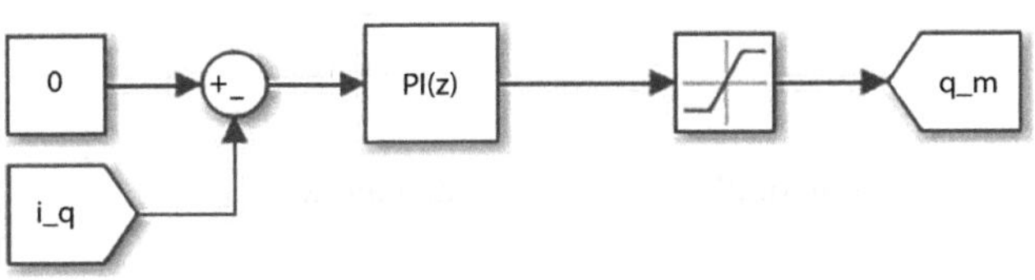

Figure 19.7 Circuit to generate q-axis modulation signal.

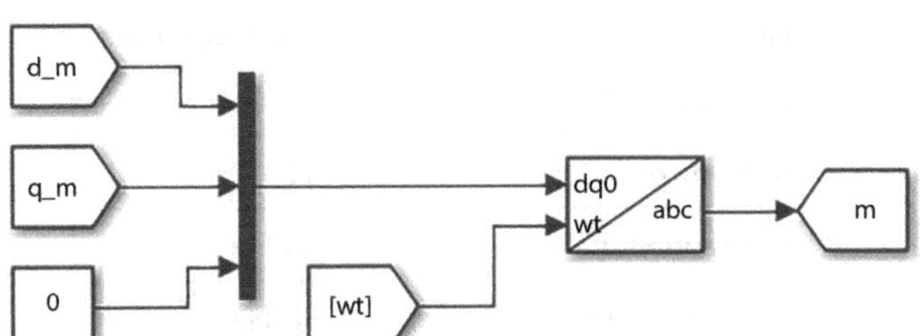

Figure 19.8 Circuit to generate final modulation signal for PWM generator.

reactive power transfer. Since we aim to transfer only active power to the grid, the q-axis reference current is taken to be 0.

The d- and q-axis current modulation signals are passed through a multiplexer and converted back to three phases in order to generate a final modulation signal for the two-level PWM generator, as shown in Figure 19.8. The final pulses generated are used to control the inverter.

19.3 Islanded Mode

Figure 19.9 shows the block diagram for a PV system working in an islanded system. Two types of loads are represented, namely, critical loads and non-critical loads. Critical loads are those which require uninterrupted power supply, whereas non-critical loads are interfaced with a circuit breaker and can be disconnected from the system as and when required. Table 19.3 gives the specifications for the model simulated on MATLAB.

An RLC filter is used for the inverter in order to generate smooth waveforms, and the filter design is governed by Equations (19.3) and (19.4), and the calculated values are shown in Table 19.4.

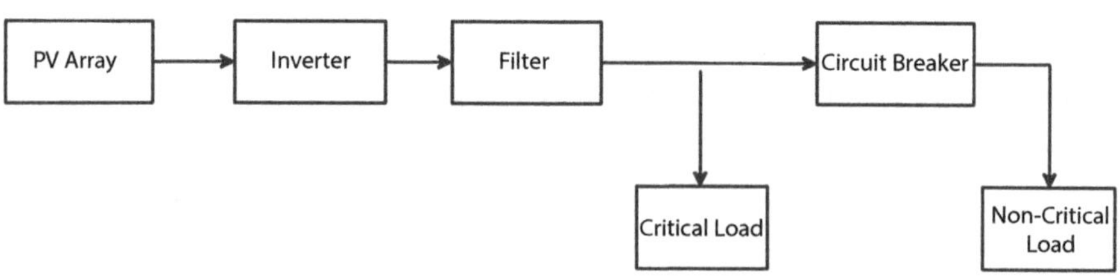

Figure 19.9 Block diagram of PV system in islanded mode.

Table 19.3 Model specifications for PV system in islanded mode.

Parameter	Specifications
Load Rated Power	10 kW
Grid Voltage	415 V
Grid Frequency	50 Hz
Switching Frequency	5k Hz
Maximum Power of PV Array	213.5 W

Table 19.4 Calculated values for filter design.

Parameter	Specification
Filter capacitance	9.2411 μF
Filter inductance	5.5 mH
Filter resistance	0.5482 ohm

$$L_f = \frac{0.1\,V^2}{2\pi fP} \tag{19.3}$$

$$C_{fmax} = \frac{0.05\,P}{2\pi fV^2} \tag{19.4}$$

where C_{fmax}= maximum capacitance, L_f= inductance, P = rated power of load frequency, and V = inverter phase to phase voltage.

The inverter voltage has to be controlled for the islanded mode of operation. The value of wt is once again obtained by using a PLL. The measured inverter voltage undergoes three-phase and two-phase transformations (abc-dq0) in order to give the peak voltage value. This peak voltage is then compared with a reference value $(380*\sqrt{2})$ in order to generate an error signal, as shown in Figure 19.10, in order to obtain a modulation index.

The modulation index is then multiplied with each phase of the inverter voltage in order to generate a modulation signal, as shown in Figure 19.11, which is then fed to a PWM generator in order to supply the gating pulses to the inverter.

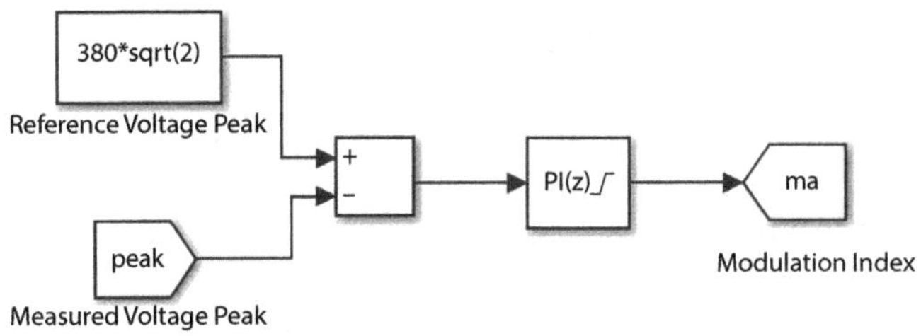

Figure 19.10 Control loop to obtain modulation index.

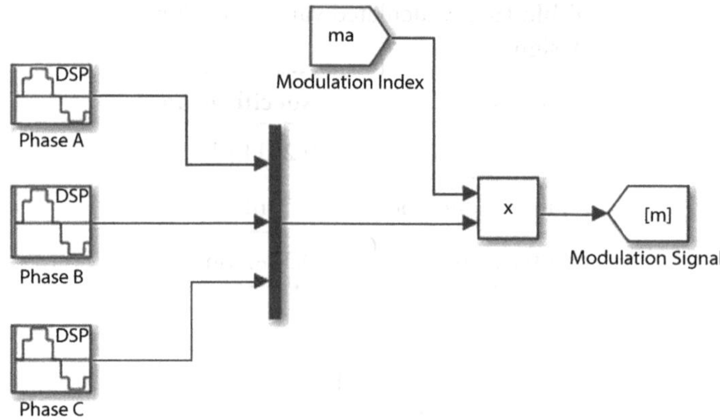

Figure 19.11 Circuit to obtain final modulation signal.

19.4 Results and Discussion

The grid currents are shown in Figure 19.12(a). It can be verified from the irradiation curve provided as input to the PV that the current flow in the inverter will only begin after a certain time interval. The DC bus voltage can be seen in Figure 19.12(c), and it is observed that it stabilizes to the reference value once power transfer starts, which ensures that our system is stable. Figure 19.12(b) shows a comparison of the power curves obtained by calculating the output power to the grid and that obtained from the MPPT algorithm. The coinciding nature of the graphs verifies that the MPPT algorithm is working well. We are only transferring active power to the grid, as can be seen in Figure 19.12(d), which shows the active and reactive power curves. The reactive power maintains a constant value of 0, as expected. Figure 19.12(e) gives the PV current, following a similar characteristic. The Total Harmonic Distortion (THD) is also represented in Figure 19.12(f), and it is seen that the THD is maximum initially when there is no power transfer, and it starts to fall as power transfer begins from the PV array to the grid, reaching its minimum value at the rated power, i.e., when there is maximum power transfer.

Further analysis of the grid currents and the DC bus voltage can be performed by observing Figures 19.13 a and b, respectively. Initially, the capacitor voltage is 0, and hence, it starts charging by drawing power from the grid. This is why there is a peak in the grid currents at the very beginning, as the capacitor draws large current from the grid during that time to charge itself. Once the power transfer from the PV array starts, the PI controller ensures that a constant DC bus voltage is maintained. This further

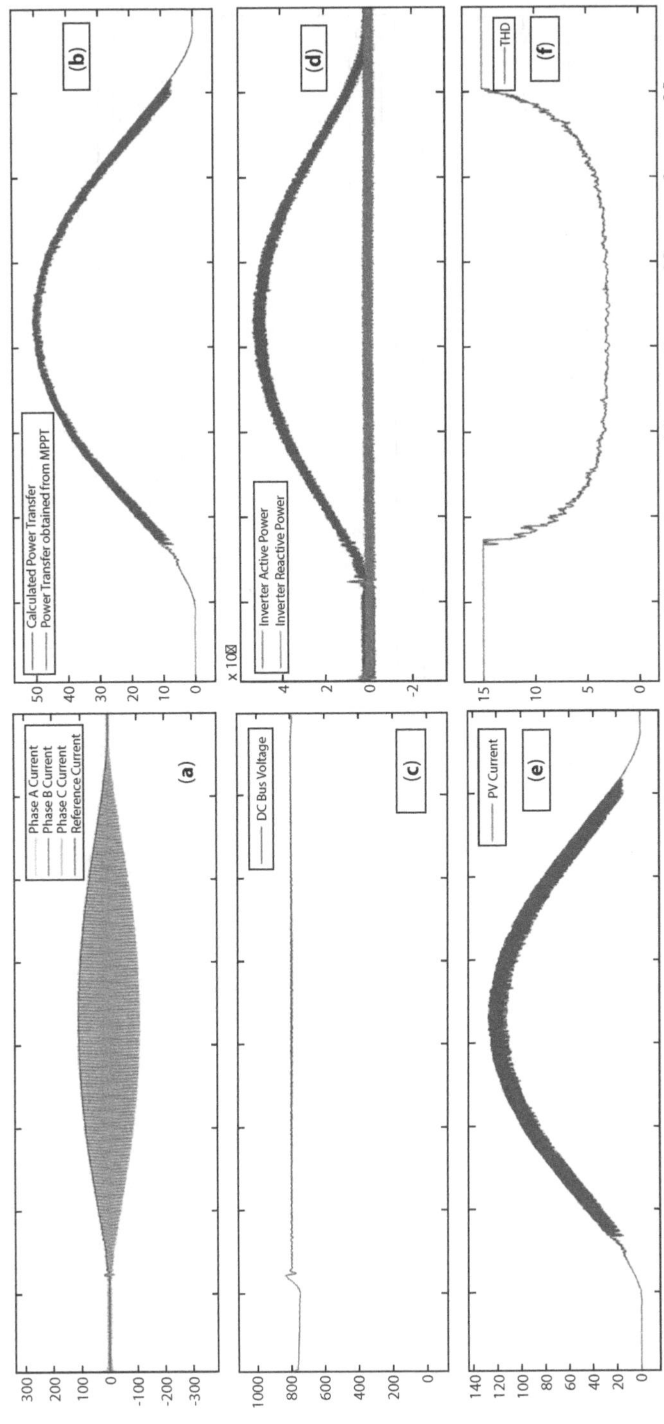

Figure 19.12 Waveforms for PV system in grid connected mode. (a) Grid currents, (b) Comparison between theoretical power and power delivered by MPPT, (c) DC bus voltage, (d) Inverter active and reactive power, (e) PV current, (f) Total harmonic distortion.

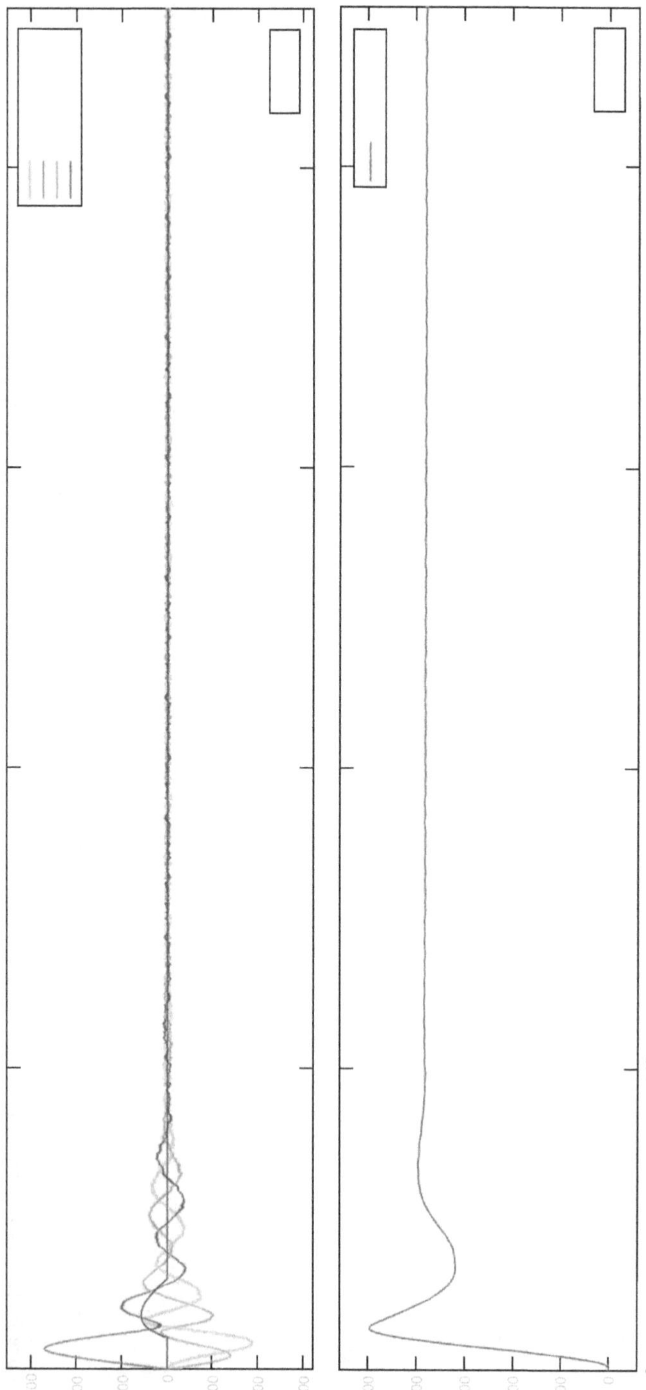

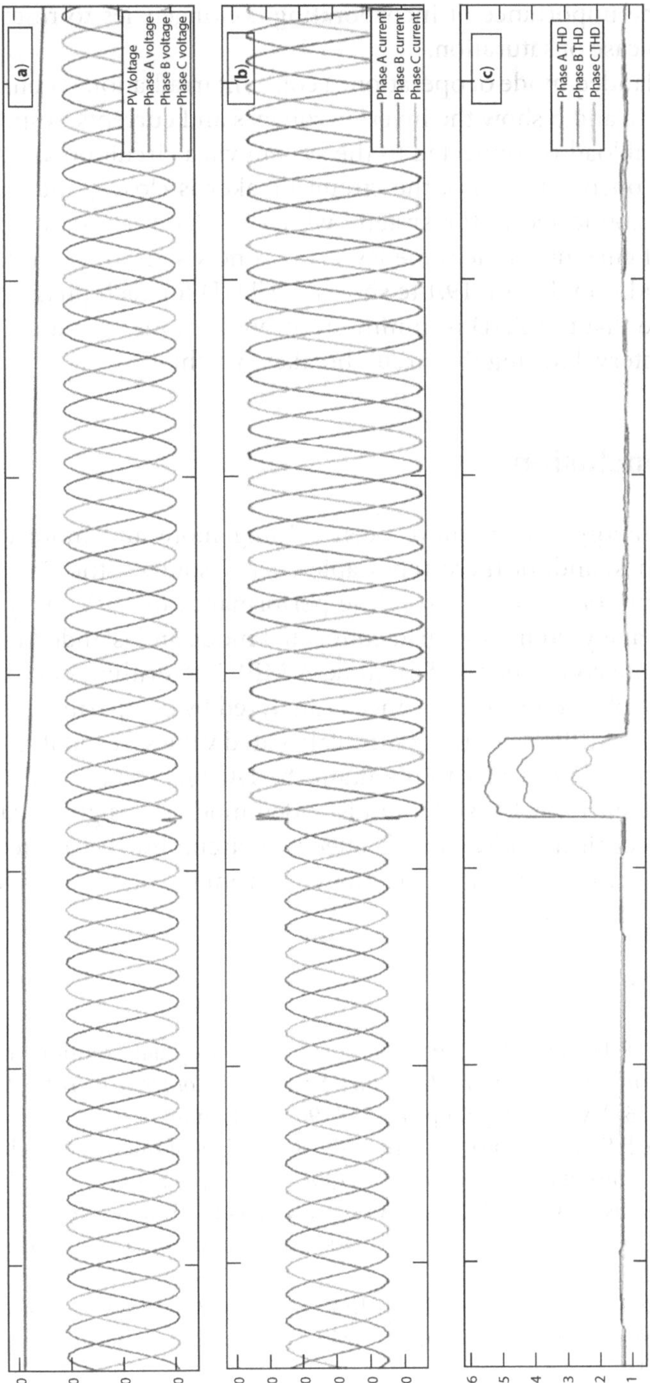

Figure 19.14 Waveforms for PV system in islanded mode. (a) Inverter voltage, (b) Inverter current, (c) THD of each phase current.

highlights the importance of incorporating contingencies to reset the PI controller in case of saturation.

For the islanded mode of operation, a constant irradiation is considered. Figures 19.14a and b show the inverter voltages and currents, respectively. A non-critical load is connected to the system via a circuit breaker, which is originally open. At t = 1 s, the circuit breaker is closed, and the non-critical load connected to the system, which results in an increase in the amplitude of current, as the same PV array is now satisfying two loads in islanded condition. Figure 19.14c shows the THD for each phase current. It is observed that the THD is minimum throughout the operation, except for a short interval during the circuit breaker switching.

19.5 Conclusion

Renewable energy systems have been gaining more and more traction in recent times, and, of these renewable energy sources, the PV array is the most commonly used source. The performance of a PV array–based microgrid is analyzed in both grid connected mode and islanded mode by simulating both conditions on Simulink. A MPPT algorithm is used for the DC side control, and its performance is analyzed by comparing the power curve from the MPPT with that of the calculated values. A suitable control system is proposed in order to maintain the DC bus voltage constant via the AC side control. A PV system in islanded mode is designed with critical and non-critical loads, and the effects of circuit breaker switching in order to add a non-critical load are analyzed from the output waveforms.

References

1. Lasseter, R.H. and Paigi, P., Microgrid: A conceptual solution, in: *2004 IEEE 35th Annual Power Electronics Specialists Conference (IEEE Cat. No. 04CH37551)*, vol. 6, IEEE, pp. 4285–4290, 2004, June.
2. Hartono, B.S., Budiyanto, Y., Setiabudy, R., Review of microgrid technology, in: *2013 International Conference on QiR*, IEEE, pp. 127–132, 2013, June.
3. Anastasiadis, A.G., Tsikalakis, A.G., Nikos, D.H., Environmental benefits from dg operation when network losses are taken into account. *Proc. 2009 DISTRES Conf.*, 2009.
4. Mohanty, A., Viswavandya, M., Mishra, D.K., Ray, P.K., Pragyan, S., Modelling & simulation of a PV based micro grid for enhanced stability. *Energy Proc.*, *109*, 94–101, 2017.

5. Khalil, A. and Ateea, K., Modelling and control of photovoltaic-based microgrid. *Int. J. Renew. Energy Res. (IJRER)*, *5*, 3, 826–835, 2015.
6. Jain, M., Gupta, S., Masand, D., Agnihotri, G., Jain, S., Real-time implementation of islanded microgrid for remote areas. *J. Control Sci. Eng.*, *2016*, 1–9, 2016.
7. Jiang, Q., Xue, M., Geng, G., Energy management of microgrid in grid-connected and stand-alone modes. *IEEE T. Power Syst.*, *28*, 3, 3380–3389, 2013.
8. Zhou, X., Guo, T., Ma, Y., An overview on microgrid technology, in: *2015 IEEE International Conference on Mechatronics and Automation (ICMA)*, IEEE, pp. 76–81, 2015, August.
9. Cai, S., Wang, S., Wang, C., Jia, G., Economic perspective of smart grid, in: *Automation of Electric Power Systems*, vol. *20*, pp. 13–87, 2009.
10. Mahmoud, M.S., Hussain, S.A., Abido, M.A., Modeling and control of microgrid: An overview. *J. Franklin Inst.*, *351*, 5, 2822–2859, 2014.
11. Chatzivasiliadis, S.J., Hatziargyriou, N.D., Dimeas, A.L., Development of an agent based intelligent control system for microgrids. *2008 IEEE Power and Energy Society General Meeting-Conversion and Delivery of Electrical Energy in the 21st Century*, IEEE, 2008.
12. Logenthiran, T. *et al.*, Intelligent control system for microgrids using multiagent system. *IEEE J. Emerg. Sel. Top. Power Electron.*, *3*, 4, 1036–1045, 2015.
13. Olivares, D.E., Cañizares, C.A., Kazerani, M., A centralized optimal energy management system for microgrids. *2011 IEEE Power and Energy Society General Meeting*, IEEE, 2011.
14. Ghafouri, A., Milimonfared, J., Gharehpetian, G.B., Fuzzy-adaptive frequency control of power system including microgrids, wind farms, and conventional power plants. *IEEE Syst. J.*, 12, 3, 2772–2781, 2017.
15. Hajiaghasi, S., Salemnia, A., Hamzeh, M., Hybrid energy storage system for microgrids applications: A review. *J. Energy Storage*, 21, 543–570, 2019.
16. Güler, N. and Irmak, E., MPPT based model predictive control of grid connected inverter for PV systems, in: *2019 8th International Conference on Renewable Energy Research and Applications (ICRERA)*, IEEE, pp. 982–986, 2019, November.

Index

Also of Interest

Check out these other related titles from Scrivener Publishing

SMART GRIDS AND MICROGRIDS: Concepts and Applications, Edited by P. Prajof, S. Mohan Krishna, J. L. Febin Daya, Umashankar Subramaniam, and P. V. Brijesh, ISBN: 9781119760559. Written and edited by a team of experts in the field, this is the most comprehensive and up to date study of smart grids and microgrids for engineers, scientists, students, and other professionals. *NOW AVAILABLE!*

MICROGRID TECHNOLOGIES, Edited by C. Sharmeela, P. Sivaraman, P. Sanjeevikumar, and Jens Bo Holm-Nielsen, ISBN 9781119710790. Covering the concepts and fundamentals of microgrid technologies, this volume, written and edited by a global team of experts, also goes into the practical applications that can be utilized across multiple industries, for both the engineer and the student. *NOW AVAILABLE!*

INTEGRATION OF RENEWABLE ENERGY SOURCES WITH SMART GRIDS, Edited by A. Mahaboob Subahani, M. Kathiresh and G. R. Kanagachidambaresan, ISBN: 9781119750420. Provides comprehensive coverage of renewable energy and its integration with smart grid technologies.

Encyclopedia of Renewable Energy, by James G. Speight, ISBN 9781119363675. Written by a highly respected engineer and prolific author in the energy sector, this is the single most comprehensive, thorough, and up to date reference work on renewable energy. *NOW AVAILABLE!*

Green Energy: Solar Energy, Photovoltaics, and Smart Cities, edited by Suman Lata Tripathi and Sanjeevikumar Padmanaban, ISBN 9781119760764. Covering the concepts and fundamentals of green energy, this volume, written and edited by a global team of experts, also goes into the practical applications that can be utilized across multiple industries, for both the engineer and the student. *NOW AVAILABLE!*

Energy Storage, edited by Umakanta Sahoo, ISBN 9781119555513. Written and edited by a team of well-known and respected experts in the field, this new volume on energy storage presents the state-of-the-art developments and challenges in the field of renewable energy systems for sustainability and scalability for engineers, researchers, academicians, industry.

Energy Storage 2nd Edition, by Ralph Zito and Haleh Ardebili, ISBN 9781119083597. A revision of the groundbreaking study of methods for storing energy on a massive scale to be used in wind, solar, and other renewable energy systems. *NOW AVAILABLE!*

Hybrid Renewable Energy Systems, edited by Umakanta Sahoo, ISBN 9781119555575. Edited and written by some of the world's top experts in renewable energy, this is the most comprehensive and in-depth volume on hybrid renewable energy systems available, a must-have for any engineer, scientist, or student. *NOW AVAILABLE!*

Progress in Solar Energy Technology and Applications, edited by Umakanta Sahoo, ISBN 9781119555605. This first volume in the new groundbreaking series, Advances in Renewable Energy, covers the latest concepts, trends, techniques, processes, and materials in solar energy, focusing on the state-of-the-art for the field and written by a group of world-renowned experts. *NOW AVAILABLE!*

A Polygeneration Process Concept for Hybrid Solar and Biomass Power Plants: Simulation, Modeling, and Optimization, by Umakanta Sahoo, ISBN 9781119536093. This is the most comprehensive and in-depth study of the theory and practical applications of a new and groundbreaking method for the energy industry to "go green" with renewable and alternative energy sources. *NOW AVAILABLE!*

DESIGN AND DEVELOPMENT OF EFFICIENT ENERGY SYSTEMS, edited by Suman Lata Tripathi, Dushyant Kumar Singh, Sanjeevikumar Padmanaban, and P. Raja, ISBN 9781119761631. Covering the concepts and fundamentals of efficient energy systems, this volume, written and edited by a global team of experts, also goes into the practical applications that can be utilized across multiple industries, for both the engineer and the student. *NOW AVAILABLE!*

INTELLIGENT RENEWABLE ENERGY SYSTEMS: Integrating Artificial Intelligence Techniques and Optimization Algorithms, edited by Neeraj

Priyadarshi, Akash Kumar Bhoi, Sanjeevikumar Padmanaban, S. Balamurugan, and Jens Bo Holm-Nielsen, ISBN 9781119786276. This collection of papers on artificial intelligence and other methods for improving renewable energy systems, written by industry experts, is a reflection of the state of the art, a must-have for engineers, maintenance personnel, students, and anyone else wanting to stay abreast with current energy systems concepts and technology. *NOW AVAILABLE!*

SMART CHARGING SOLUTIONS FOR HYBRID AND ELECTRIC VEHICLES, edited by Sulabh Sachan, Sanjeevikumar Padmanaban, and Sanchari Deb, ISBN 9781119768951. Written and edited by a team of experts in the field, this is the most comprehensive and up to date study of smart charging solutions for hybrid and electric vehicles for engineers, scientists, students, and other professionals. *NOW AVAILABLE!*

Printed in the USA/Agawam, MA
April 28, 2026

904343.001